Textiles and Their Use in Microbial Protection

Textile Institute Professional Publications

Series Editor: Helen D. Rowe for The Textile Institute

Care and Maintenance of Textile Products Including Apparel and Protective Clothing
Rajkishore Nayak, Saminathan Ratnapandian

Radio Frequency Identification (RFID) Technology and Application in Fashion and Textile Supply Chain
Rajkishore Nayak

The Grammar of Pattern
Michael Hann

Standard Methods for Thermal Comfort Assessment of Clothing
Ivana Špelić, Alka Mihelić Bogdanić, Anica Hursa Sajatovic

Fibres to Smart Textiles: Advances in Manufacturing, Technologies, and Applications
Asis Patnaik and Sweta Patnaik

Flame Retardants for Textile Materials
Asim Kumar Roy Choudhury

Textile Design: Products and Processes
Michael Hann

Science in Design: Solidifying Design with Science and Technology
Tarun Grover and Mugdha Thareja

Textiles and Their Use in Microbial Protection: Focus on COVID-19 and Other Viruses
Jiri Militký, Aravin Prince Periyasamy, Mohanapriya Venkataraman

For more information about this series, please visit: www.crcpress.com/Textile-Institute-Professional-Publications/book-series/TIPP

The aim of the ***Textile Institute Professional Publications*** is to provide support to textile professionals in their work and to help emerging professionals, such as final year or Master's students, by providing the information needed to gain a sound understanding of key and emerging topics relating to textile, clothing and footwear technology, textile chemistry, materials science, and engineering. The books are written by experienced authors with expertise in the topic and all texts are independently reviewed by textile professionals or textile academics.

The textile industry has a history of being both an innovator and an early adopter of a wide variety of technologies. There are textile businesses of some kind operating in all counties across the world. At any one time, there is an enormous breadth of sophistication in how such companies might function. In some places where the industry serves only its own local market, design, development, and production may continue to be based on traditional techniques, but companies that aspire to operate globally find themselves in an intensely competitive environment, some driven by the need to appeal to followers of fast-moving fashion, while others by demands for high performance and unprecedented levels of reliability. Textile professionals working within such organizations are subjected to a continued pressing need to introduce new materials and technologies, not only to improve production efficiency and reduce costs, but also to enhance the attractiveness and performance of their existing products and to bring new products into being. As a consequence, textile academics and professionals find themselves having to continuously improve their understanding of a wide range of new materials and emerging technologies to keep pace with competitors.

The Textile Institute was formed in 1910 to provide professional support to textile practitioners and academics undertaking research and teaching in the field of textiles. The Institute quickly established itself as the professional body for textiles worldwide and now has individual and corporate members in over 80 countries. The Institute works to provide sources of reliable and up-to-date information to support textile professionals through its research journals, the *Journal of the Textile Institute*[1] and *Textile Progress*[2], definitive descriptions of textiles and their components through its online publication *Textile Terms and Definitions*[3], and contextual treatments of important topics within the field of textiles in the form of self-contained books such as the *Textile Institute Professional Publications*.

REFERENCES

1. http://www.tandfonline.com/action/journalInformation?show=aimsScope&journalCode=tjti20
2. http://www.tandfonline.com/action/journalInformation?show=aimsScope&journalCode=ttpr20
3. http://www.ttandd.org

Textiles and Their Use in Microbial Protection

Focus on COVID-19 and Other Viruses

Jiri Militký, Aravin Prince Periyasamy
and Mohanapriya Venkataraman

CRC Press
Taylor & Francis Group
Boca Raton London New York

CRC Press is an imprint of the
Taylor & Francis Group, an **informa** business

First edition published 2021
by CRC Press
6000 Broken Sound Parkway NW, Suite 300, Boca Raton, FL 33487-2742

and by CRC Press
2 Park Square, Milton Park, Abingdon, Oxon, OX14 4RN

CRC Press is an imprint of Taylor & Francis Group, LLC

Library of Congress Cataloging-in-Publication Data
Names: Militky, Jiri (Material engineer), editor. | Periyasamy, Aravin Prince, editor. | Venkataraman, Mohanapriya, editor.
Title: Textiles and their use in microbial protection : focus on COVID-19 and other viruses / edited by Jiri Militky, Aravin Prince Periyasamy, Mohanapriya Venkataraman.
Description: First edition. | Boca Raton, FL : CRC Press, 2021. | Series: Textile Institute professional publications | Includes bibliographical references and index. | Summary: "Textiles and Their Use in Microbial Protection: Focus on COVID-19 and Other Viruses provides readers with vital information about disinfection mechanisms used in textile applications in the fight against dangerous microbes and viruses. It is aimed at textile and materials engineers as well as readers in medical fields and offers a comprehensive view of fundamentals and solutions"-- Provided by publisher.
Identifiers: LCCN 2021004336 (print) | LCCN 2021004337 (ebook) | ISBN 9780367691059 (hardback) | ISBN 9780367691059 (paperback) | ISBN 9781003140436 (ebook)
Subjects: LCSH: Biomedical materials. | Industrial fabrics. | Anti-infective agents. | Disinfection and disinfectants. | Virus diseases--Prevention.
Classification: LCC R857.M3 T49 2021 (print) | LCC R857.M3 (ebook) | DDC 610.28/4--dc23
LC record available at https://lccn.loc.gov/2021004336
LC ebook record available at https://lccn.loc.gov/2021004337

ISBN: 978-0-367-69108-0 (hbk)
ISBN: 978-0-367-69105-9 (pbk)
ISBN: 978-1-003-14043-6 (ebk)

Typeset in Times
by SPi Global, India

Contents

SECTION III Textile Applications

Foreword

The COVID-19 pandemic is ravishing human lives and economies of many countries. To this day and across the globe, the virus has affected millions of patients and has killed over a million. It is important to provide researchers across the world with as much information as possible.

Editor Prof. Jiri Militký is known to me for decades and he is one of the foremost experts in the field of Textiles. He has written multiple books and published numerous articles in peer-reviewed journals. He also has a long history of cutting-edge research in Textile materials. He has been the past Dean and Head of Faculty of Textile Engineering at the Technical University of Liberec which is known for its academic excellence and research in Textiles. He and his protégés Dr. Mohanapriya Venkataraman and Dr. Aravin Prince Periyasamy have edited this book.

This book, edited and authored by the members of the Department of Textile Engineering, Faculty of Textile Engineering, Technical University of Liberec, contains a detailed literature review of the pandemic-causing viruses and possible solutions available to develop textile structures that will protect humans.

Various chapters of this book cover a broad spectrum of topics about the virus and methods that can be used on Textiles to passivate the virus. Initial chapters provide foundational knowledge about the SARS-CoV-2 coronavirus and usage of textile materials used in health care. The state-of-the-art information about the usage of textile materials for antivirus is very relevant to the current problem. The latter chapters discuss various methods available to characterize, indicate, and passivate the virus. The chapter about moisture harvesting is particularly useful to design protective clothing with the right properties that ensure the safety of the wearers of protective clothing.

I strongly believe this book will help textile researchers with its contemporary and relevant scientific information. I wish all the best to the Editors and Authors of the book.

Prof. Jaroslav Sestak, M.Eng., D.Sc., Dr.h.c.
Holder of the State Award granted by the Czech President 2017
University of West Bohemia in Pilsen
Plzen, Czech Republic

Preface

Through the pages of history, we can read about pandemics caused by viruses. Viruses continue to emerge as a serious threat to the health of humans and animals. The latest virus to cause chaos across the world is COVID-19. COVID-19 is a disease caused by a new coronavirus called SARS-CoV-2. It is a novel infectious disease caused by the virus family of Severe Acute Respiratory Syndrome (SARS). First infections of COVID-19 were diagnosed at Huanan Seafood Wholesale Market, located in Wuhan, Hubei, China. Due to global spread, the World Health Organization declared the outbreak a Public Health Emergency of International Concern in January 2020 and a pandemic in March 2020. As of November 2020, more than 60 million cases have been confirmed, with more than 1.4 million deaths attributed to COVID-19.

With such a massive impact of COVID-19 on mankind, it is important to consider all possible means to save and protect humans from such a tragedy. In this book, we strive to explore the devices used to protect against the transmission of the novel COVID-19 disease. Notably, most protective devices are more or less related to textile structures. It is imperative that the textile structures considered for protection of human life is evaluated logically and extensively.

This book provides a detailed review of the literature necessary for textile professionals to develop textile structures that have necessary properties for effective use by humans. The biggest challenge to Textile researchers is the Design and development of textile materials and structures that is capable of capturing and passivating the virus while not impacting the well-being of the wearer. The book has been logically organized to provide a background of the problem and navigate through topics that can provide information to develop possible solutions.

The literature review begins with an Introduction to textile materials used in Health care and provides State-of-the-Art information about available protection against microbes. In the following chapters, the Virology and Characterization, Indication, and Passivation of COVID-19 are discussed in detail. Different disinfection methods like UV light and ozonization; Photocatalysis; Photo-oxidation; Carbonization; Usage of metals such as Titania and Copper have been reviewed. Detailed reviews of antiviral finishes for protection against SARS-CoV-2 have been provided. To provide information about the construction of face masks, a chapter about the fundamental Principles for moisture harvesting systems and its design of fabric has been included.

This book provides timely and relevant information to jumpstart the efforts of textile researchers to develop textile materials and structures that will result in saving millions of lives not just during the current COVID-19 pandemic but also as protection during any such future unfortunate calamity.

Prof. Jiri Militký, M.Sc., Ph.D.
Dr. Aravin Prince Periyasamy, M.Tech., Ph.D.
Dr. Mohanapriya Venkataraman, M.Tech., M.F.Tech., Ph.D.
Technical University of Liberec, Czech Republic

Editor Biographies

Prof. Ing. Jiri Militký CSc. is a university professor in the Department of Material Engineering, Faculty of Textile Engineering, Technical University of Liberec, Czech Republic. He is a renowned contributor in the field of Textile Sciences. His scientific activities are mainly in the areas of textile physics, textile material engineering, nanocomposites, and statistical data treatment. In this field, he has published 24 books besides 450+ scientific papers and 400+ conference presentations. He started to work in the field of modeling the kinetic processes in a solid phase. He was engaged in the State Textile Research Institute in the Department of the Mathematical Modeling of the Textile Structures from 1973 to 1976. He started with research in the field of statistical data analysis and quality control there. On these themes, he published 4 books and about 100 scientific papers. From 1976 to 1989, he was engaged in the Research Institute of Textile Finishing in Dvůr Králové, in many positions, from the head of the research department till scientific secretary. There, he worked in the field of textile dyeing kinetics, physics of the fibers, mathematical modeling in the textile branch, and control of dyeing and drying processes. In collaboration with University Pardubice, he is working in the field of chemometry in analytical laboratories. The two-volume monographs published in England were finished in 1994 and 1996, respectively. In 1982, he defended a Ph.D. degree concerning the properties of the modified polyester fibers. Since 1989, he is employed at the Technical University of LIBEREC (TUL). In 1995, he was appointed Academician of the Ukraine Academy of Engineering Sciences. In 1996, he obtained the professional title EURING. From 1991 to 1993, he was held the position of vice chancellor for foreign relations, and from 1994 to 2000, he was dean of the Faculty of Textile Engineering. He has held multiple positions of Dean, Vice-rector, and Head of Faculty of Textile Engineering and is now responsible for research, supervising Ph.D. students, and teaching. He is the leader of multiple research projects focusing on cutting-edge research. In 2018, he was recognized at AUTEX in recognition of his decades of high-quality contribution to textile sciences, famously known as "TEXTILE OSCAR."

Dr. Mohanapriya Venkataraman is a passionate textile material scientist, working as an assistant professor at the Department of Material Engineering, Faculty of Textile Engineering, Technical University of Liberec, Czech Republic. Hailing from Chennai, India, she is a holder of a Ph.D. and multiple post-graduations in Textile Material Engineering, Fashion Technology, and Garment Manufacturing Technology. Her teaching and research areas include Textile Materials, Thermodynamic Analysis, Micro- and Nanoporous Materials, Heat Transfer,

Polymers, and Composites. She is a leader and team member of multiple international research projects funded by the *EU*, the *Technology Agency of the Czech Republic* (TA ČR), and the *Czech Science Foundation* (GA ČR). She has authored and co-authored over 50 scientific papers in peer-reviewed journals; 70+ conference publications; 10+ keynote speeches; and 25+ book chapters. She has won international recognition as "Outstanding Researcher" in multiple forums like SGS, TBIS, etc.; prior to endeavoring into academics and research, she worked as an executive in Material Quality Assurance in an International Textile behemoth. She is certified in ISO, Lean Six Sigma, 5S, Kaizen, and Silverplus Limited brand testing. She was recently profiled in *TA.DI* magazine of *TA ČR* as one of three female researchers as an example for breaking the stereotype of a traditional scientist. She is an ambassador for INOMICS and "Study in the Czech Republic" initiatives. She is passionate about woman empowerment and the environment.

Dr. Aravin Prince Periyasamy is working as a researcher in the Department of Material Engineering, Faculty of Textile Engineering at the Technical University of Liberec (TUL), Czech Republic. His graduation and post-graduation are from Anna University, Chennai, India, in Textile Technology. He also completed Ph.D. in *Textile Techniques and Material Engineering* from TUL. His research areas include chromic materials (photochromic and thermochromic materials), metal coating (copper, nickel, silver) on various textile structures, chemical vapor deposition, sol-gel chemistry on the kinetics of chromic materials, synthesis of conductive silanes, surface modification by plasma, sustainable chemical processing on textiles, recycling of polymers, quantification of microplastics, and life cycle analysis of various textile materials. Currently, he is working on various projects funded by the Czech government and European Union. He has published 40+ research papers in international refereed journals, 20+ conference proceedings, 10+ keynote speeches at national and international level conferences, and 50+ visiting lectures. In addition, he has contributed chapters in 16 edited books, published by reputed publishers, such as CRC Press, Woodhead, Elsevier, Springer, and Apple Academic Press. He authored one book entitled *Chromic Materials Fundamentals, Measurements, and Applications* by CRC Press and one monograph on *A review of Photochromism in Textiles and Its Measurement* (Textile Progress) published by Taylor & Francis group.

Contributors

Saeed Ahmad
Center of Biotechnology
University of Peshawar
Peshawar, Pakistan

Azam Ali
Department of Material Engineering, Faculty of Textile Engineering
Technical University of Liberec
Liberec, Czech Republic

Divan Coetzee
Department of Material Engineering, Faculty of Textile Engineering
Technical University of Liberec
Liberec, Czech Republic

Sajid Faheem
Department of Material Engineering, Faculty of Textile Engineering
Technical University of Liberec
Liberec, Czech Republic

Shi Hu
Department of Material Engineering, Faculty of Textile Engineering
Technical University of Liberec
Liberec, Czech Republic

Daniel Karthik
Department of Material Engineering, Faculty of Textile Engineering
Technical University of Liberec
Liberec, Czech Republic

Hira Khaleeq
Department of Chemistry
Government College University
Lahore, Pakistan

Dana Kremenakova
Department of Material Engineering, Faculty of Textile Engineering
Technical University of Liberec
Liberec, Czech Republic

Aamir Mahmood
Department of Material Engineering, Faculty of Textile Engineering
Technical University of Liberec
Liberec, Czech Republic

Jiri Militký
Department of Material Engineering, Faculty of Textile Engineering
Technical University of Liberec
Liberec, Czech Republic

Nazia Nahid
Department of Bioinformatics and Biotechnology
Government College University Faisalabad
Faisalabad, Pakistan

Muhammad Shah Nawaz ul Rehman
Virology Lab, Centre of Agricultural Biochemistry and Biotechnology
University of Agriculture Faisalabad
Faisalabad, Pakistan

Miroslava Pechociakova
Department of Material Engineering, Faculty of Textile Engineering
Technical University of Liberec
Liberec, Czech Republic

Qingyan Peng
Department of Material Engineering, Faculty of Textile Engineering
Technical University of Liberec
Liberec, Czech Republic

Aravin Prince Periyasamy
Department of Material Engineering, Faculty of Textile Engineering
Technical University of Liberec
Liberec, Czech Republic

Jana Saskova
Department of Material Engineering, Faculty of Textile Engineering
Technical University of Liberec
Liberec, Czech Republic

Xiaodong Tan
Department of Material Engineering, Faculty of Textile Engineering
Technical University of Liberec
Liberec, Czech Republic

Mohanapriya Venkataraman
Department of Material Engineering, Faculty of Textile Engineering
Technical University of Liberec
Liberec, Czech Republic

Dan Wang
Department of Material Engineering, Faculty of Textile Engineering
Technical University of Liberec
Liberec, Czech Republic

Yuanfeng Wang
Department of Material Engineering, Faculty of Textile Engineering
Technical University of Liberec
Liberec, Czech Republic

Jakub Wiener
Department of Material Engineering, Faculty of Textile Engineering
Technical University of Liberec
Liberec, Czech Republic

Kai Yang
Department of Material Engineering, Faculty of Textile Engineering
Technical University of Liberec
Liberec, Czech Republic

Muhammad Zaman Khan
Department of Material Engineering, Faculty of Textile Engineering
Technical University of Liberec
Liberec, Czech Republic

Section I

Generalities

1 Introduction to Textile Materials Used in Health Care

Jiri Militký, Mohanapriya Venkataraman and Aravin Prince Periyasamy
Technical University of Liberec, Czech Republic

CONTENTS

1.1 INTRODUCTION

The utilization and development of textile structures are closely connected with the development of polymer chemistry, materials engineering, mechanical engineering, electronics, and other industries. Most conventional textile structures composed of standard textile fibers used for clothing purposes can be applied for modern technical textiles including medical ones without major problems. For special applications such as protective or barrier fabrics, the desired effects can be achieved by grafting, coating, laminating, and layering techniques combined with a suitable construction of standard fiber-based fabrics. To achieve especially high performance (e.g., strength and modulus) or high functionality (antimicrobial protection, electric conductivity, flame retardancy, etc.), it is often beneficial to use special fibers. These special fibers already have several required properties (mechanical, thermal, electrical, biochemical, chemical, etc.), so the fabrics made from them do not require special modifications.

On the other hand, however, problems often arise with the construction of textile structures (many special fibers are brittle, low in tension, etc.) and with possible finishing and dyeing (e.g., for technical clothing textiles). This, together with the relatively high price of special fibers, leads to the fact that even today, over 90% of all technical textiles are made from conventional fibers. A special area of technical textiles are textiles for use in medicine and textiles that have only a certain medical effect and contribute in some way to the health of the wearers. For medical purposes, it is possible to use several standard textiles and textile structures, in particular for ordinary clothing purposes, bed linen, and ordinary home textiles. However, the emphasis is on special conditions of use (higher incidence of dangerous bacteria, the possibility of influencing treatment processes, etc.). This chapter is a survey of advanced types of materials used in textile structures with special emphasis on textiles for health care. Antimicrobial textiles are described in Chapter 2.

1.2 DEVELOPMENT OF TEXTILES

The objective causes of textile development are closely related to both the human factor and the influence of civilization. The effects related to the human factor can be divided into the following groups:

- *Earth's population growth:* the planet's population is expected to increase to 8.9 billion by 2050. With an expected consumption of 20 kg of textiles per person per year, this amounts to a total of 178 billion tons of textiles per year in 2050.
- *Prolongation of life expectancy:* It was found (Tuljapurkar et al. 2000) that life expectancy in 2050 will be in the range of 80–83 years and 83–91 years in USA and Japan, respectively (resp.). This is also related to the increase in the relative share of seniors in society. The category of seniors will have other

requirements for several textiles related in particular to ensure their safety with limited mobility (e.g., improved visibility of objects, and identifiable edges).

- Growth of the share of free time that can also be spent on activities requiring special textiles (fitness, wellness).
- A lifestyle that significantly changes the size range of clothing textiles and also affects the way they are purchased.
- Civilizational influences on a person, which usually have a negative effect on his health and require the provision of special barrier functions (allergies against allergies, environmental pollution, etc.).
- Protection and prevention of health (fitness, sports, rehabilitation), which again require special textiles both in clothing and in some products (especially composites).

Closely related to the human factor, there are many changes in the availability and acquisition of information, virtualization of practically everything (including the supply of goods), and the globalization of society. Civilization factors are directly related to the development of the level of human society and the corresponding consumption. In addition to the objective factors of textile development, subjective factors are of course also manifested. Subjective factors of textile development are related to two basic approaches. Typical in particular for Asian countries there is a desire for a new (nontraditional). Examples are textiles with new functions, intelligent structures, and special fibers enabling, for example, the conversion of solar (light) energy into electrical energy. In many western countries, there is a tendency to believe in traditional (i.e., old) techniques and products. An example is the popularity of materials made of viscose, where the raw material is bamboo. These fibers are available under the misleading name bamboo fibers and are attributed to therapeutic and health-promoting effects. Subjective factors often cause manufacturers to return to old techniques and processes (e.g., dyeing with natural dyes and use of milk casein fibers), which they sometimes combine with modern techniques to ensure practical applicability. Prospectively, the predominant long-term trend will be the integration of development results in the field of chemistry, physics, and engineering, for the production of new textile structures, for the production of clothing textiles capable of adapting to changes in environmental conditions, and special technical textiles with unique properties required for their applications. The traditional aspect of fashion, style, and comfort prevailed in textiles for clothing purposes. In the future, it can be expected that the following will need to be ensured for clothing textiles:

- Optimal humidity control.
- Heat flow control.
- Air breathability control.
- Improved thermal insulation properties.
- Water vapor permeability (size 0.4 nm) but liquid water drop impermeability (size 100 μm).
- Protection against dangerous influences from the environment (microorganisms, UV radiation).
- Ecological production and disposal (biodegradability).

- Self-cleaning effects and dust repellency.
- Improved wear resistance (abrasion).
- Support for health care (vital functions, healing processes).
- Support of cosmetic manifestations (regenerative processes on the skin).
- Easy maintenance including cleaning and ironing.
- Improved hand, aesthetic sensations, and appearance even after several cycles of use and maintenance.
- Controlled active identifiability of textiles in conditions of limited visibility.

There are already partial solutions enabling the implementation of some of these requirements. In the future, multifunctional effects and solutions to problems associated with limited durability can be expected. In the field of technical textiles, the situation is usually simpler, as the requirements for their properties can often be precisely specified according to the intended purpose of use.

1.3 TEXTILE STRUCTURES

Textile structures are unique materials composed of fibers, which are responsible for their behavior and geometry. Fibers are thin, long, strong, light, anisotropic, tough, stiff, flexible, and viscoelastic rod-like materials with typical fibrous structure (Figure 1.1). Fiber is a generic name for long (length $l = 10^{-2}$ to 10^{-1} m) thin (diameter $d = 10^{-6}$ to 10^{-4} m) rod-like formations prepared from polymeric or nonpolymeric substances. The typical length to diameter ratio l/d is about 10^3 (Militký et al. 2013).

Polymeric fibers have a typical fibrous structure characterized by the hierarchy of long, thin element bundles (molecular chains, microfibrils, macrofibrils) oriented preferably in the fiber axis direction and having more or less ordered three-dimensional arrangements (semicrystalline state). The basic element of fibrous structures is thin (diameter ≈ 10 nm), long (length ≈ 1 μm) microfibril with a regular arrangement of crystalline and amorphous portions. The length of crystalline parts is 6–10 nm and the distance between two crystalline parts (long period) is equal to $L_p = 15$ nm (Figure 1.2).

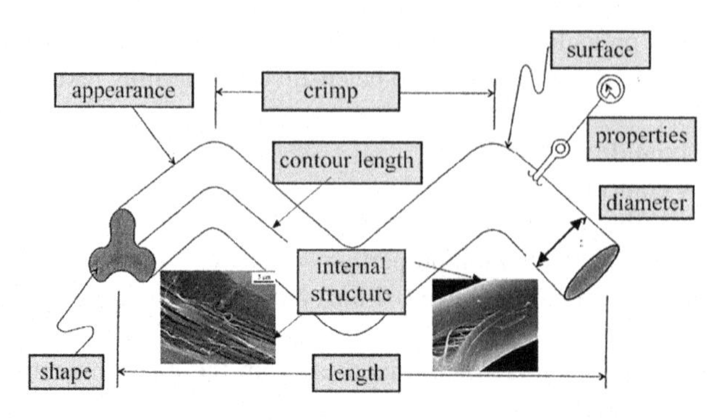

FIGURE 1.1 Fiber characteristics.

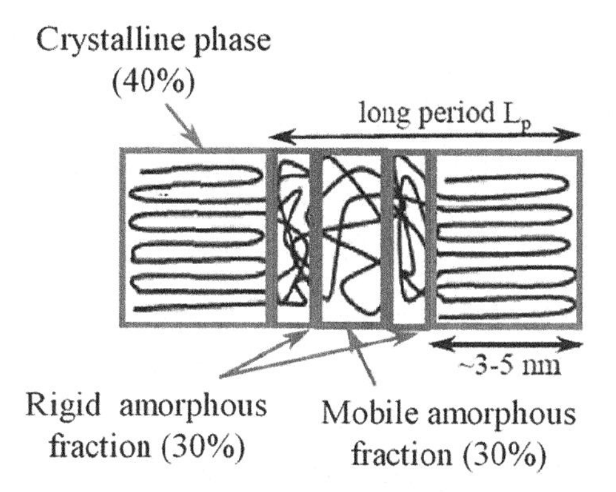

FIGURE 1.2 Structure of microfibril.

Parallel microfibrillar bundles connected by tie molecules and embedded in interfibrillar amorphous phase form well-defined fibril. This fibrillar structure is common for natural, chemical, and synthetic fibers from organic polymers (Militký et al. 2013). Due to this special structure, fibers have high anisotropy of physical and mechanical properties (Militký 2019). For example, the typical modulus of crystalline phase in the longitudinal (fiber axis) direction E_{KL} = 150 GPa modulus of crystalline phase in the transversal (perpendicular to fiber axis) direction is E_{KT} = 4 GPa and modulus of the amorphous phase is E_A = 0.6 GPa. Typical is the cooperative character of deformation where the deformation process acts on the group of elements together.

Natural fibers such as *cotton*, *bast*, and *wool fibers* are products of nature, not for textile purposes. Their structure hierarchy is the same as the structure of chemical and synthetic fibers (*viscose*, *polyamides*, *polyesters*, *acrylic fibers*, *polyolefines*, etc.) which are prepared artificially. Properties of man-made fibers can be simply changed by the variation of fiber spinning, drawing, and heat setting conditions. It is simple to change markedly the majority of properties by the selection of fiber geometries (denier, cross-section profile, texturing) and spinning conditions (rate of production, drawing degree, and temperature treatment). Textile fibers have special organoleptic properties (luster, hand), technological properties (length, strength, crimp, surface roughness, etc.), and utility properties (sorption, ability to set, abrasion resistance, etc.) the details are given in this book (Militký 2019). Many characteristics of fibrous textile structures are similar as for all kinds of, generally 3D objects but some of them are quite different due to limitations of the thickness of fibers, yarns, and plane structures (woven, knitted fabrics and, to some extent, nonwovens). Standard characterization of the relative mass of 3D objects is density ρ [kg/m^3],

$$\rho = m / V = T_T / S = M_P / h \tag{1.1}$$

Density is the ratio between object mass m [kg] and object volume V [m^3]. For *linear textile materials* (Figure 1.3) the density is defined as fineness T_T divided by cross-section area S (see Equation 1.2) and for *planar textiles* (Figure 1.6) density is defined as planar mass M_p divided by planar textiles thickness h (Figure 1.4).

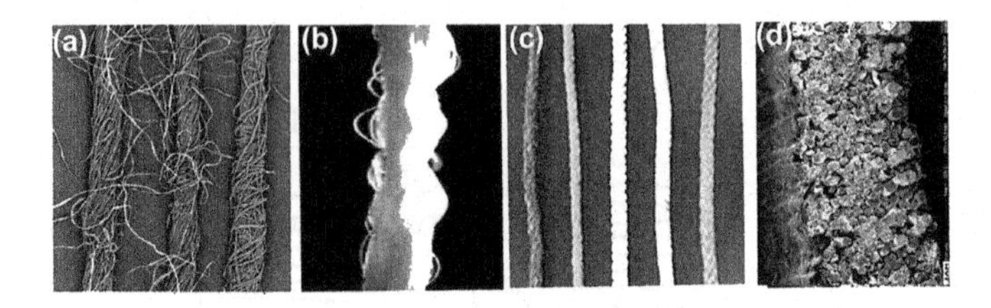

FIGURE 1.3 Some linear textile materials: a) cotton yarn prepared by various techniques, b) textured multifilament, c) ropes, and d) hybrid tape prepreg.

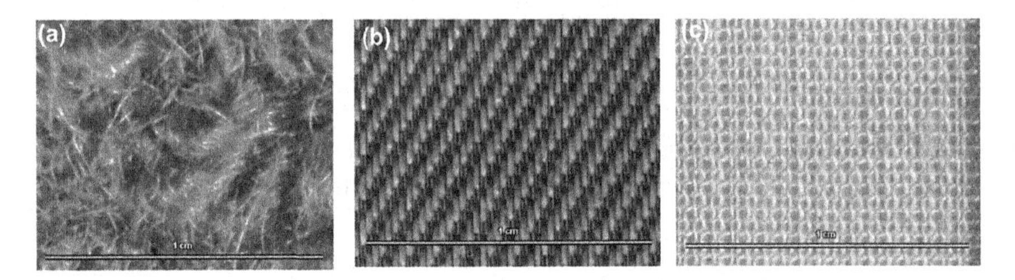

FIGURE 1.4 Some woven textile materials: a) Krull weave with surface loops, b) twill weave, and c) plain weave (low sett).

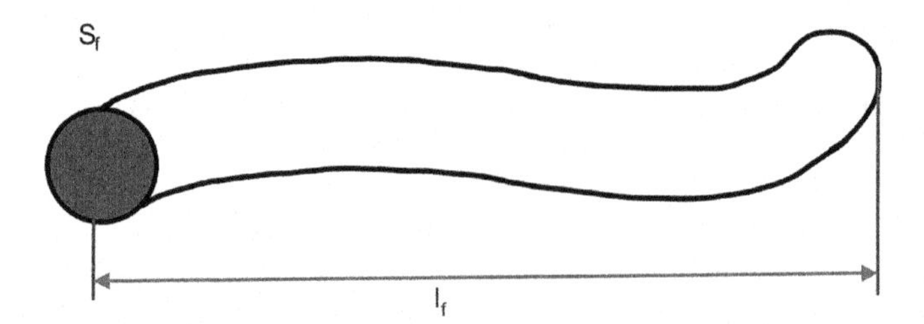

FIGURE 1.5 Fiber geometry.

The density of common textile fibers is from 980 kg/m^3 for polypropylene to 1,530 kg/m^3 for cellulosic fibers. For metallic fibers, the density is usually much higher (for steel it is 7,800 kg/m^3). In the case of fibers and other linear textile structures (e.g., yarns) there exists real limitation due to restricted thickness/diameter (see Figure 1.5).

The thickness of *fibers* cannot be selected arbitrarily but it should be in the restricted range from approximately 5 to 35 μm due to fiber flexibility required for subsequent operations. Instead of density, it is beneficial to define relative mass expressed as fineness T_T usually in tex units (mass in grams per length in kilometers). It can be simply shown that

$$T_T\left[\text{tex}\right] = \frac{m\left[\text{g}\right]}{l\left[\text{km}\right]} = 10^{-6}\rho\left[\text{kg/m}^3\right]S_f\left[\mu\text{m}^2\right] \quad (1.2)$$

Fineness can be directly used for the calculation of the mass of linear textiles by multiplying them by their length. It is visible that the fineness is dependent on the material characteristics, i.e., density ρ and geometric characteristics, i.e., cross-section area S_f. Higher value T_T is for more coarse fibers with higher thickness. Standard cotton fiber fineness is in the range 1.1 (fine) to 2.3 (coarse) dtex, and wool fiber fineness is in the range 2.8 (fine) to 15 (coarse) dtex. Man-made fibers (chemical and synthetic) have a fineness of about 1–5 dtex depending on spinning technology. For cotton spinning technology, the fineness of these fibers is 1.3 or 1.75 dtex and for woolen spinning technology, the fineness is usually 3.3 dtex. The fibers are commonly referred to as fine if they have a fineness of about 1 dtex, extrafine if they have a fineness of about 0.5 dtex, and superfine if they have a fineness of about 0.1 dtex. Finer fibers are generally more flexible and have a larger surface area (cohesion) in yarns. This is especially true in fibers from metals, which have a high density but can be very thin and fine. For example, for fibers of fineness 1 dtex from polyethylene (POE) (900 kg/m^3) diameter is 14.5 μm and for the same fineness of steel fibers (density 7,800 kg/m^3) diameter is only 5.1 μm. Fineness can be used as well for another linear textile structure as yarns from staple fibers and filaments or multifilaments. Yarns' fineness is usually in the range above 15 tex. In the case of planar textile structures, there are real limitations due to restricted thickness *h* only (see Figure 1.6).

Better to express relative mass as planar mass M_P usually in *GSM* (gram per square meter) units. For this case it is,

$$M_P\left[g/m^2\right] = \rho\left[kg/m^3\right] h\left[mm\right] \tag{1.3}$$

The planar mass of woven textiles is about 120 g/m^2 and for heavier textiles, it is about 350 g/m^2.

Relative units of concentration C_m [mg/g] as amount of substance [mg] per material mass [g] are for plates and planar textiles better expressed as concentration C_s [mg/m^2], i.e., amount of substance [mg] per material surface area [m^2]. Multiplication by real surface area leads here directly to the amount of substance. It is simple to derive interrelation,

$$C_m = C_s M_p \tag{1.4}$$

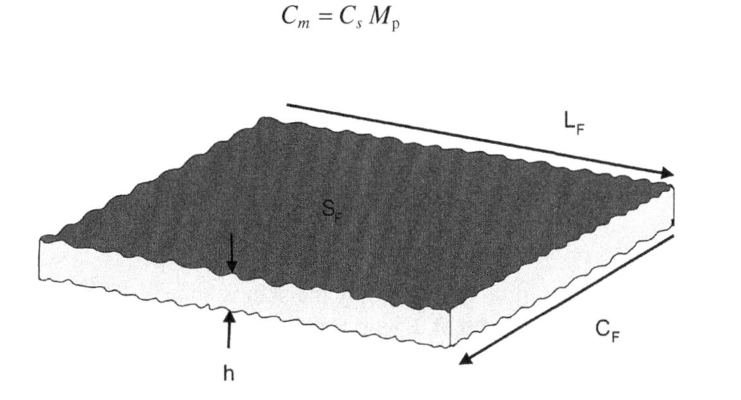

FIGURE 1.6 Geometry of planar textiles.

Equation 1.4 is useful for the cases of filtration (real amount of captured materials) or medical drug delivery by planar structures (real amount of drug). The textile structures are therefore limited by their thickness starting from fibers and ending by planar structures. In the case of necessity, it can be possible to enlarge textiles thickness by layering but there are necessities to join individual layers by special techniques (sewing, gluing, using adhesive intermediate layers, etc.). Textiles are increasingly being used as special flexible construction materials and composite structures. For these purposes and special applications (light-conducting systems, electrically conductive systems, etc.) polymers, ceramic materials, and metals are used. A separate problem in these cases is the possibility of creating fibrous structures by textile techniques. Many special materials can be converted into the form of thin wires, but for weaving purposes resp. knitting, these structures often do not have the required stiffness (inverse to flexibility) expressed in terms of flexural rigidity *FR* [N mm^2]. This quantity is related to the specific initial modulus of elasticity in tension E_s [N/tex], fineness T_T [tex], and fiber density ρ [kg/m^3]. For fibers of circular cross section,

$$FR = \frac{E_s T_T^2}{4\pi\rho} \tag{1.5}$$

Specific initial modulus of elasticity in tension E_s [N/tex] is directly related to initial modulus of elasticity in tension E_s [GPa],

$$E\left[GPa\right] = 10^{-3}\rho\left[kg/m^3\right]E_s\left[N/tex\right] \tag{1.6}$$

Then, finer fibers allow for textile processing of materials with increased initial tensile modulus E. Reducing the fineness is therefore necessary especially for fibers with a high modulus E. It can be easily determined that if we replace the wire (monofilament) of fineness T_M and flexural rigidity FR_M with a bundle of *N* finer wires (multifilament) of fineness $T_V = T_M/N$, the flexural rigidity of this bundle FR_Y will decrease to the value $FR_Y = FR_M/N$. Thus, it can be seen that replacing a thicker monofilament with *N* thinner ones will reduce the flexural rigidity and improve processability by textile techniques while maintaining the same external geometry. Reducing the fineness is, therefore, necessary especially for fibers with a high modulus *E*. At the same time, the number of fibers in a unit of textile structure is increasing. The number of fibers N_u [m^{-2}] per unit surface area of fabric with planar mass M_P [g/m^2] composed from staple fibers with length l_f [cm] and fineness T_T [tex] is equal to,

$$N_u = \frac{10^5 M_P}{T_T\, l_f} \tag{1.7}$$

The surface of the adult body covered by the skin is about 1.5 m^2. The weight of typical cotton fiber with a length of 2.5 cm and a fineness of 1.5 dtex is only 3.75 µg. In one square meter of finer cotton fabric with a planar mass of 150 g/m^2, there are 40 million fibers. During the preparation for spinning and the actual production of the yarn, this amount of fibers is processed (loosening, cleaning, mixing, carding, drawing, and twisting). Similarly, in weaving, a relatively large length of yarn must be interwoven. It is thus clear that textile machinery will have to be relatively complicated and, above all, able to produce a large number of textiles from an extraordinary big number of fibers in a short time. Thus, even here, reducing the fineness of the fibers is advantageous. The increase in the number of fibers is reflected in an increase in the relative surface area of the yarn and fabrics, which plays a significant role, for example, in the construction of composites or the application of the surface coating. These simple considerations demonstrate that fibrous structures have a unique behavior due to their geometry (long thin objects) and can be used to advantage in technical applications. Many other unique properties result from the fact that the polymer fibers are drawn and the polymer chains are relatively well oriented in the direction of their axis, for example, to an increase of up to 100 times the tensile modulus E and thus the stiffness compared to rods of the same material and the same geometry. Typical characteristics of fabrics are their porosity, drape, and hand. In particular, the drape of staple fiber fabrics is unique and necessary for clothing purposes. The textile structures of the present and the future have a dual role. They are not only typical consumer goods (clothing purpose) but also a special construction material with specific manifestations. Basic features of standard fibrous structures are:

a. Drape ability, i.e., adaptation to external fields changes or formability
b. Low initial stiffness (good initial deformability)
c. High porosity, more than 75% (wide pores distribution from nano to micro)
d. A partially random hierarchical structure
e. Surface roughness (huge surface area: 10^2–10^4 m^2/g)
f. Planar unevenness (periodic, random)
g. Unique hand and appearance
h. Washing ability
i. Abrasion resistance
j. Slow aging during use (influence of moisture, heat, UV, temperature, friction).

The majority of these features are negatively influenced by using special coatings, membranes, high-performance fibers, resins, etc., necessary for the achievement of protections and smartness. Negative changes of standard fibrous structure features can cause loss of comfort, bad appearance, skin irritation/sores, and nonacceptance of clothing purposes. A typical feature of classic textile structures is that their usefulness is proven only in practical use. The user's experience with similar products therefore plays an important role here. Classical principles of production persist here, and innovations are manifested primarily in productivity and economics. The need for market penetration often leads to the use of rapid innovations without a detailed analysis of the real benefits or the importance of the new effect. There is still a significant difference between laboratory implementation and results from industrial production. This leads to a situation where inaccurate and often distorted information is only used to attract customers due to advertising and sales-promoting campaigns. On the other hand, several breakthrough solutions are slowly being implemented because customers are not ready for them. It is, therefore, necessary to ensure both the preparation of structures with new or more advantageous features, as well as to

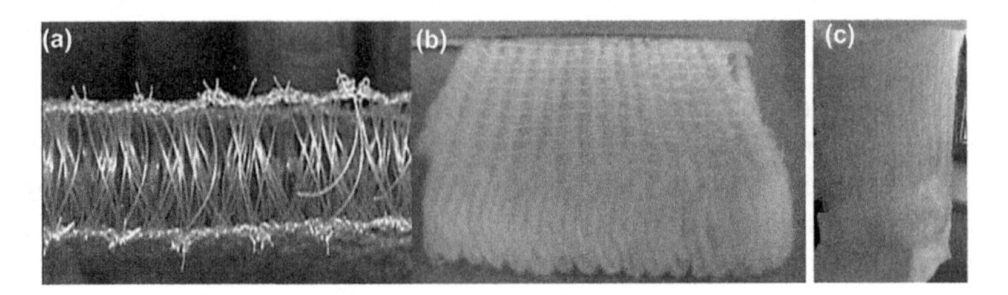

FIGURE 1.7 3D textile structures: a) spacer fabric, b) perpendicular laid nonwoven ROTIS, and c) distance fabric for inflated structures (DIFA loom VUTS Liberec).

verify their applicability in practice and ensure their acceptance by the customer community. In the production of so-called "high-tech" textile structures for special applications, it will also be necessary to use:

- New structures including 3D structures are spacer fabrics, 3D structures, structures with controlled surface relief, porosity, density, etc. (see Figure 1.7).
- Special types of energy (plasma, laser) focused on surface resp. localized action.
- New techniques for fixing substances (encapsulation, molecular traps, nanolayers).
- New types of materials (aerogels, nanoparticles, shape, and temperature-sensitive materials, electrically conductive, thermo-electrical, opto-electrical materials, etc.).

Some of these possibilities are already verified mainly on a laboratory scale, but practical applications are nonsystemic and sporadic.

When developing new products, a computer-aided system approach will be used as a standard. For these purposes, software systems will be gradually created combining existing software focused on the classic construction of textiles (visual design) with systems predicting the properties of textiles depending on their construction (analogy of CAD systems in mechanical engineering). For clothing and protective clothing fabrics, it will be necessary to include comfort in the design (permeability to water vapor and impermeability to liquid water, thermal resistance, and ventilation).

1.4 TEXTILES FOR ADVANCED APPLICATIONS

Textiles for advanced application are a broad class of structures comprising textiles for technical purposes, high functional textiles, and textiles as information systems. All these groups of textiles can be adopted for health care and medical purposes. These chapters are an overview of classical results and recent developments in these fields.

1.4.1 Textiles for Technical Aims

Interestingly, more than 90% of technical textiles are made from conventional fibers and nonwoven fabrics use a greater proportion of textile technologies (Holmes 1998). The predominant technology for processing technical fibers into fabrics is the weaving of yarns made of technical fibers. Generally speaking, although weaving means relatively low added value, it is a highly competitive sector still having a significant market position. An example where the growth in fabric consumption is relatively significant are applications that take advantage of their unique strength, stability, and characteristic directional orientation, typical of yarn-based products. These are, for example, the construction sector and structures (membrane structures, etc.), geotextiles, especially for applications related to reinforcement, protective clothing, and sports textiles. Knitwear has not previously found greater use in technical applications due to its relatively low strength and low dimensional stability. The main use of traditional knitting technologies has therefore been in applications where extensibility and shape adaptability were required, such as in medical fabrics or backing fabrics for application in artificial leather. A new development that can significantly increase the need for knitted fabrics is the technique of inserting a weft into the warp knit, which makes it possible to guide the strength characteristics in the construction of the fabric and to create complex knitted patterns and shapes. These fabrics, which are close to nets, offer various advantages, such as material reduction and thus lower production costs, as well as applications ranging from composite reinforcement to car seat covers. Modern computer patterning technology allows controlling the construction and shaping of knitted fabrics. At present, knitted technical textiles represent 3%–5% of the total production of technical textiles. One of the limitations of their growth is the development of modern technologies for the production of nonwovens, which are more successful at the expense of established textiles, where high strength and stability are not essential and where otherwise knitting would be an ideal alternative.

The end use of products based on special yarns can be divided into two groups in terms of the complexity of the environment for which they are intended. One group consists of products with specific utility properties, the design of which is standard and they are not subject to special requirements in terms of their physical and mechanical properties. Such products are, for example, bedding with antibacterial effects, nonflammable curtain fabrics, antistatic socks. The second group includes products that have specific useful properties and are designed for more demanding conditions. The physical and mechanical properties of the materials are used to reach a higher level of products. This category includes special work and protective clothing and clothing for sports. The quality and useful properties of such products are determined by the fibrous material used, the strength of the yarn, and the construction of the fabric. Fibers with high-performance properties are used for special protective functions (e.g., nonflammable Basofil fiber, carbon fibers, and fibers with antistatic effects). One of the promising techniques for finishing technical textiles is coating. Technologically, the application is realized in two phases:

a. Applying a sufficiently viscous agent to the surface of the fabric,
b. Drying with possible subsequent curing.

For the application of polymer solution, melt, or resins either application rollers (less viscous agents) or deposition with different types of doctor blades (viscous agents) is used. The use of new forms of deposition (foam, fog, scattering of particles in an electric field, etc.) is as well as attractive. Depending on the type of finishing agent, either crosslinking or only solvent removal is performed. A special problem is the deposition of metals (metallization). Laminating of metal foils, spraying, or coating in an electric field is used as standard. It is quite interesting to use the technique of physical vapor deposition (PVD), which allows the formation of a film of defined structure on the surface of the fabric. Depending on the type of metal (titanium, zirconium) and the atmosphere (nitrogen, acetylene), nitrides are formed, resp. carbides. A suitable industrial device for PVD is the HTC 1000 from Hauser (Holmes 1998). Using this technique, it is possible to prepare textiles with antistatic properties, electrical conductivity, protection against electromagnetic and thermal radiation, resp. against bacteria, etc. The deposition of metallic nanoparticles can be realized by the technology of electroless plating. This method is successfully used for surface deposition of copper nanoparticles on the surface of polyester light nonwoven MILIFE. The principle is the autocatalytic reduction of cuprous salts deposited from the liquid phase on the fabric surface (Militký et al. 2020). The image and structure of copper-coated MILIFE are shown in Figure 1.8. Some coatings are used separately only to impart the required properties of textiles and some are used to attach other active substances (pigments, powders, particles, etc.). The basic types of polymers that are used for coatings on technical textiles include:

Polyvinyl chloride (PVC), which is produced by radical polymerization of vinyl chloride. It is a hard solid which is softened (plasticized) for applications such as coating, for example, by cyclohexyl isoctyl phthalates. The plasticizer is completely mixed with the polymer at 120°C, and upon cooling, a flexible polymer is formed. The properties of the coating are controlled by the amount of plasticizer (commonly used plasticizer concentration is up to 50%). Plasticized PVC has good abrasion resistance and low permeability. It is usually incorporated into color pigments, flame retardant agents, and many other special additives. PVC coatings are relatively well resistant to acids and alkalis, but chemical solvents can extract the plasticizer, leading to embrittlement and cracking. The advantage of PVC is high dipole moment and dielectric constant, which allows the use of a high-frequency field for their connection. Polyvinylidene chloride has similar properties, which is used mainly to achieve a nonflammable effect due to the higher price.

Polytetrafluoroethylene (*PTFE*) is produced by the addition polymerization of tetrafluoroethylene. Its typical feature is very low surface energy, which prevents

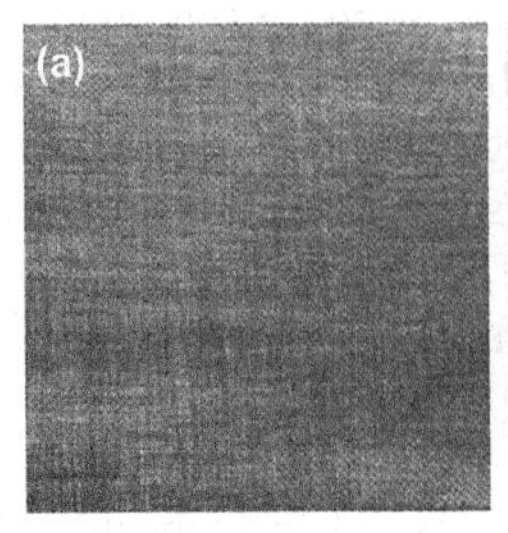

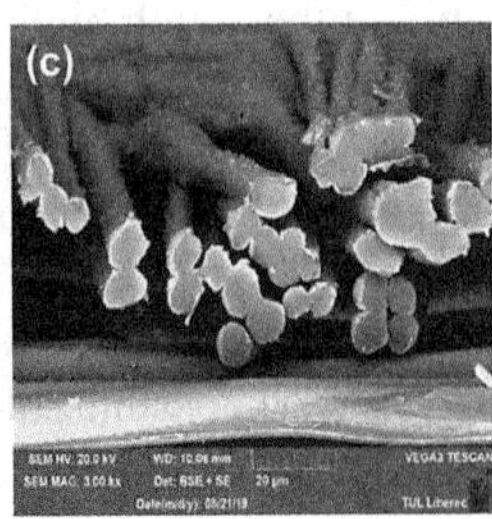

FIGURE 1.8 Copper-coated MILIFE fabrics: a) macro image, b) cross section, and c) longitudinal view.

wetting of both hydrophilic and hydrophobic liquids. Thus, the application of PTFE causes both water-repellency and oil-repellency. Besides, PTFE is thermally stable up to 250°C and resistant to solvents or chemicals. Due to strong etchants, surface etching occurs, which, however, improves adhesion to textiles. It is well resistant to sunlight and weather. The main disadvantage is the hitherto high price and harmfulness due to the presence of toxic perfluorooctanoic acid (C8).

Polyurethanes (*PURs*) are condensation products of diols and diisocyanates. Polycondensation is supported by increasing the temperature. The problem is that crosslinking occurs immediately after mixing the components. Stable pre-polymers usually contain diisocyanate products. The advantage of PUR is their flexibility accompanied by resistance to abrasion and water or solvents. With the help of diols, the permeability to water vapor as well as the elasticity can be influenced to a large extent. Segmented PURs containing "soft" and "hard" blocks can be used as elastomers or shape-memory materials. As these materials pass through the glass transition temperature, the modulus is significantly reduced and the segmental mobility of the soft blocks increases. This results in a significant increase in permeability to both air and water vapor, but impermeability to liquid water is maintained. It also returns to its original shape (which was delivered to the material at a temperature above T_g and then fixed below this temperature). Depending on the application conditions, it is, therefore, possible to change both the transition temperatures and the dependence of the breathability and permeability of water vapor on the temperature, resp. reversible shape changes. The resulting coatings are nonporous but allow air exchange and transport of water vapor.

In the construction of technical textiles, all techniques of achieving special effects can be advantageously used, which are described in Section 1.6. Less traditional are technical textiles for desalination of seawater (based on intelligent gels), ion-exchange textile membranes, textile pipes for aggressive environments, and textile ropes for anchoring drilling rigs (at sea). The field of textiles for electronic applications is developing rapidly. Common to all these applications is that it uses textile techniques for the preparation of special structures (from spinning, weaving, knitting, and production of nonwovens to wrapping and 3D spatial weaving). The resulting textile structures are characterized by relatively easy formability, simple methods of joining and separating, and low weight compared to conventional materials such as metals.

1.4.2 High Functional Textiles

Highly functional textiles represent a wide class of products providing a range of functions required to improve their practical applicability or for applications in special conditions. Several special functions can also be partially achieved by modifying standard textile production procedures and selecting suitable fibers resp. processing procedures. In the case of highly functional textiles, the aim is to use adaptive techniques, i.e., they react in a positive direction only when conditions change. Lamination, coating, and layering are often used to achieve the special properties of high-performance textiles. For lamination, cotton fabrics have traditionally been used, which has been gradually replaced by polyamide-based fabrics. Polyamides

(PAD 6) have several practical advantages such as high flexibility, resistance to wear, abrasion, rot, and cyclic deformations. On the other hand, they have higher creep under long-term loading, sensitivity to acids, higher thermal shrinkage (can be reduced by thermal fixation before lamination), low modulus, low torsional resistance, higher moisture uptake, and lower softening temperature. More preferred is the use of polyester fibers, which are generally stronger and stiffer (higher initial modulus), more heat resistant, and have very low creep under long-term loading. Their slightly worse adhesion can easily be improved, for example, by controlled surface alkaline degradation. Although polyester fibers are not as resistant to abrasion and cyclic stress as polyamide fibers, they are more resistant to weathering and UV radiation, making them standard for lamination in technical textiles. The basic areas of future use of high-performance textiles include:

Active dosing systems – enabling the production of textiles in the clothing sector releasing various substances according to needs, from cosmetics and medicines, through vitamin supplements to insect/bacteria repellents, etc. In nongarment applications, these will be mainly intelligent filters, separators, and sieves.

Monitoring – it will be monitoring both the human vital functions' signs and the environment, which will be usable for both clothing and technical textiles.

Intelligent sensors – in addition to conventional sensors, odor, gas, and bacterial sensors in the air or water are also developed. Motion and mechanical field sensors will be part of the control system, especially for sick people, resp. athletes.

Information technology – textiles will have to, in addition to electronic functions (replacement of keyboards, touchpads, printed circuit boards, displays, and consumer electronics carriers), serve for data storage, storage of electricity (electric batteries), and as carriers of micromechanical systems (capable of cleaning from dust, repairs, etc.).

Adaptive materials – which will change the structure, shade, hand, or other characteristics. Many of these areas of use are already at least partially feasible. Problems associated with ergonomics, comfort, durability, maintenance, production, testing, and finally disposal will still need to be addressed for practical application.

1.4.3 Textiles as an Information System

The use of clothing textiles as an interface for the transmission of information is natural because clothing forms an integral part of man and accompanies him during most activities. In 1996, the US military commissioned the development of an intelligent T-shirt for soldiers that would be able to report on their condition in the event of missile injuries or grenade fragments. Information was requested on the extent of injury (depth of penetration into the body) or vital functions of the wounded soldier (Park and Jayaraman 2001). It was specified that it must be a breathable, lightweight material that meets the requirements of comfort when wearing and using, including maintenance (cleaning). A woven structure was chosen to meet these requirements. A technique was used to create a T-shirt without the need for cutting and sewing. A plastic optical fiber was integrated into the structure of the T-shirt. The following materials were used (Park and Jayaraman 2001):

- Polypropylene fiber (for specific gravity),
- Polymer optical fiber for detecting the depth of penetration of the projectile or fragments,
- POE-coated copper fiber and polyamide fibers with inorganic particles for electrically conductive connections,
- PUR elastomer (spandex) to ensure mechanical comfort and snugness,
- Nega-stat to ensure dissipation of electrostatic charge.

Basic information about a person's condition is obtained from temperature, heart rate, and respiratory rate sensors. These sensors are complemented by a microphone for voice information (Park and Jayaraman 2001). This information is collected in a small electronic device, which is also part of the T-shirt. A similar project was solved in Finland for the design of a snowmobile driver's coverall, which has the following functions (Park and Jayaraman 2001):

- Protects against extreme cold and humidity
- Provides information on the driver's status
- Provides position and orientation information (via global GPS positioning system)
- Allows you to enter information via a textile display

One of the problems associated with the use of textiles for "clothing electronics" is the provision of a suitable power supply. The classic solution consists of placing electrically charged batteries resp. batteries into the garment, which somewhat limits some functions of the garment and besides requires recharging. It is also possible to use energy sources that are available from the environment (e.g., solar cells) or energy generated by the movement of the wearer.

1.5 SMART TEXTILES

In the context of new materials and structures, terms such as "smart" are used to indicate the difference between traditional materials and structures (sometimes referred to as "stupid"). A very detailed discussion of these concepts is in the book (Srinivasan and Mc Farland 2001). There, it is stated that the materials themselves cannot be smart. These have only special properties usable for the construction of smart/intelligent structures. Structures that can independently evaluate the state of the environment and respond appropriately to it are then referred to as smart. They are not, in the true sense of the word, cognitive systems, because they are not chosen from various reaction options. Smart textiles will probably be one of the fastest-growing segments in the textile industry. Smart textiles are textile structures that are sensitive to external stimuli (various types of radiation, pH, mechanical, magnetic, or electric field) and, depending on changes in these stimuli, react reversibly (usually by changing shape). According to the way of reaction to external stimuli, these textile structures are further divided into two basic categories:

Passive intelligent textiles – that are only sensitive to external stimuli and indicate changes (perception). This includes several textiles that act as sensors and indicators of the state of the environment. Examples are optical fibers, which not only transmit a light signal but also are sensitive to deformation, chemical concentration, pressure, acceleration, electric current, magnetic field, etc.

Active intelligent textiles – which are able not only to identify a change in external stimulus but also to respond to this change in a way that leads to their passivation (perception and reaction). Examples are fabrics that change color depending on temperature (chameleon), heat-regulating fabrics (capable of storing or releasing thermal energy according to changes in ambient temperature), fabrics with shape memory (with reversible shape changes during heating or cooling), fabrics with variable breathability and permeability to water vapor and temperature stabilizing fabrics.

In the case of active smart textiles, a group of highly active textiles is further divided, which, in addition to indicating and reacting to changes in ambient conditions, are also able to adapt to changed conditions. It is interesting that the passive intelligent textiles also include components, the so-called textile/clothing (wearable) electronics. In the first generation of clothing, electronics are standard fabrics with built-in common miniaturized electronics. In the second generation, the functions are already integrated directly into textiles, i.e., it is textile electronics (textile displays, keyboards, switches, etc.), and in the third generation, it is based on fibers with electronic functions (textile computers). In many cases, these fabrics are used as clothing fabrics providing top comfort (protecting against temperature fluctuations, adjusting the conditions of air, and water vapor ventilation) or facilitating communication, resp. use of common electronic devices (mobile phones, human position, and status indicators, computers). They are also important for military and medical purposes. These especially include protection against extreme climatic conditions, making visual identification difficult (camouflage), and protection against toxic gases, bacteria, and viruses. Their use is also in the field of nongarment applications, whether they are materials serving as barriers to mechanical, electric, magnetic fields, and radiation of various lengths (from ultraviolet to infrared) or materials acting as intelligent filters and separators (seawater desalination plants), respectively special energy converters (chemo-mechanical responses). In the field of medicine, in addition to their barrier abilities, they can also serve as materials for intelligent drug dosing (depending on the patient's condition) and the diagnosis of disorders of human vital function.

Thus, intelligent textile structures are generally capable of positive reactions to changes in external stimuli (biological, mechanical, thermal, magnetic, optical, chemical, electrical, pH, etc.). Positive reactions/responses are typically mechanical, electrical, optical, and biological. It is easy to classify intelligent structures according to the type of stimulus. Physical stimuli are temperature, electric field, light, pressure, etc. Chemical and biochemical stimuli include pH and specific molecular resolution. Materials for intelligent structures make it possible to use conversions between different types of energies resp. structural changes for the implementation of sensors or actuators. The basic types of materials include:

a. *Piezoelectric:* (PZT: lead + zirconium + titanium, PVDF, silicon). Conversion of mechanical energy to electrical and vice versa.
b. *Electrochromic:* (transition metal oxides, polymers, nanocrystals). Conversion of radiation to electricity and vice versa.
c. *Electrostrictive:* (polyacetylene, elastomers, lead + magnesium + niobium) – nonlinear and heat sensitive. The conversion from mechanical energy to electrical and vice versa.
d. *Magnetostrictive:* (Terphenol D, organometallics). Conversion of mechanical energy to magnetic and vice versa.
e. *Rheological:* (suspension elastomers, silicone oil + barium titanyl oxalate). Conversion of electrical (magnetic) energy to change of state.
f. *Shape memory:* (Nitinol, alloys, PURs). Conversion of thermal energy to mechanical by the change of structure.
g. *Intelligent gels*: (hydrogels, poly(*N*-isopropyl acrylamide) cellulose ethers – xerogel is brittle, PEG – xerogel is rubbery). Conversion of energy to precipitation (swelling).
h. *Thermoelectric I:* (two metals, semiconductors, or conductive polymers on cellulose). Seebeck effect: an electric current caused by a thermal difference in a joint between two materials. Peltier effect: thermal difference caused by an electric current in a joint between two materials.
i. *Thermoelectric II:* (metals, conductive polymers). Thompson effect: the passage of an electric current through a conductor causes heating (proportional to the square of the voltage).
j. *Combined:* examples are shape memory electroactive materials (PURs filled with 5% carbon nanotubes).

Using these principles, it is possible to construct several smart textiles. New types of smart textiles for technical applications also include self-repairing and self-adapting structures. To ensure the efficient use of smart textiles, it is also necessary to provide a suitable method of power supply. For this purpose, structures enabling the transformation of light energy (elastic photocells in a suitable form) or energy induced by motion to electrical energy are developed. Solar energy conversion systems using photocells in the form of strips and fibers also seem promising. The use of kinetic and thermal energy produced by people seems to be promising, especially for powering mobile electronics (Donelan et al. 2008). When people move (high amplitude and low frequency), energy up to 67 W is generated in each step. The thermal energy of the human body can also be used for conversion to electrical energy (Seebeck effect). Based on the analysis of a special Seiko watch, it was found that at a thermal difference of 5°C the power density of electricity is up to 0.14 $\mu W/mm^2$.

At Georgia Tech, a special fiber covered with ZnO nanowires has been developed that can convert mechanical energy into electrical energy. It responds to all types of vibrations and movements. Gold-plated 3.5-micron ZnO nanowires rub against uncoated ZnO nanowires, which bend and generate an electric charge (Wilson 2011). Also interesting is a piezoelectric generator using nanofibers of polyvinylidene fluoride, which can convert mechanical deformations into electrical energy (Chang et al. 2010). Several special oxides such as ZnO, doped In_2O_3, and SnO_2 are used to construct transparent conductive layers suitable for optoelectronics (Chopra et al. 1983). These layers can also be formed on fibrous substrates or with the use of coating or printing technology.

1.6 SELECTED MATERIALS FOR THE PREPARATION OF TEXTILE STRUCTURES

Textile structures with high functional effects for special applications are composed of fibrous polymers and other special materials including nanomaterials, aerogels, shape-memory materials, and stimuli-sensitive materials. All these materials can be adopted for the preparation of textiles with some medical effects or promoting human health. This chapter is an overview of classical results and recent developments in these fields.

1.6.1 Fibrous Materials

Only polymers that meet some basic requirements may be used for the production of fibers:

a. The sufficiently high and uniform average degree of polymerization (PPS) or molecular mass. If the PPS is low, the processing properties of the fibers deteriorate. High PPS causes deterioration of spinning ability.
b. The linear shape of a macromolecule without bulky side chains allowing "approximately parallel" arrangement of macromolecular segments. The statistical segment contains typically 10–20 units.
c. Regularly repeating polar groups in chains allowing the formation of strong one's inter-chain bonds (not necessary, the strongest POE fibers use only weak van der Waals forces).
d. Spatially regular structure allowing at least partial crystallization and formation of fibrous structures.
e. Sufficient stiffness of the chain to allow the formation of crystals with folded chains. On the contrary, some types of fibers (liquid crystalline polymers) require high stiffness leading to taut chain crystallites.
f. The ability to melt or dissolve so that a fiber can be prepared by spinning. Other fiber production methods can also be used for special polymers.

The distance between the polymeric chain ends in the melt is about 15 nm. The length of a fully extended polymeric chain is about 1,500 nm and therefore chains in melt are in the form of coils. Polymers have some typical features:

a. High viscosity increasing with decreasing temperature. Only regular linear structures without side groups can crystallize. Some polymers are after solidification in an amorphous state.
b. Polymer structures are amorphous regions, crystallites, and para-crystallites
c. There exists a possibility to prepare polymeric monocrystals (from solution).
d. The majority of fibers are semicrystalline polymers having crystallinity degree *X*.

Mean polymeric chains molecular mass M_p is simply monomer molecular mass M_M multiplied by mean polymerization degree *PPS*,

$$M_p = PPS\, M_M \tag{1.8}$$

Mechanical and thermodynamic properties of polymers can be described by the relation

$$X = X_\infty - A / M_n \tag{1.9}$$

where X is polymer property, X_∞ is limit value for infinite long chains, and A is constant. Other polymer properties are divided into two groups:

1. Properties directly related to polymer chemical composition and polydispersity (cohesive forces, density, molecular mobility, phase behavior).
2. Properties are indirectly related to polymer chemical composition and polydispersity (morphology and relaxation phenomena).

Details about the main properties of individual fibrous polymers are presented in the book (Militký et al. 2013). Focus on the sustainability and environmental acceptability of textile structures leads to the use of biopolymers and controlled degradable polymers. Both of these factors are important for medical purposes. One of the basic mechanisms of degradation is hydrolytic degradation. Degradation, where the rate of water penetration is rate-determining and the polymers are first converted to water-soluble products, is referred to as *volume erosion.* During volume erosion, degradation occurs throughout the volume and the rate of water penetration is higher than mass loss. Most polymers used in vivo degrade by this mechanism. The second type of biodegradation, i.e., *surface erosion*, occurs in cases where the rate of penetration of water into the polymer is lower than the rate of disruption of polymer chains to water-soluble products. The result is thinning while maintaining shape (Figure 1.9).

Surface erosion occurs mainly in hydrophobic materials, but the bonds in the main chain are sensitive to hydrolytic action. Examples are polyanhydrides and polyorthoesters. Weight loss is faster here and water penetrates into the polymer structure. The time sequence of changes in polymer properties during in vivo degradation is shown in Figure 1.10.

Degradation in the organism, i.e., in vivo, is mainly evaluated for medical applications. Usually, analogous degradation processes also occur in storage conditions in the soil, compost, or landfills. It is understood that biodegradation must not

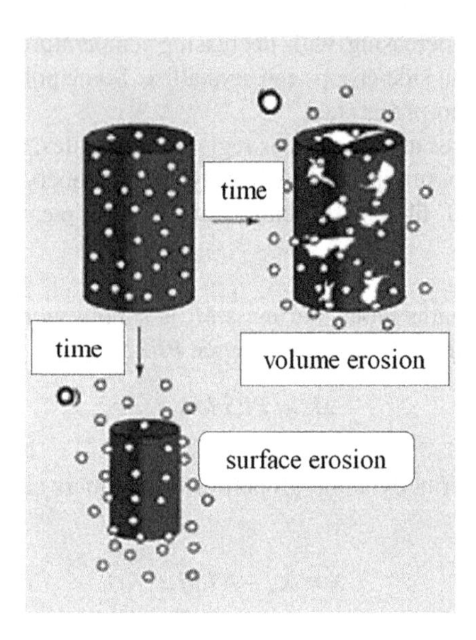

FIGURE 1.9 Basic types of fiber erosion.

significantly affect materials during their use. Only after the end of the cycle of their use should biodegradation begin, leading to their decomposition into simple compounds, which are easily removed, resp. excreted from the organism. The majority of biopolymers are degradable under some environmental condition or in vivo. Biopolymer is a special polymer using products of natural origin in the synthesis (standard ASTM D6866-06). Biopolymers are often extracted directly from biomass (via polysaccharides), which are further modified resp. unmodified. Examples are polymers of modified starch or cellulose. Another group consists of polymers produced directly by microorganisms in their natural or genetically modified state. Examples are Pullulan (1,6-linked maltotriose) and polyhydroxyalkanoates (PHA). Polymers can also be obtained from biointermediates from renewable raw materials. Examples are polylactic acid (PLA); biopolyethylene (BPE) obtained by polymerizing bioethanol; biopolyamides from diacids obtained from biomass and bio-PURs containing polyols from natural sources. Biopolymers are characterized by their low energy consumption required to prepare the raw materials and the polymer. The basic group of biopolymers for medical purposes consists of aliphatic polyesters. Natural sources are glycolic and lactic acid. These acids can also be produced synthetically. The preparation of aliphatic polyesters based on simple aliphatic hydroxy acids was already described by Carothers in 1932. However, their low molecular weight and poor mechanical properties did not allow the production of fibers. The production of high molecular weight aliphatic polyesters based on lactic acid was patented in 1954, but due to the high hydrolytic degradation, the development was not continued. In 1972, fiber-based copolymers of lactic acid and glycolic acid was introduced. This fiber was intended for absorbable bandages. However, the higher price of the polymer prevented wider development. It was not until the late 1980s that a technique for bacterial fermentation of D-glucose derived from corn starch was discovered, which

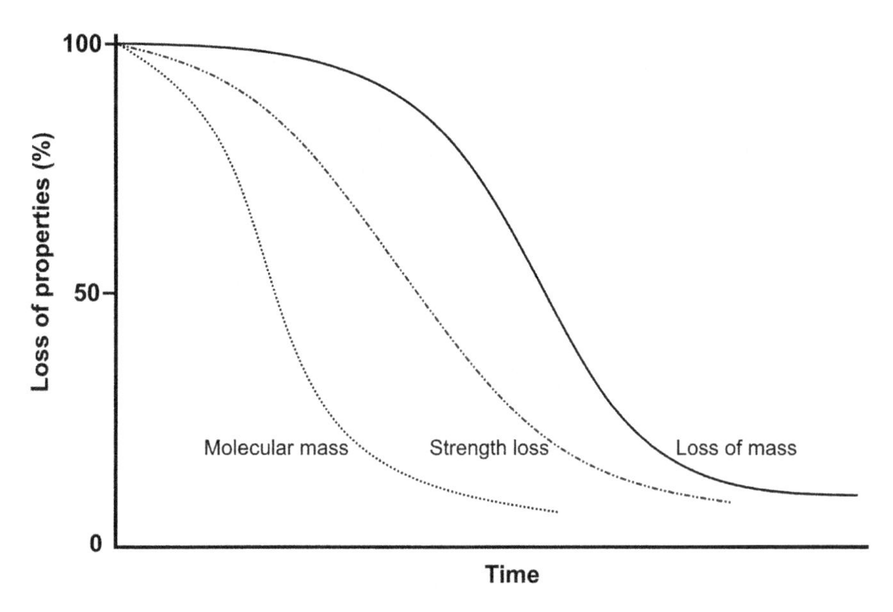

FIGURE 1.10 Time sequence of changes in polymer properties during in vivo degradation (Pietrzak et al. 1997).

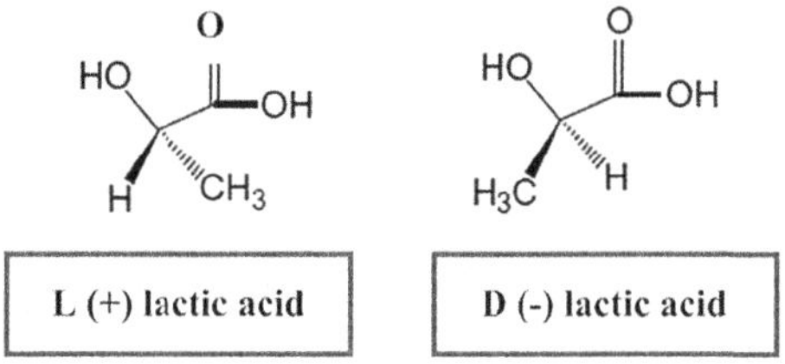

L (+) lactic acid

D (-) lactic acid

FIGURE 1.11 Stereoisomers of lactic acid.

made it possible to produce these biodegradable polymers on a larger scale (Sin and Tueen 2019). During their biodegradation, the polymer chains are first shortened and then decomposed into water and CO_2. Thus, they are biopolymers that are not only biodegradable but also bioabsorbable and biocompatible.

Polyglycolic acid (PGA) has a high crystallinity (46%–52%) and a high melting point. It is relatively hydrophilic. Glycolic acid-based fibers are too brittle and have a higher rate of hydrolytic degradation for many applications. Lactic acid exists in two basic stereoisomers, D and L (Figure 1.11).

These isomers affect both the degree of crystallinity and the melting point. Lactic acid obtained by fermentation is 99.5% only L-form. These are colorless, strongly hygroscopic crystals. The melting point is 18°C and for the mixed form, it is 53°C. The boiling point is 85°C. Poly-(L) lactic acid has a melting point of 178°C. In the presence of the D-form, the melting point drops to 130°C. The glass transition temperature of these polymers is T_g = 60°C. The average molecular weight of PLA is about 300,000. The methyl group of PLA increases hydrophobicity (low water

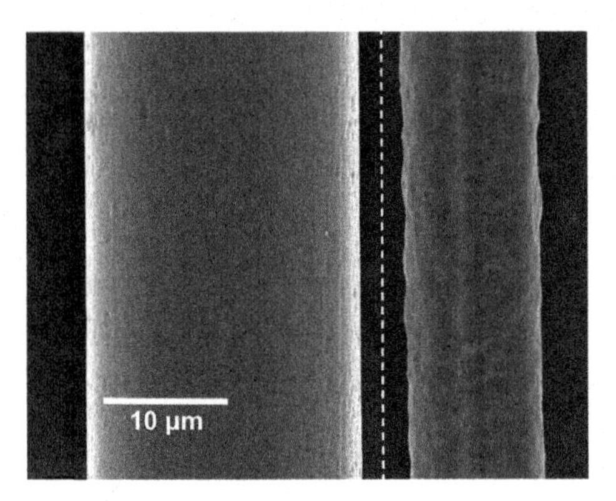

FIGURE 1.12 Alkali etching of PLA (Sun et al., 2009).

sorption), and the rate of hydrolysis is significantly lower than that of PGA. The strength of conventional fibers ranges from 0.5 to 0.8 GPa and the modulus ranges from 7 to 10 GPa. A strength of 2.3 GPa can be achieved. Break deformation is 18%–25% (Yang 2011). The practical use of PLA is in the technical sector mainly as packaging materials, hygiene products, and products for medical purposes (Avinc and Khoddami *2009*).

High molecular weight poly(L-lactic acid) fibers are produced by melt spinning (190°C–240°C) or wet spinning from solution (chloroform/toluene). Subsequently, stretching takes place at higher temperatures (160°C). The fibers are crystalline (degree of crystallinity 60%) and their biological resistance depends on the level of mechanical properties (Drumright et al. 2000). During spinning and drawing, the molecular weight of the fibers is reduced to 1/3. The moisture content of PLA fibers is 0.6% and the specific weight is 1,250 kg.m^{-3}. PLA fibers are resistant to UV radiation and their melting point can be varied within wide limits. Swelling in water is around 2% (Avinc and Khoddami *2009*). At 98% relative humidity above 60°C, PLA fibers are hydrolyzed. Only the hydrolyzed form is very rapidly microbiologically degraded. PLA fibers are nonflammable and have good recovery properties. It is possible to dye them with disperse dyes (Farrington 2015). Fixation of staple PLA fiber yarns at 130°C for 15–60 s does not cause a change in strength and elongation. Pre-cleaning (washing in the water at 60°C for 10 min) reduces the initial tensile modulus. The modulus decreases by approximately 25% during washing. After fixation at 130°C for more than 45 s, the modulus is stabilized. Alkaline action on PLA fibers leads to their refinement. In the article (Sun et al., 2009), the degradation of PLA was monitored depending on the concentration of NaOH (0.25–3 mol/L), time (0–1.5 h), and temperature (25°C–80°C). It was found to be a surface hydrolysis mechanism (Figure 1.12).

The disadvantage of PLA is acidic decomposition products. During degradation, harmful residues are formed in the human body. PLA fibers decompose completely in the soil within 20–30 months and compost within 30–40 days (cellulose fibers in compost already after 15–20 days). Initially, hydrolysis occurs and then microbial degradation (Drumright et al. 2000). PLA fibers were produced in 1994 by Kanebo (Japan) and in 1997 by Cargill Dow Polymers (USA). In practice, copolymers of

PLA with PGA are also used, which are significantly more hydrophilic (Vicryl™ and Polyglactin 910™).

Varying the PLA and PGA proportions in copolymer fibers can be used to control the degradation rate and strength retention of medical implants and adopting their properties according to the requirements of specific medical applications. During the process of degradation, the degraded fibrous implant is replaced by new connective tissues (Sin and Tueen 2019). Bacterial polyesters are products of various types of bacteria (e.g., *Alcaligenes eutrophus*, *Alcaligenes vinelandii*, or *Rhodobacter spheroids*) growing on alcohols, molasses, and glucose. Typical are polyhydroxy butyrate PHB (Biopol) and its copolymer with hydroxy valerate PHV. By adding different aliphatic acids, different copolymers can be obtained. In nature, bacteria accumulate 3%–20% of PHB of their weight. Under laboratory conditions, yields of up to 70%–80% of PHB by weight can be achieved. The high crystallinity of PHB causes their fragility. Thermodegradation occurs near the melting point. The initial modulus is comparable to PP. Tensile deformation till break is 5%, i.e., very low. Microbial polyesters are piezoelectric and optically active. PHA decomposes better in the wet (e.g., seawater) than in the air and soil. Enzymatic degradation must occur first.

1.6.2 Nanomaterials

In recent years, nanotechnologies have shifted significantly from research to industrial applications. It will significantly influence fields like medicine, computer technology, optoelectronics, materials engineering, biotechnology, etc. From a physical point of view (special effects), nanomaterials are considered to have substances that have at least one dimension smaller than 100 nm. A comprehensive description of recent applications of nanotechnology in the textile sector is published in the book (Mishra and Militký 2018). Nanomaterials are usually divided into four groups:

- Materials with one dimension in the order of nanometers, with the other two dimensions being larger (e.g., layered clays, i.e., silica-based materials) with a layer thickness of several nanometers but other dimensions around 1,000 nm.
- Materials with two dimensions (usually cross section) in the order of nanometers, with the remaining dimension being larger (e.g., carbon nanotubes and nanofibers).
- Materials with three dimensions in the nanometer range (e.g., different types of nanoparticles).
- Nanostructured materials, where at least one structural element has a significant dimension in nanometers. These are, for example, aerogels, nanoporous membranes, nanocrystalline structures, and block copolymers.

In many cases, the textile sector was worked on at the nanolevel before this concept was extended. Examples are disperse dyeing, thermal fixation, and several other operations where changes begin at the level of tens and hundreds of nanometers. Many of these effects are indeed reflected at the micro and macro level, and the nanolevel structures have not been changed in a controlled manner. Nanotechnologies are mainly used in the textile field as follows:

- Nanofibers, nanoyarns, and nanowebs are produced mainly by electrostatic spinning technology.
- Nanoparticles with an enhanced intensity of effect in comparison with bigger particles (antimicrobial treatments, abrasion resistance, self-cleaning, lotus effect).
- Nanoporous materials (aerogels).
- Nanocomposite (improvement of mechanical properties and heat resistance of fibers).
- Carbon and other types of nanotubes (improvement of mechanical and electrical properties).

1.6.2.1 Nanoparticles

The behavior of nanoparticles depends critically on its size. The 100 nm particles contain about a million atoms. Particles with a radius of 1 nm already contain only about 25 atoms. The monomolecular layer has a thickness of about 2–3 nm. It can be easily deduced that the number N of smaller spherical particles of diameter d formed by dividing one spherical particle of diameter D is equal to $N = (D/d)^3$. The number of smaller particles of the same total volume (mass) is thus significantly higher (see Figure 1.13), and there is a reduction in the size of free spaces in a filling volume of the same size.

A common mistake is the assumption that nanoparticles have better mechanical properties in comparison with bigger particles. The energy of all atoms binding in particle (cohesion energy) per atom (diameter d) is dependent on the diameter of particles D.

$$E = E_b\left(1 - d / D\right) = E_b\left(1 - 1 / L\right) \tag{1.10}$$

where E_b is cohesive energy for bulk material and L is the particle to diameter ratio. For nanoparticles, it is ratio $L = D/d$ from 10 till 100 and cohesive energy is increasing with particle diameter. Interestingly, starting from the diameter of nanoparticles above 30 nm,

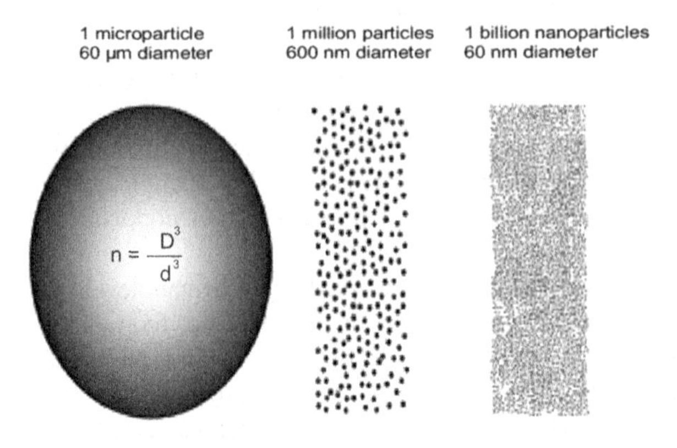

FIGURE 1.13 Influence of particle size on covering of space.

E is practically constant. This is one natural definition of nanorange. Gravitational and electromagnetic forces exist between particles. Gravitational force is a function of particle mass, and inter particle distance is very weak between nanoparticles. The electromagnetic force is a function of particle charge and inter particle's distance is not affected by particle mass, and it can be very strong even when we have nanoparticles. The nanoparticle liquid dispersions are very stable due to very small settling velocity v_s which is proportional to the square of particles diameter $v_s \sim d^2$. Suppose we have spherical particles of volume V_e and diameter d_p dispersed evenly in prescribed volume V_a. Then the volume fraction of particles ϕ (proportional to their concentration) and the number of particles N are given by relations (Mishra and Militký 2018).

$$\phi = \frac{\text{particles volume} = \text{N}\,\text{V}_e}{\text{total volume}\,\text{V}_a} \quad \text{N} = \frac{6\,\phi\,\text{V}_a}{\pi\,\text{d}_p^3} \tag{1.11}$$

In Figure 1.14 the growth of the number of particles with a given volume concentration in the volume of 50,000 μm³ depending on their diameter is shown.

A high increase is evident for particle diameters below 0.3 μm. This indicates a high occupancy of the entire available volume and very small inter-particle distances. Therefore, it can be expected that significantly lower concentrations will be sufficient to achieve the same effects. The size of the nanopart generally produces interesting effects:

- Extremely large specific surface contact area per particle mass is associated with the fact that most atoms are on the surface (Figure 1.15).
- Dimensional similarity with the length of UV and visible radiation.

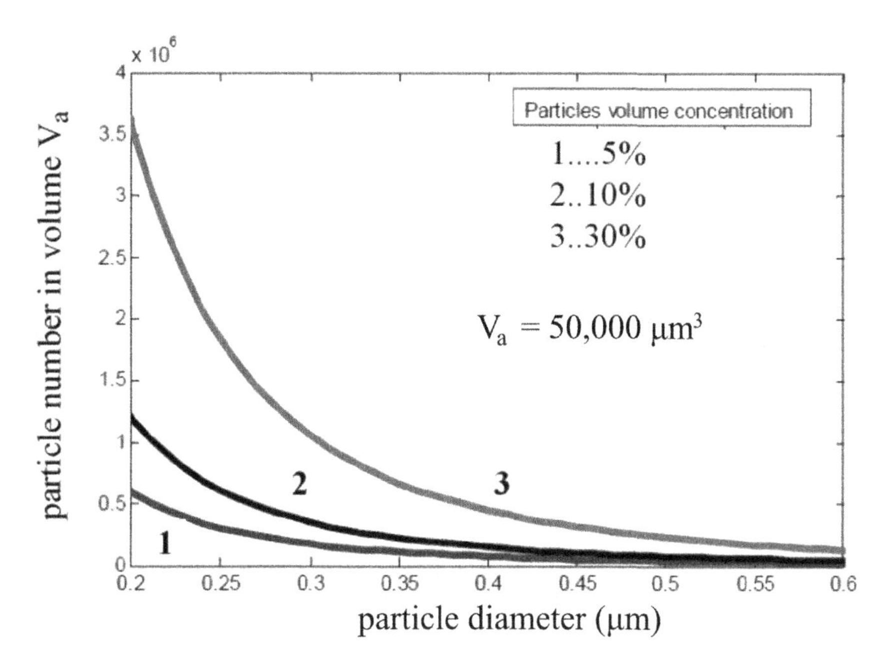

FIGURE 1.14 Dependence of the number of particles in given volume V_a on their diameter.

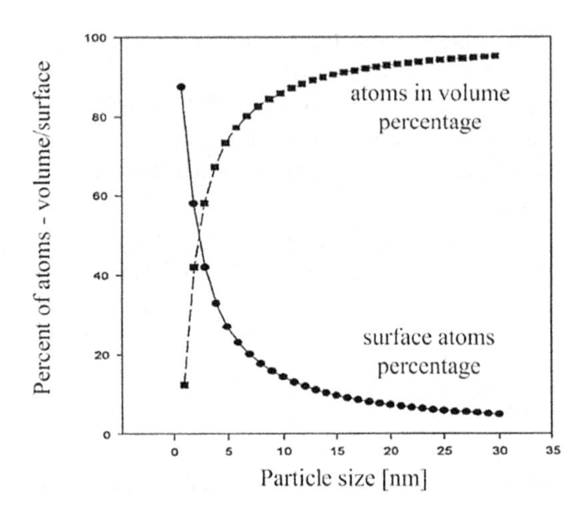

FIGURE 1.15 Percentage of atoms on surface and volume for different particle sizes.

- Radiation absorption, scattering, and color then depend on the particle size.
- Critical lengths (mean free length of motion) for transport properties such as diffusion are comparable to or higher than the nanoparticle size.
- Particle toxicity increases with decreasing particle size.

Also, particles larger than 100 nm still have a relatively high relative surface area per their mass and are used to enhance various effects. Such particles are commonly referred to as submicrons.

The disadvantage of all particles smaller than 1 μm is their relatively high price compared to coarser particles (approximately, the price of 10 nm particles is four times over the price of 1 mm particles). Nanoparticles have several size-dependent properties:

- Lower melting point and lower sintering temperature.
- A shift of the optical spectrum to small wavelengths.
- Larger bandwidth for semiconductors.
- Change in magnetic properties (magnetic moment).
- Higher catalytic activity.
- Weaker cohesion.
- Higher surface energy and reactivity.
- Higher electric and thermal conductivity.
- Higher bioactivity.
- Higher tendency to aggregation and degradation.

Nanoparticles are commonly applied from dispersions in a suitable medium (usually liquid). A key problem is an appropriate stabilization to prevent aggregation (Mishra and Militký 2018).

Electrostatic stabilization involves Coulomb repulsion between particles caused by an electrical bilayer formed by ions adsorbed on the surface of the particles (e.g., sodium citrate) and corresponding counterions. An example is a gold nanopowder prepared by reduction of [$AuCl^{4-}$] with sodium citrate.

Spherical stabilization is ensured through arranged organic molecules forming a protective layer on the surface of the particles. The nanoparticles themselves are separated from each other and their agglomeration is prevented. The basic types of the protective layer forming molecules are polymers and copolymers, phosphines, amines, thioethers, solvents, higher aliphatic alcohols, surfactants, and organometallics.

Nanoparticles offer several new or improved options: electrical conductivity, antistatic properties, surface structure (hydrophobic/hydrophilic), UV protection, gas penetration, electromagnetic smog, combustion, microbes, improved dyeability, abrasion resistance, and mechanical properties. It is known that the wettability of surfaces is affected by free surface energy and surface roughness (i.e., lotus effect). The surface free energy can be controlled by chemical modification, for example, by a coating of agents containing fluoropolymers. Nanoparticles enable the preparation of surfaces, which in combination with chemical modification, can ensure superhydrophobic behavior (Ramaratnam et al. 2008). Note that surfaces with a water contact angle greater than 150° are referred to as superhydrophobic. On smooth surfaces, a maximum water contact angle of 120° for PTFE can be achieved. Interestingly, the superhydrophobicity of surfaces is mainly influenced by the size and shape of particles, which allows the use of nanoparticles, for example, from waste materials (fly ash, etc.). An overview of current approaches to the formation of superhydrophobic surfaces (nanoparticles, controlled destruction, etching, phase separation, crystal growth, particle aggregation, nanolithography, pattern embossing, etc.) is given in (Roach et al. 2008). Also, the layer of hydrophobic nanofibers on the surface of textile substrates causes an increase in hydrophobicity (Ma et al. 2005). At the same time, these layers are more resistant to oils, dirt, and dust. They are also in the category of self-cleaning surfaces, where dust particles are easily removed by packing them into drops of water. One of the most preferred polymers for this purpose is trifluoroethoxy polyphosphazene (Singh et al. 2005). Nanoparticles are often advantageously used as fillers in polymers and coatings, where they show several special manifestations, such as:

- Interactions between particles arising at low concentrations (<0.1%).
- Low percolation threshold (1%).
- A high number of particles (10^{30} in 1 cm^3).
- High interfacial surface in contact with the polymer matrix.
- Small distances between particles.
- Optical transmittance.
- Special dielectric properties.
- Enhanced electric conductivity.
- Enhanced biological activity.
- Photocatalytic activity (TiO_2, MgO).
- UV blocking.

In the future, intensive research related to particle size optimization can be expected concerning their effects, cost, and possible toxicity. These particles will be partly prepared from industrial wastes (waste glass, waste from the production of carbon fibers and composites, fly ash, fiber wastes) and partly obtained by reducing the dimensions of common particles intended for special treatments. Particles of optimal sizes will be suitably combined with textile means for special treatments to use synergistic effects. Activation of particles will be solved mainly by controlled physicochemical degradation and surface modification. The research will focus on the possibilities of anchoring particles and their combinations in coatings and the use of dyeing techniques with the possibility of subsequent stabilization.

1.6.2.2 Nanofibers

A special class of nanomaterials consists of *nanofibers*. Usually, there are layers of entangled submicron fibers. Electrospinning technology is mostly used for its production. It is a technology known for more than 100 years. The advantage is the possibility of preparing an entangled assembly from segments of fibers with a diameter of 5–500 nm with *very low porosity*. In a standard arrangement, the polymer is fed through a metal needle metering device about 1 mm in diameter at a rate of 0.5–10 mL/h. There is a high voltage field of 5–65 kV between this device and the collector plate with opposite charge. In this field, the polymer droplets originally form the so-called Taylor cones and then a continuous liquid jet which solidifies and forms fibers during passage through the gap between the electrodes (10–30 cm). A liquid jet narrows and branches and then solidifies (Figure 1.16).

In reality, there are chaotic movements of the polymer jet caused by repulsive forces between charged particles. The placement of the fibers on the collecting plate is chaotic, which is advantageous for some applications (filters, membranes). On the other hand, it is a discontinuous production with the uneven placement of the fibrous assembly. In the case of low viscosity of the polymer solution (low molecular weight) and its high surface tension, the liquid jet is divided into smaller droplets (electrospraying). At higher viscosities (high molecular weight), the droplet stretches (Taylor cone formation), and a fibrous structure is formed. More than 50 different polymers (in many cases with different particles) have already been used for electrospinning (Mishra and Militký 2018). Water and organic solvents are standardly used to prepare polymer solutions. It is also possible to use polymer melts (in the air or an inert

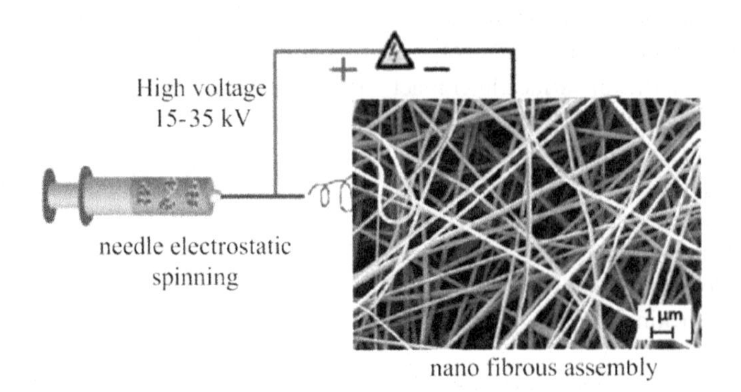

FIGURE 1.16 Classical needle electrospinning.

atmosphere). Additional processing (crosslinking, thermal stabilization, and mechanical reinforcement) is also advantageous. Today, injectable dosing devices are grouped into frames with computer-controlled movements in a direction perpendicular to the direction of the machine axis, which ensures uniform deposition of the fibrous web in an endless belt. Another possibility is replacing the classic needle-type nozzles with other principles (Li et al. 2004). It is also possible to purposefully change the arrangement of the collecting electrode and thus obtain various forms of nanofiber entanglement. There are also electrospinning systems using a combination of centrifugal force in a container and an electrostatic field and other principles (Song et al. 2020). The industrially realized NANOSPIDER device based on the patent of Jirsák et al. from the Faculty of Textiles of the Technical University of Liberec uses a rotating roller and modification for a smaller amount of spinning solution (Figure 1.17).

Controlled deposition of the fibrous web can be realized, for example, by changing the direction of the electric field between two parallel plates, wing electrodes causing the fibers to converge on the collector, and a rotating cylindrical collector enabling the formation of a multifilament. Procedures using special collectors are described (grid, frame, disk, divided frame) or combination with other principles (auxiliary air blowing, spinning on the surface of the liquid, auxiliary electric fields, mechanical vibrations), which allow controlled deposition or tuning web orientation. Noninjection techniques also include the use of magnetic fluids and direct spinning from the molten film. Methods for forming a hybrid yarn using air vortex to secure a nanofiber surface layer to staple yarn are described as well (USP 6106913 of 2000). The use of these yarns is expected for the construction of wrapped structures, fabrics, and knitwear. It is therefore clear that there are already several possibilities for the production of textiles containing nanofibers. An overview of methods for the production of nanofibers on an industrial scale is available (Yamashita 2008). There are also several methods for the production of nanofibers, such as special drawing (USP 2578899 of 1949), matrix synthesis, spontaneous aggregation, direct spinning, phase separation, and crystal separation. For the time being, these methods are used rather marginally (Mishra and Militký 2018). A number of both standard and special polymers in the form of solutions or melts are used for electrospinning. It is also possible to spin the precursors and form the final polymer only after processing. Thus, for example, ceramic, metal, and carbon nanofibers can be prepared. It is also possible to use polymer mixtures and nanocomposites (Ramakrishna et al. 2005). An interesting possibility is the use of ionic liquids as solvents for electrostatic spinning

FIGURE 1.17 NANOSPIDER System – Rotating roller electrode, single wire electrode, and a rotating wire electrode (from left to right) (Venkataraman et al. 2018).

of biopolymers (dissolving, e.g., cellulose), which are nonvolatile (Meli et al. 2010). By default, polymers with a sufficiently large chain length and concentration are required for spinning purposes so that electrostatic field chain arrangement can be used. However, it has also been possible to use for electrospinning low-molecular compounds that form overlapping cylindrical micelles. One of these substances is cyclodextrins (specifically methyl-β-cyclodextrin), which are capable of self-assembly and the formation of hydrogen-bonded aggregates (Celebioglu and Uyar 2011). Very interesting is the use of the cyclodextrin triclosan complex (triclosan is one of the most effective antibacterial agents in the past, today its use is forbidden) in a ratio of 1:1, when relatively high-quality nanofibrous assemblies are formed.

Nanofibrous assemblies are also useful for obtaining very special effects. An example is a structure called Janus (after the god Janus with two faces), which is hydrophilic on the one side and hydrophobic on the other side. It is produced by the electrostatic spinning of PAN and tetraethyl orthosilicate in dimethylformamide on aluminum foil. This is followed by heat treatment (in the air at 200°C), which causes surface hydrolysis (COOH groups are formed). This layer has practically zero water contact angle. Another layer is applied to this structure by electrospinning the same solution again. It is superhydrophobic with a water contact angle of 151.2° (Lim et al. 2010). The major limitations of nanofibrous assemblies till yet are their high sensitivity to mechanical field action (especially abrasion), instability in maintenance processes, and limited durability in their use. It can therefore be expected to apply where these disadvantages do not manifest themselves, for example, in the field of tissue engineering. The use of nanofiber webs for controlled drug delivery and sensor design is also interesting. Thanks to the low dimensions and specific surface area (nanoscale), the response of the sensors is greater and the sensitivity threshold is lower. Some promising results are already known today when using special nanofibers to create composite structures.

1.6.3 Aerogels

A very interesting class of materials with a nanoporous structure are aerogels, which were discovered in the years 1929–1930 by Kistler. Aerogels are highly porous materials with very low solid content (usually below 10%) formed by removing liquid from silica gels while maintaining their structure (Venkataraman et al. 2016). Supercritical drying techniques are used as standard. These materials have extremely low thermal conductivity, making them suitable as heat insulators. The appearance of aerogel and its major exceptional characteristics are shown in Figure 1.18.

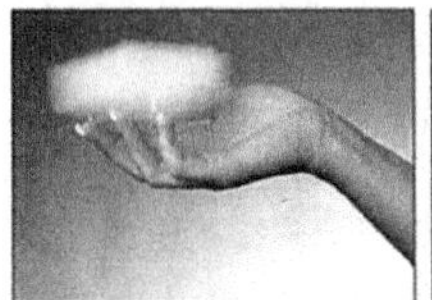

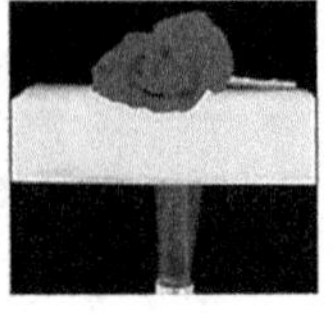

low density   low index of refraction   low thermal conductivity

FIGURE 1.18 Various features of aerogel.

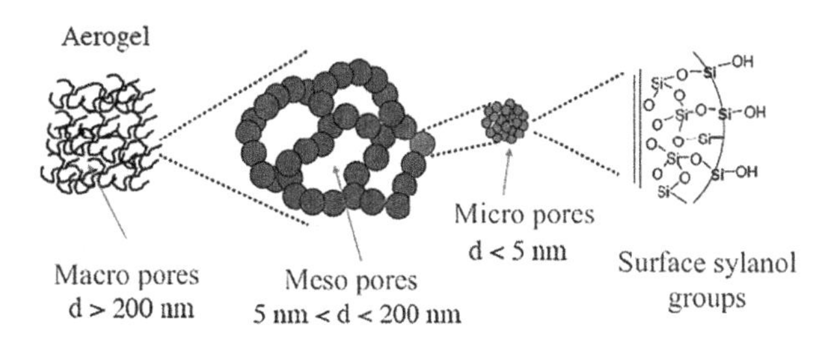

FIGURE 1.19 Aerogel porous structure (Venkataraman et al. 2016).

The practical expansion of aerogels was hindered by the high price and risky production technology (Venkataraman et al. 2016). In 2003, Cabot carried out the industrial production of a silicon aerogel called Nanogel®. The name is somewhat misleading because the classic nanogel is composed of nanoparticles placed in a gel. This material is supplied in the form of particles with a size of 5 μm to 4 mm. Its thermal conductivity is only 9–12 mW/(m K), the porosity is around 95% and the pore size is 20–40 nm. The specific heat capacity is 0.7–1.15 kJ/(kg K). The relative surface area is extremely high, around 750 m^2/g, and the density is extremely low, around 30–100 kg/m^3 (http://www.cabot-corp.com/aerogel). The typical hierarchical porous structure of aerogel responsible for extreme insulation is shown in Figure 1.19. It can be expected that aerogels especially in the form of coating additives can be used for many applications in the textile field. Also interesting is the high oil absorption capacity (due to the hydrophobic nature of the silica aerogel), which could be used, for example, for cleaning oil spots or as part of special filters.

1.6.4 Shape Memory Materials

Shape memory is the ability to "remember" a shape, i.e., to return to its original shape, after strong deformation when heated above a certain transition temperature. If shape memory is manifested during heating, there are materials with unidirectional shape memory. If the shape changes in a targeted manner during cooling, these are materials with bidirectional shape memory (Santhosh et al. 2018). Classic textiles also have a partial shape memory. Shape memory materials are characterized by being stable in several temperature ranges (states). In the individual states separated by the transition temperature T_R, they can have different shapes. Below T_R, these materials are easily deformable. At T_R, the material becomes rigid and a force is released to return it to its original state. The T_R varies according to the material composition. The transition temperature suitable for applications in protective textiles is T_R around 45°C. It has been found that good protection against a heat flux of 15 kW/m^2, which is a dangerous dose for humans, can be achieved by using these materials. The best-known shape memory material is NITINOL, i.e., an alloy of nickel and titanium, where this effect is caused by a phase transformation in the solid phase. The principle of NITINOL operation is the rearrangement of the atomic lattice due to the effect of temperature (Ziolkowski 2015). At the Tas temperature, the formation of an austenitic structure, which is cubic, begins to form. This formation is terminated at the temperature Tak (Tak > Tas). At lower temperatures, the formation of a martensitic structure begins from the temperature Tma, which is characterized by a strong shift of the lattice atoms (compared to the cubic lattice). This formation is terminated at Tmk (Tmk < Tas). Martensite has a stiffness (initial tensile modulus) of 28 GPa and austenite 75 GPa. The differences in the mechanical behavior of the two structures are evident from Figure 1.20.

It can be seen from Figure 1.12 that when the martensitic structure (temperature Tmk) is stretched due to the mass H, deformation to the value of eM occurs. If the material is heated to a temperature, it will shorten to deformation eA. Thus, cooling and heating reversibly change the length of the loaded material. During the periodic alternation of cooling and heating, the material then oscillates. NITINOL has a melting point of around 1,300°C and a density of 6,450 kg/m^3. The electrical resistance of the austenitic phase is about 1 μΩ m, and the martensitic structure (phase) is about 0.70 μΩ m. The thermal conductivity of the austenitic structure is 18 W/(m K) and the martensitic structure is 8.5 W/(m K). The phase transition temperature varies from −200°C to 110°C, depending on the composition. The deformation caused by the transformation is 8.4%. NITINOL has excellent biocompatibility and corrosion resistance. For these reasons, it is used primarily in medicine from surgical instruments to stable implants, including those placed in the bloodstream. The use under the name NITEK for anchoring tendons to bones and Simon NITINOL filter for capturing blood clots are successful. In addition to being used in the form of a straight wire, NITINOL is also used as a helical spring. Depending on the heat treatment method, different effects can be achieved. Temperatures of at least 400°C and a time of about 1–2 min can be used to stabilize the shape of the spring, although it is better to choose 500°C and more than 5 min. It is advantageous to choose fast cooling. As a standard, the spring is fixed in shape during heat treatment. Another possibility is to wind the spring a little thicker than the final size and then heat it without limiting the shape. Thanks to the shape memory effect, the diameter of the spring changes by up to 25% during heat treatment. Copper and zinc alloys are capable of double-sided transitions and can therefore reversibly change their shape according to the ambient

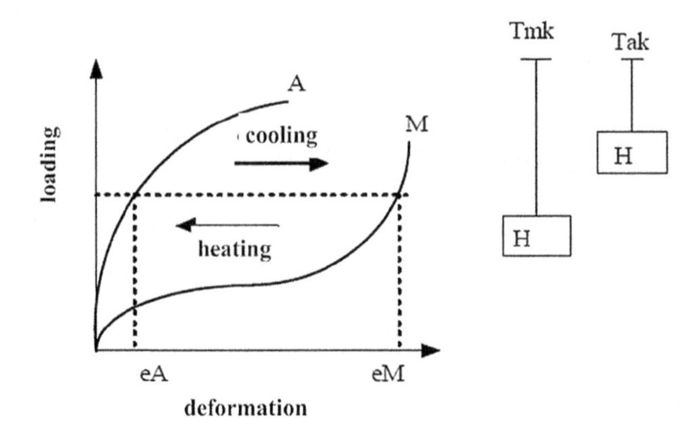

FIGURE 1.20 Changes in NITILON length due to the heating to temperature Tak (deformation eA), resp. cooling to temperature Tmk (deformation eA).

temperature. Usually, these materials are used in the form of a flat spring, which expands, respectively shrinks, according to ambient temperature. When inserting this spring between two textile layers, insulation against extreme temperatures can be achieved due to the widening of the air gap between the layers. Shape memory is implemented in three phases:

- Setting – the material in the required state is heat-treated (500°C–540°C; 2–5 min) and the shape is fixed using an austenitic structure.
- Deformation – use under conditions of martensitic structure (deformable), i.e., at normal temperature.
- Recovery – return to the original (required) state by heating to the transformation temperature (into the austenitic structure).

An interesting manifestation of shape memory materials is super-elasticity, i.e., 100% elastic recovery after tensile deformation but accompanied by hysteresis due to temperature changes (Ziolkowski 2015). A typical manifestation of the super-elasticity of shape memory materials is shown in Figure. 1.21.

Shape memory polymers are an interesting alternative to alloys because they have a low cost, are easy to process, and have a significant ability to return to their original shape. Above the glass transition temperature T_g, these materials are in a rubbery state, so that they can be easily deformed without significant relaxation if the time of action of the external stress is shorter than the time of relaxation. If this material is cooled below T_g, the deformation of the material is stabilized (Parameswaranpillai et al. 2020).

The material returns to its original (undeformed shape) simply by heating to a temperature above T_g. Thus, for significant shape memory, a rapid transition from the glassy to the rubbery state, and a large difference between the modulus in the glassy and rubbery state are necessary. Shape memory polymers usually consist of two phases. One is thermally reversible, which maintains the transition state, and the other is a rigid structure allowing a return to the original state. Crystals, entanglements, or crosslinks are used as the rigid structure. The thermal return phase is selected so that there is a significant decrease in modulus when the temperature rises above T_R.

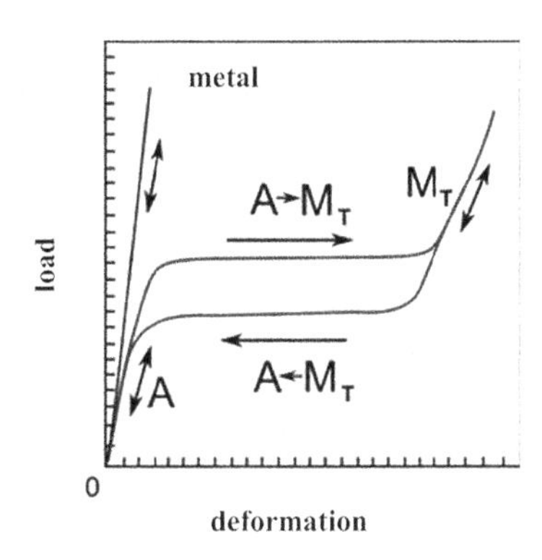

FIGURE 1.21 Effect of superelasticity.

The temperature T_R here is usually either the melting point of the crystalline segments or the glass transition temperature T_g of the amorphous segments. It is obvious that, unlike alloys, the elasticity decreases with increasing temperature. The glass transition temperature is directly proportional to the weight fraction of phases. For shape memory polymers, the time constant of relaxation and retardation processes also plays an important role, which usually differs properly at temperatures below and above T_R (Moukman 2000). Segmented PURs, polyester ethers, polynorbornylenes, styrene–butadiene copolymers, etc., are used as shape memory polymers. Segmented PURs are particularly preferred because they have a large deformation recovery (more than 95%) and easy influence of the softening temperature (T_g or crystallization temperature T_c) in the range from −30°C to 70°C (Parameswaranpillai et al. 2020). These polymers are commonly used as coatings.

1.6.5 Thermally Adaptive Textiles

The human body is very sensitive to temperature changes. The temperature of human skin on the body is 35°C and on the head 34.4°C. The temperature of the hands is approximately 31.6°C and the temperature of the feet is 30.8°C. The average temperature of human skin that is considered comfortable is 33.3°C. If this temperature drops to 31°C it occurs feeling cold and at 29°C usually the body becomes hypothermic. On the other hand, an increase in human skin temperature to 35.5°C causes excessive sweating, and when it rises to 40°C, the human cells die. To some extent, the human body is capable of self-regulation. When cold, the capillaries contract, which limits blood flow and thus heat transfer (Parsons 2014).

Thermoregulatory textiles must therefore be able to either absorb or release heat, depending on the state of the environment and the human body. As the temperature rises, even ordinary fibers, or fabric absorbed heat and this heat is released during cooling. However, the actual effect for the preparation of thermally adaptive textiles is negligible (heat absorption is about 1 J/g of fabric at a temperature change of 1°C). The study of thermally adaptive fibers and textiles began in the 1980s (Vigo and Frost 1989). Either cavities in hollow fibers or the surface of cotton/PES fabrics were used as carriers. Polyethylene glycol (PEG) was chosen as the heat absorber. Currently, two thermal energy storage options are used:

a. *Thermosensitive materials* – these materials absorb heat during heating and release it during cooling. An example is water and ceramics. Water is usable from 1°C to 99°C. When the water temperature is increased by 1°C, the heat of 4.18 J/g is absorbed. An example of the use of water for temperature control is NASA spacesuits containing a system of forced circulation tubes filled with water.
b. *Phase change materials (PCM)* – here, the fact is used that during the melting of materials it is necessary to supply latent heat of melting, and during solidification, the heat is released (often as crystallization heat).

The latent heat of PCM fusion is significantly higher than the heat absorbed by heating, and therefore PCMs are preferably used to control the temperature of garment fabrics. With regard to thermal comfort 33.4°C ± 4.5°C, materials are used where melting occurs in the temperature range of 20°C–40°C and solidification (crystallization) in the temperature range of 30°C–10°C. Among the basic types of PCM belongs (Mehling and Cabeza 2008):

1.6.5.1 Hydrated Inorganic Salts

These salts (containing bound water molecules) were historically the first PCM for textile applications. Among the most effective is lithium nitrate $LiNO_3.3H_2O$, which has a melting point of 30°C and a melting heat of 296 J/g. A certain disadvantage is that the effect of storing thermal energy is not permanent (it lasts only 25 heating/cooling cycles).

1.6.5.2 Polyethylene Glycol

PEG is one of the most widely used PCM. Melting and crystallization temperatures of PEG are strongly influenced by molecular weight. For PEG molecular weight 600, the melting point is 18°C, the melting heat is 121 J/g, the crystallization temperature is 6.9°C and the crystallization heat is 116 J/g. For PEG molecular weights of 1,000, the melting point is 35°C, the melting heat is 137 J/g, the crystallization temperature is 12.8°C and the crystallization heat is 134.5 J/g. PEG with a molecular weight of 800–1,500 is usually used, when the melting point is over 33°C. At these molecular weights, PEG is soluble in water. One possibility of using PEG as a PCM is a resin treatment based on dimethylol ethylene urea combined with PEG crosslinking. Approximately 50% PEG 600 solution and conventional treatment (e.g., noncreasing) are used. However, the heat capacity for polyester fabrics is only 54 J/g, which is slightly more than the heat capacity of untreated polyester (approximately 40 J/g). Bacteriostatic properties are also an advantage of textiles treated in this way.

1.6.5.3 Block Copolymers

PEG/polyester block copolymers are used for the production of elastomers or fibers with improved dyeability (dyeable with disperse dyes at boiling without a carrier). If the molecular weight of PEG is 4,000 and its content in the fiber is above 50%, the PEG segments crystallize independently, their melting point is 33°C and the melting enthalpy is 30.6 J/g.

1.6.5.4 Higher Hydrocarbons

Depending on the number of carbon atoms Nc in hydrocarbons, the melting point, and crystallization change. For $Nc = 17$ (*n*-heptadecane), the melting point is 21.7°C, the melting heat is 171.4 J/g, and the crystallization temperature is 21.5°C. For $Nc = 20$ (eicosane), the melting point is 36.6°C, the melting heat is 246.6 J/g, and the crystallization temperature is 30.6°C. The advantage of these hydrocarbons is their low cost, nontoxicity, and the possibility of an easy combination of different hydrocarbons (adjustment of melting and crystallization temperatures). For fixing PCM to textiles, hollow fibers (usually viscose), or surface coating in combination with resin treatment is commonly used. Recently, there has been a shift to microcapsule techniques that reduce PCM leakage during fabric use and maintenance. It is a special encapsulation technique in which PCM is stored in a capsule with a diameter of several micrometers. These 1–10 μm microcapsules containing PCM can be dispersed into solution-spun polyacrylonitrile fibers (active substance content is 6–10%). An example is the Outlast fiber (see Figure 1.22). Another option is to incorporate microcapsules into PUR foam, resp. application to the surface of textile structures. Frisky supplies TERMABSORB microcapsules in sizes ranging from units to hundreds of micrometers (typically 15–40 μm), where the core containing PCM is coated with an impermeable membrane less than 1 μm thick.

Various PCMs are available with a phase change temperature from 6°C to 100°C with a latent heat capacity from 160 to 200 J/g. The possibilities of preparing PCM-containing blended fibers were also investigated. The problem here is the very low viscosity of all PCMs at spinning temperatures so that thickeners must be used. In this way, it is possible to produce fibers with a core containing PEG and a polypropylene sheath. The disadvantage of PCM materials is that after a certain time interval (real-time of action is of the order of 6–10 min) their storage capacity, resp. heat release is exhausted. It is therefore particularly suitable for conditions of rapidly changing temperatures. For real textiles, the maximum heat content is around 50 J/g. It is therefore advantageous to combine PCM with a suitable fabric construction containing closed air pores. One way to obtain thermal energy is to take advantage of the fact that metal carbides of the VI transition group convert near-infrared, light, and UV radiation into heat (Park and Jayaraman 2001). Zirconium carbide particles ZrC added to the surface by resin coating are commonly used. ZrC reflects electromagnetic waves of lengths greater than 2 μm. However, it absorbs at wavelengths below

FIGURE 1.22 Schematic cross section of Outlast fiber.

2 μm (especially in the light and UV range) and converts light into heat (increases the wavelength). It is also interesting to note that some copolymers of polyacrylonitrile and acrylic acid absorb moisture and release heat. Thus, lamination of textiles from such polymers with a thermoregulatory layer of PCM and ZrC for clothing intended, for example, for winter sports is offered. For active thermal insulation, ZrC microparticles are dispersed into the core of polyamide, resp. polyester fibers (an example is SOLAR α fiber). Thermocath W fiber – containing this carbide in a PAN matrix with a fineness of 3.3 dtex, is a photo, thermo, electro, and conductive material. Even if direct sunlight is limited by clouds, the temperature of the fabric increases by 2°C–8°C. MASONIC M (Kanebo) is a PA 6.6 filament with ceramic microparticles, and Lovwave is a hollow polyester fiber blended with ceramic microparticles (Park and Jayaraman 2001). Even more advantageous are the so-called thermo regulating textiles, where the electrical resistance changes due to changes in the volume of fibers caused by changes in ambient temperatures. The fibers contain graphite particles. Heating enlarges fibers, i.e., the interparticle distances grow, and the electrical resistance increases. During cooling, the interparticle distances are shorter due to macroshrinkage, and the electrical resistance decreases. An example is the GORIX fabric made of carbonized PAN fibers. The carbonization takes place in an inert atmosphere at 1,000°C. The difference in the fineness of the warp and weft is also important. This fabric has the advantage over PCM in that the temperature of fabric maintains the same at a given ambient temperature (Park and Jayaraman 2001). The systems of active temperature control employing a garment containing tubes filled with water, cavities with the possibility of regulating changes in thickness employing the amount of air contained or regulating changes in thickness using the "shape memory" effect are applied as well.

1.6.6 Materials Sensitive to External Stimuli

Stimulus-sensitive materials (SSMs) can change some characteristics depending on changes in environmental conditions (temperature, pH, mechanical pressure, light, or electrical energy). These are usually gel materials capable of resizing shape or other characteristics. A gel is a cohesive mass containing a liquid in which the particles are either dispersed or arranged in a network covering the entire volume. The gel can be sufficiently elastic and jelly (gelatin) or solid and rigid (silica gel). Polymer gels contain a crosslinked polymer network swollen with a suitable liquid (Fernandez-Nieves et al. 2011). They can reversibly swell and precipitate (up to 1,000 times the volume) due to small changes in ambient conditions (pH, temperature, electric field) (Smeets and Hoare 2013). Gel fibers change shape in milliseconds, but thick polymer layers take minutes to hours. They can transmit relatively high voltages. The 3D network created by crosslinking polymer chains can swell strongly in liquids. However, there is no dissolution and the gel retains its shape and integrity. The gel's response to the outer subject is to either swell or collapse. The reason may be hydrogen cation transfer, ion exchange, redox reactions, phase changes, electro-kinetic processes, etc. A special group consists of so-called hydrogels, which form a water-swollen 3D soft elastic network with a controllable arrangement. The water content of the hydrogels varies widely (greater than 20% to 1,000%). There are several polymeric hydrogels from various sources:

Natural polymers

- Anionic: hyaluronic acid, alginic acid, pectins, and dextran sulfate
- Cationic: chitosan
- Amphipathic: collagen (gelatin), carboxymethyl chitin, and fibrin
- Neutral: dextran, agarose, and pullulan.

Synthetic polymers

- Aliphatic polyesters: a combination of PLA, polyglycolic acids (PEG), and polycaprolactone (PCL).
- Other polymers: PEG-bis-(PLA-acrylates), poly(*N*-isopropylacrylamide) – PAAm.

Combination of polymers

- (PEG-co-peptides), collagen–acrylate, alginate–acrylate, and block copolymers composed of polyoxyethylene and polyoxypropylene.

The basic advantages of hydrogels include an aqueous medium that is biologically advantageous (peptides, proteins, DNA), good transport properties, easy modification, and biocompatibility (Das et al. 2020). The disadvantages are handling problems, lower mechanical resistance, difficulty filling with active substances, and difficult sterilization. Gel eye lenses are an example of successful hydrogel application.

Physical gels are generally formed mainly by the following methods:

- Heating of polymer solution (polyethylene oxide-PEO based block copolymers)
- Cooling of the polymer solution (agarose or gelation)
- Freezing – melting cycles forming polymeric microcrystals (polyvinyl alcohol – PVA)
- Lowering pH and formation of H-bridges (PEO and poly(acrylic acid) – PAAc)

Chemical gels are formed by the following procedures:

- Crosslinking of polymers in solid-phase or solution (PEO irradiation)
- Chemical crosslinking (collagen with glutaraldehyde)
- Multifunctional reactive substances (PEG and diisocyanate)
- Copolymerization of monomer and crosslinker in solution
- Copolymerization of monomer and multifunctional macromer (PLA-PEO-PLA and photosensitizer and light irradiation)

- Polymerization of a monomer with solid polymers for mutually permeable gels.

Many polymer gels can respond to external stimuli (physical or chemical) by either isotropic or anisotropic reversible swelling (Fernandez-Nieves et al. 2011). Chemical stimuli reversible swelling occurs, for example:

- When replacing the bad solvent (low ionic strength), with a good solvent (high ionic strength). An example is a gel based on polyvinyl alcohol.
- When the pH changes, for example, in the alkaline range, gels based on the acrylic acid swell, and in the acidic range, they shrink.
- When the solubility changes depending on the temperature. Polyvinyl methyl ether-based gels swell very strongly at low temperatures.

After exceeding the critical temperature T_c (30°C–40°C) there is a significant shrinkage. In all these cases, isotropic swelling occurs. In the case of physical stimuli, reversible swelling occurs:

a. *Due to irradiation in the UV range.* An example is PVC containing spyrobenzopyran.
b. *Due to the electric field.* Examples are polyelectrolyte gels based on polyacrylic acid and nonionic polymer gels based on polyvinyl alcohol crosslinked with glutaraldehyde with dimethyl sulfoxide (solvent).
c. *Under the influence of a magnetic field.* Here, a superparamagnetic liquid (containing, e.g., iron) is added to the gels.

Due to these physical stimuli, anisotropic reversible deformation occurs depending on the direction of radiation or field. Thus, these gels induce a kind of motion according to the polarity of the electric field and its direction. The polyvinyl alcohol/dimethyl sulfoxide system in particular has excellent properties, with a 7% elongation in the direction of the applied electric field. The rate of change is relatively high (about 0.1 s) and the field is 350 V/mm at 1 mA. The magnitude of the elongation increases with the square of the magnitude of the electric field. There are no significant losses due to heating (which is typical of polyelectrolyte gels of the Nafion). A similar effect can also be achieved with the use of PUR elastomers (without gel formation). In general, SSM responds to a certain stimulus (chemical, electrical, pH, biological, mechanical, pressure, temperature, optical, or magnetic by a measurable response (mechanical, electrical, optical, bioactivity, etc.). The majority of SSMs are temperature sensitive, i.e., the response is phase transformation. At low temperatures, SSM gels are highly swollen (containing a large amount of water in equilibrium). Above the transition temperature, the gel collapses and the water separates. A typical manifestation here is LCST = low critical dissolution temperature. Examples of LCST polymers are polyethylene oxide, hydroxypropyl acrylate, and polymethacrylic acid. A typical representative of LCST polymers is poly(*N*-isopropyl acrylamide) (PIPAAm) and its copolymers (Jeong and Gutowska 2002).

This material is soluble in the water below 32°C and precipitates above 32°C. The cause is the transition from an extended spiral to a shrink ball shape. PIPAAm treated cotton exhibits hydrophilic behavior below 32°C and hydrophobic behavior above 32°C. PIPAAm hydrogels allow active substances to diffuse at temperatures below 32°C. Above this temperature, the release of active substances is blocked due to the formation of "skin" on the surface of the gel (switching on/off mechanism). The porosity of textile structures can also be controlled by the amount of applied LCST polymer (Figure 1.23)

Sulfobetaine-(*N*-morpholino) ethyl methacrylate changes its character (hydrophilic/hydrophobic) in the temperature range 30–40°C (Armes et al. 2002). BST-Gel™ is a chitosan thermosensitive gel that is liquid at normal temperature but solidifies at human body temperature (37°C). In general, there are four types of interactions with smart gels:

a. Electrically charged ionic regions may be attracted or repelled.
b. Nonpolar areas may repel water.
c. Hydrogen bridges can connect chains.
d. Dipole–dipole interactions can attract or repel chains.

Another SSM is ionic polymers with weak acidic groups (e.g., carboxyl). These materials are highly swollen in the alkaline region because the carboxyl groups are ionized and separate the polymer chains. This allows the sorption of large amounts of water. At low pH (acidic region), the carboxyl groups lose charge and the gel collapses. These gels are sensitive to both pH and electrolytes. One of the most interesting groups of SSM is polyelectrolyte gels. Because they are ionic, they respond to a variety of stimuli from pH, through electrolyte content to electric field potential. The polyelectrolyte gel contains both positive and negative charges, the mutual positions of which are stabilized due to the crosslinking of polymer molecules. Due to external electrical stimuli, the charges either come near or move away, leading to either

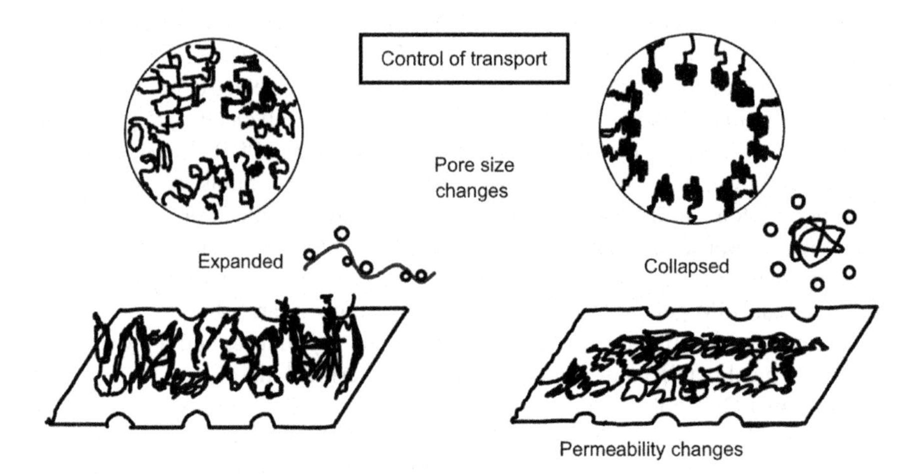

FIGURE 1.23 Porosity control by LCST polymer.

swelling or collapse of the gel. The presence of various salts promotes this effect (low molecular weight salts such as NaCl and KCl are suitable). These gels are used in the desalination of water and the creation of artificial muscle fibers for robots. They can also serve as sources of mechanical energy in space or underwater. Polymer gels bond to textile structures using various grafting and coating techniques. Acrylic acid polymerization initiated by γ beam irradiation (cobalt bomb) or UV radiation is used to attach the gels to the fibers. When using UV radiation, a solution of uncross-linked SSM, a crosslinking monomer, and a photoionicizer are first applied to the surface. UV irradiation produces free radicals by decomposing photoinitiators and subsequently polymerizing crosslinkers. Usually, acrylic acid or vinylpyrrolidone is chosen. Difunctional oligomers with vinyl groups at both ends can also be used.

An example of the preparation of a pH-sensitive gel is the crosslinking and polymerization of a system consisting of polyacrylic acid and carboxymethyl acrylate (polymer solution), acrylate urethane (crosslinking agent), and a photoinitiator. This gel precipitates by up to 50% at pH 7.5. At pH 10, there are no dimensional changes, and at pH 12, swelling occurs by up to 90%. An aqueous NaCl solution with a concentration of about 1 M is suitable as the swelling liquid. *N*-vinylpyrrolidone-based gels behave similarly, but only water is sufficient for swelling. These gels are also sensitive to temperature changes. They can be used as high-comfort textile materials that slowly release water in hot conditions and retain water in cold conditions. They can also be used in medical applications for sensitive drug dosing. Depending on the preparation conditions (amount of water added to the solution before crosslinking), these gels can absorb from 150% to 600% of water. These hydrogels are also sensitive to ambient humidity. If the material is in a humid environment, it is soft and flexible. In a dry environment, it is rigid and inflexible. Materials sensitive to external stimuli are usually only in the laboratory research phase and are used only to a limited extent by industry.

1.7 TEXTILES FOR MEDICAL PURPOSES

Textiles are used in medical applications either alone (medical textiles) or as part of various structures (composites) and systems (textile sensors, parts of devices and equipment, etc.). Textiles in medicine use properties such as strength, flexibility, formability, flexibility, breathability for gases, and permeability to liquids (Rajendran and Anand 2002). For medical applications, special properties are relatively easily supplied in the form of binding of active substances in fibers or coating on the surface of textiles (Bartels 2011). The use of textiles ranges from clothing fabrics for hospital staff, bed linen, surgical sutures, and bandages through barrier fabrics, laminated, and sandwich structures to complex composite structures to replace human organs, bones, and skin (Arundhathi 2015).

1.7.1 General Topics

Textiles for medicine are generally divided into true medical textiles approved for use in medicine and supporting medical textiles that have only a certain supporting medical effect (e.g., antibacterial) or contribute in some way to the health of the

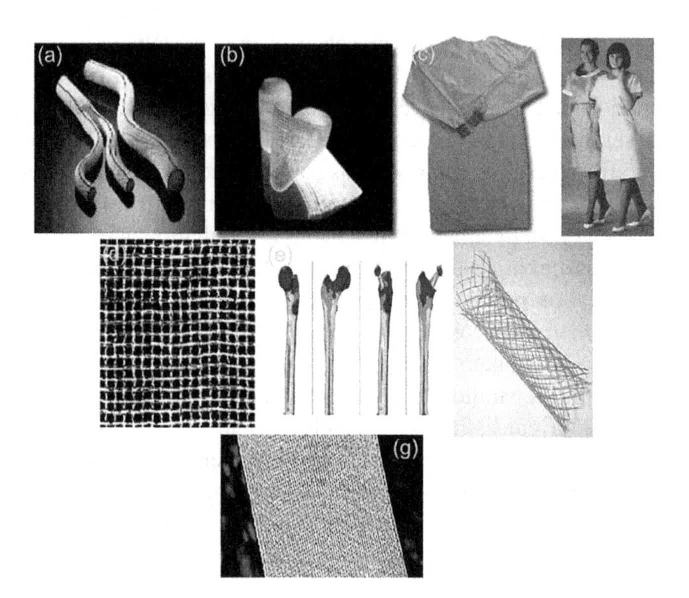

FIGURE 1.24 Some medical textiles.

wearers. For medical purposes, it is of course possible to use several standard textiles and textile structures, in particular for ordinary clothing purposes, bed linen, and ordinary home textiles. Some examples of medical textiles are given in Figure 1.24

Even here, however, the emphasis is placed on special conditions of use (higher incidence of dangerous bacteria, the possibility of influencing treatment processes, etc.). Relatively special demands are placed on textiles that come into contact with the wound and cover the wound during treatment. About the speed of healing and the ease of removing textiles from wounds, it is advisable to keep the area around the wound moist. Therefore, fibers with a high absorption capacity are used, which can form a swollen gel when wet (Qin 2011). Such fibers include, for example, alginates (based on glucuronic acid) or crosslinked structures based on acrylic acid or various cellulose-based superabsorbents (microcrystalline cellulose, etc.). Chitin and chitosan are also substances that accelerate the healing and growth processes of new tissues. In particular, chitosan is easily processed into fibers. Polyacrylonitrile and polyvinyl alcohol fibers are mainly used for the incorporation of various medical devices. Common requirements for medical textiles include:

- Nontoxicity of textiles and nontoxicity of their decomposition products,
- Inability to cause allergic reactions, resp. promotes the development of malignant cell proliferation (cancer),
- Possibility of sterilization without deterioration of mechanical and other properties.

Many textiles for medical purposes require biocompatibility (compatibility with human tissue) and sometimes biodegradability over time (e.g., surgical sutures).

The use of textiles for medical purposes ranges from clothing textiles for hospital staff, bed linen, surgical sutures, and bandages through barrier fabrics, laminated, and sandwich structures to complex composite structures for the replacement of human organs, resp. bones and artificial skin. Medical textiles can be classified into four basic categories:

- Nonimplant materials such as bandages, patches, and orthoses.
- Organ replacements – artificial kidneys, lungs, and liver.
- Implant materials – surgical sutures, venous transplants, artificial joints, atria, skin, etc.
- Clothing and protective materials – bedding, clothing, drapes, towels, and coats.

Most textiles from the group of clothing protective materials are intended for single use. In the case of multiple-use, in addition to the removal of impurities, sterilization must also be ensured (usually with hot air or steam or radiation irradiation). Of the natural materials, cotton and natural silk are very often used. Chemical fibers are mainly regenerated cellulose fibers (viscose). Of the synthetic fibers, PES, PAD, PAN, POP, and PVA fibers are mainly used. In special cases, PTFE, glass, carbon, alginate, collagen, chitin, chitosan, and many others are also used (Qin 2011).

1.7.2 Nonimplant Materials

These materials are intended for external use and can, resp. may not come into contact with the human body. Examples of nonimplant materials is shown in Figure 1.24b. Some materials are intended to absorb liquids and some to repel (Qin 2011). Bandages, towels, and bath towels should have the ability to absorb. This ability is related to:

- The liquid absorption rate,
- The absorption capacity of the material,
- Retention of liquids under pressure,
- Rise rate,
- Wettability.

It is obvious that in addition to the type of fiber (which directly affects wetting and wicking), it also depends on the construction of the fabric (porosity and pore volumes). In some textiles (viscose), the structure collapses in the wet state (due to a strong swelling and reduction of resistance to deformation in the wet state). The ease of penetration of liquids through these materials is directly proportional to the wetting angle and the surface tension of the liquid and indirectly proportional to the pore radius. Materials used in wound care are classified as:

- *Primary (which are in direct contact with the injury),*
- *Secondary (forming outer layers).*

In some cases, these materials have both functions.

Primary materials (absorption pads, contact layers of images) have the main task of protection against infection and absorption of blood or body fluids. Advantageously, they can also actively contribute to the wound healing process such as:

- Carriers for topical treatment (antibiotics, antiseptics, growth substances),
- Substances that are involved in the healing process (collagen, alginates, chitin fibers),
- Odor absorbents, resp. inflammatory products (activated carbon, superabsorbents),
- Substances that maintain the conditions for healing (especially moisture).

It is not considered appropriate in the medical community to use topical antibiotics without the advice of a specialist. Nevertheless, there are several patents and products with these properties or effects. In general, textiles that come into contact with injuries are required to have the following properties:

- Required capacity for liquid absorption,
- Limited possibility of moisture transport,
- No release of fibrous or other particles,
- Slight adhesion to healed skin (this can be reduced, e.g., by using viscose rayon).

The composition of the primary wound healing layer depends on the extent of the injury.

Secondary materials include some textile structures with sufficient absorption capacity, flexibility, warmth, and possibly impermeability to liquids. A typical wound care material is usually coated with an acrylate-based adhesive to allow it to adhere. An interesting possibility is the use of bandages with bacterial sensors working on the principle of color change. A separate group consists of gauze, which is used mainly for burns and scalds. It is a thin cotton fabric often covered with paraffin wax. In surgical operations, gauze is used for blood aspiration, where a high absorption capacity is required. In these applications, instead of waxing, the fibers are labeled with $BaSO_4$ to identify the gauze by X-rays. Chemically modified cotton gauze is known, which absorbs the enzyme "elastase" and thus breaks the proteins that make up connective tissue. To improve the absorbency, superabsorbents such as alginates are grafted onto the gauze or the surface of the cotton gauze is covered by carboxymethyl cellulose (action of chloroacetic acid in an alkaline medium). It is also possible to use alginate fibers with a chitosan layer on the surface for the construction of gauze. An activated polyester fabric (ethylenediamine: formation of amino and carboxyl groups) with the bound antibiotic ciprofloxacin (CIPRO-staining technology) and the enzyme thrombin (promotes blood clotting) can be used for the production of intelligent gauze. Another possibility is to use a three-block pluronic copolymer (BASF) composed of hydrophobic polypropylene oxide (PPO) and a pair of hydrophilic PEO segments. In an aqueous environment, the soluble PEO blocks become bulky and form a spatially stable brush-like surface. The result is the prevention of adhesion of bacterial and animal cells. The material adsorbs on strongly hydrophobic surfaces and reduces the formation of deposits. Several bandages and orthopedic pads are designed especially for:

- Fastening of materials that are in direct contact with the injury,
- Protection resp. stabilizing the position of parts of the human body,
- Induction of sufficient pressure (pressure therapy).

A variety of textile structures are used for this purpose, especially knitted fabrics, woven fabrics, and nonwovens. Sufficient elasticity is achieved either by a special textile production technique (crepe bandages) or by using elastic fibers (see Figure 1.24g). Depending on the pressure they can exert, pressure bandages are divided into categories with extra high, high, medium, and low effect. The pressure under the bandage is directly proportional to the tension in the bandage and indirectly proportional to the radius of curvature of the bandaged limb. The elasticity and extensibility of the dressings are evaluated as the stress required to stretch by 20% (not to be higher than 14 N/cm) and to recover from this deformation (not less than 95%).

1.7.3 Organ Replacements

Textiles are used in many cases as replacements for parts of the human body. Selected options for using textiles for these purposes include:

- Abdominal wall – PES fibers (woven fabrics)
- Vessels –PUR, Polylactones, and PES fibers (nonwovens, knitwear, braided, fabrics)
- Vascular grafts – PES, PTFE, PUR, and polyglycolid acid (PGA) (fabric, knit)
- Vases – PES, carbon, aramids, and glass (braided, fabric, knitted fabric)
- Skin – chitin fibers (fabric, knitwear, nonwovens), polyglycols, PAD, and collagen fibers (foam)
- Tendons –POE, PES, silk, and PAD fibers (yarn, braided)
- Trachea – PGA fibers (nonwovens)
- Bones – PGA fibers and polyhydroxyapatite (knitwear, foams)
- Cartilage – PGA, polylactone fibers (nonwovens), POE, PES, PTFE fibers (braided, fabric), and Heart flap – PGA and PES fibers (fabrics, knitwear)
- Liver – PGA and polyanhydride fibers (braided, knitwear, fabrics).

With a slight exaggeration, it can be stated that at present there are already means for the production of artificial human parts, the vast majority of which use at least partially fibrous structures or directly textiles. In many cases, the fabrics only form supporting grids allowing tissue growth. Mechanical organs made using textile structures are also available. Usually, textiles are only part of a replacement human organ. In particular, the possibility of easy formability and hollow fibers enabling blood purification processes, resp. blood and air transport.

Artificial kidneys: remove waste products from the blood plasma. These are membranes or bundles of hollow fibers (hollow viscose and PES fibers, polycarbonates,

PAN, chitin, chitosan, polysulfone). It is a multilayer filter, where there is a different filling density in each layer.

Artificial liver: remove (detoxify) waste products from the blood and allow hematopoiesis (hollow viscose fibers, carbon fibers, polyether urethanes, special filters).

Mechanical lungs: remove CO_2 from the blood and allow the supply of fresh oxygen. These are microporous membranes or bundles of hollow fibers (POP, silicones). High air permeability but low liquid permeability is required.

In the field of medical textiles, the whole spectrum of fibers is used, from natural (especially cotton and natural silk) through viscose fibers (where medically active substances can be added during spinning) to synthetic fibers (both classic and special). The fiber is usually chosen with regard to biodegradability in the body, biocompatibility, intactness to body fluids, microorganisms, and bacteria, and last but not least, the possibility of sterilization by simple means (steam). Relatively special demands are placed on textiles that come into contact with the wound and cover the wound during treatment. With regard to the speed of healing and the ease of removing textiles from wounds, it is advisable to keep the area around the wound moist. Therefore, fibers with a high absorption capacity are used, which can form a swollen gel when wet. Such fibers include, for example, alginates (based on glucuronic acid) or crosslinked structures based on acrylic acid or various cellulose-based superabsorbents (microcrystalline cellulose, etc.). Chitin and chitosan also accelerate the healing and growth processes of new tissues. In particular, chitosan is easily processed into fibers. Polyacrylonitrile and polyvinyl alcohol fibers are mainly used for the incorporation of various medical devices. From the point of view of biocompatibility, microporous carbon fibers are suitable. Textiles containing intelligent alloys such as NITINOL are also used.

1.7.4 Implant Materials

These materials are used as replacements for human body parts, fasteners, resp. tendon, cartilage, and skin replacements. The basic requirement is biocompatibility. Sufficient porosity is required to ensure that the implants grow through new tissue. It is also verified that thin circular fibers encapsulate better than coarse fibers with an irregular cross section. The biodegradability of the fibers is also largely determined by the success of the implantation (King 2013). Of the synthetic fibers, polyamide fibers are relatively significantly biodegradable, which lose their strength almost completely after about 2 years. PTFE, on the other hand, is practically nondegradable.

Surgical sewing threads are monofilament or multifilament. Biodegradable surgical sewing threads are made from collagen, polyglycolic acid, and polylactone fibers. The problem is that these fibers often lose strength before the healing process is completed; so there is an increased risk of seam cracking. Nonbiodegradable sewing threads include polyamide, polyester, and polypropylene materials.

Tendon replacements are constructed from PTFE, polyester, and polyamide fiber wovens or wrapped tapes coated with silicone resin. Wrapped composites containing carbon fibers and polyester silk are also used for some tendons.

Soft cartilage is usually replaced by POE nonwovens.

Hard cartilage is created from carbon-fiber-reinforced composite structures.

Orthopedic implants, i.e., artificial joints and bones, are constructed as fiber-reinforced composites to achieve both biocompatibility and sufficient strength, resp. abrasion resistance. To ensure the growth of tissues around the implant, a nonwoven web containing carbon or polytetrafluorethylene fibers is used.

Venous implants are from knitted tubes made of polyester or PTFE with a diameters of 6.8 and 10 mm. Polyester venous implants are often fixed in a crimped shape so that they are easily shaped while maintaining roundness. To reduce the porosity of knitted venous implants the collagen or gelatin impregnation is used. These structures decompose after about 14 days. Artificial veins with an inner diameter of about 1.5 mm are made of porous PTFE tubes. There is a layer of collagen and heparin on their inner wall and a layer of collagen on their outer wall.

1.7.5 Clothing and Protective Materials

Cotton fabrics were used as a traditional material for medical clothing textiles (hospital gowns, trousers, etc.). However, these materials had virtually no protective effect against viruses, bacteria, and microorganisms. In recent years, protection against microorganisms from both air and liquids has been a requirement. Therefore, disposable garment fabrics with water-repellent and oleophobic finishes are used. These fabrics are produced by spun-blown or melt-blown techniques. Laminating is also often used, which limits the penetration of microorganisms. In addition to disposable medical clothing fabrics, fabrics that are washed or cleaned are still used. Examples are PES/cotton hospital gowns with hydrophobic and oleophobic treatment and low porosity. The general requirements for medical clothing textiles include:

- Low wettability;
- Resistance to the penetration of blood, fluids, and microorganisms;
- Antistatic properties and nonflammability;
- Low release of particles (fibers);
- Sufficient comfort.

In many cases, it is a problem to achieve comfort while maintaining resistance to the penetration of microorganisms. This is the same problem as for cleanroom textiles. Wipes and pads are composed of materials used to clean the skin and wounds, respectively, or to absorption of body fluids during surgery. Cellulose fiber nonwovens are commonly used. The requirements are fast absorption, large sorption capacity, retention of liquids under pressure, and detectability under X-rays. The structure of typical wipes is shown in Figure 1.24d. A special group consists of single-use diapers for patients. It is a layered structure comprising an inner layer, an absorbent layer (usually foam), and an outer layer (PVC foil). The inner layer is made of PES fabric, resp. polypropylene nonwovens with hydrophilic treatment. The general problem is to achieve the desired effects at a low cost (especially for disposable textiles). Therefore, nonwoven fabrics produced by the simplest possible process (punching, etc.) are used in particular. For protective medical textiles, the size of the bacteria or viruses against which it is intended to protect is often decisive. This is the case, for example, with the SARS virus and virus responsible for COVID 19, which has a size of 80–120 nm. To protect against this virus, special protective clothing based on PTFE/PUR membrane laminated on PES textured fabric (containing conductive fibers) with antimicrobial and water-repellent treatment has been proposed (Hao et al. 2004). PTFE is a porous membrane formed by stretching at 200°C successively in two directions. The average pore size is 250 nm. PUR is a nonporous membrane enabling the transport of sweat.

1.7.6 Development of Medical Textiles

In general, progress in the field of medical textiles is influenced by:

- Development of new materials and principles (Qin 2011),
- Imitation of nature (biomimetics),
- Development of new textile preparation technologies,
- Transfer of materials and technologies from other fields.

At first glance, it seems that the development of medical textiles is based primarily on special materials and technologies, which requires large investments, long-term research, and is very time-consuming. Truly new comprehensive solutions are the result of long-term intensive activities, often on an international scale. On the other hand, several successful solutions are the result of simple ideas. In the field of medicine, it is the possibility of controlled dosing depending on changes in the volume of injuries and bandages that do not exert excessive pressure on the swelling. The bandage itself consists of auxetic yarns encapsulating the healing substance. During infection, swelling occurs and the active substance is released. After healing, shrinkage occurs and further release of the active substance is prevented (controlled release of active substances).

1.8 CONCLUSION

There are already several interesting ideas and procedures that will probably be used on an industrial scale for medical applications in the future. On the other hand, it cannot be expected that all solutions will be automatically successful simply because they do not ensure their trouble-free use, long-term applicability, and durability under the conditions common to medical standards or requirements. It is possible to expect the expansion of some results even outside the area of their originally intended use. The use of various combinations of materials and structures with the possibility of obtaining synergistic effects resp. more comprehensive problem-solving.

ACKNOWLEDGEMENT

This work was supported by the Ministry of Education, Youth and Sports of the Czech Republic and the European Union – European Structural and Investment

Funds in the frames of Operational Programme Research, Development and Education under project Hybrid Materials for Hierarchical Structures [HyHi, Reg. No. CZ.02.1.01/0.0/0.0/16_019/0000843].

REFERENCES

Armes, T. et al. 2002 Synthesis and aqueous solution properties of a well-defined thermo-responsive schizophrenic diblock copolymer, *Chem. Commun.*: 18 2122–2123.

Arundhathi, G. 2015 *Emerging Research Trends in Medical Textiles*. Singapore: Springer Verlag.

Avinc, O. and Khoddami, A. 2009 Overview of PLA fibre, *Fibre Chem.*: 41, (6), 391–401.

Bartels, V. T. 2011 *Handbook of Medical Textiles*. Cambridge: Textile Institute & Woodhead.

Celebioglu, A. and Uyar, T. 2011 Electrospinning of polymer-free nanofibers from cyclodextrin inclusion complexes, *Langmuir*: 27, 6218–6226.

Das, S. S. et al. 2020 Stimuli-responsive polymeric nanocarriers for drug delivery, imaging, and theragnosis, *Polymers*: 12, 1397–1432.

Donelan, J. M. et al. 2008 Biomechanical energy harvesting: generating electricity during walking with minimal user effort, *Science*: 319, 807–810.

Drumright, R. E. et al. 2000 Polylactic acid technology, *Adv. Mater.*: 12, 1841–1846.

Farrington, D. W. et al. 2015 Poly(lactic acid) fibers. In *Innovative Biofibers from Renewable Resources*, eds. Reddy, N. and Yang Y., 191–220: Heidelberg: Springer Verlag.

Fernandez-Nieves, A. et al. 2011 *Microgel Suspensions: Fundamentals and Applications*. Weinheim: Wiley-VCH Verlag.

Hao, X. et al. 2004 Study of new protective clothing against SARS using semi-permeable PTFE/PU membrane, *Eur. Polym. J.*: 40, 673–678.

Holmes, I. 1998 Recent advances in chemical processing, *Colourage*: 45, 41–56.

Chang, Ch. et al. 2010 Direct-write piezoelectric polymeric nanogenerator with high energy conversion efficiency, *Nano Lett.*: 10, 726–731.

Chopra, K. L. et al. 1983 Transparent conductors – A status review, *Thin Solid Films*: 102, 1–46.

Jeong, B. and Gutowska, A. 2002 Lessons from nature: stimuli-responsive polymers and their biomedical applications, *Trends Biotechnol.*: 20, 305–311.

King, M. W. 2013 *Biotextiles as Medical Implants*. Duxford: Woodhead Publishing.

Li, D. et al. 2004 Electrospinning of nanofibers: reinventing the wheel? *Adv. Mater.*: 16, 1151–1170.

Lim, H. S. et al. 2010 Superamphiphilic Janus fabric, *Langmuir*: 26, 19159–19162.

Ma, M. et al. 2005 Electrospun poly(styrene-co-dimethylsiloxane) block copolymer fibers exhibiting microphase separation and superhydrophobicity, *Langmuir*: 21, 5549–5554.

Mehling, H. and Cabeza, L. F. 2008 *Heat and Cold Storage with PCM- An up to Date Introduction into Basics and Applications*. Berlin: Springer Verlag.

Meli, L. et al. 2010 Electrospinning from room temperature ionic liquids for biopolymer fiber formation, *Green Chem.*: 12, 1883–1892.

Militký, J. et al. 2020 Exceptional electromagnetic shielding properties of lightweight and porous multifunctional layers, *ACS Appl. Electron. Mater.*: 2, 1138–1144.

Militký, J. 2019 Tensile failure of polyester fibers. In *Handbook of Properties of Textile and Technical Fibres* 2nd ed., ed. Bunsel, A. R., 421–514. Duxford: Elsevier & Woodhead Publishing.

Militký, J. et al. 2013 Fibres and dyeing. In *Textile Dyeing-Theory and Applications*. eds. Kryštůfek, J. et al., 39–114. Liberec: TUL Press.

Mishra, R. and Militký J. 2018 *Nanotechnology in Textiles*. Duxford: Woodhead Publishing.

Moukman, G. J. 2000 Advances in shape memory polymer actuation, *Mechatronics*: 10, 489–498.

Parameswaranpillai J. et al. 2020 *Shape Memory Polymers, Blends and Composites- Advances and Applications*. Singapore: Springer.

Park, S. and Jayaraman, S. 2001 Adaptive and responsive textile structures. In *Smart Fibers, Fabrics and Clothing*, ed. Tao, X., 226–245. Boca Raton Boston: CRC Press.

Parsons, K. 2014 *Human Thermal Environments*. Boca Raton: CRC Press.

Pietrzak, W. S. et al. 1997 Bioabsorbable polymer science for the practicing surgeon, *J Craniofac Surg.*: 8, 87–91.

Qin, Y. 2011 *Medical Textile Materials*. Duxford: Elsevier & Woodhead Publishing.

Rajendran, S., and Anand, S.C. 2002: Development in Medical Textiles, *Text. Prog.*: 32, 1–42.

Ramakrishna, S. et al. 2005 *An Introduction to Electrospinning and Nanofibers*. Singapore: World Scientific Publishing Company.

Ramaratnam, K. et al. 2008 Ultrahydrophobic Textiles Using Nanoparticles: Lotus Approach, *J. Eng. Fibers Fabrics*: 3, 1–14 08.

Roach, P. et al. 2008 Progress in superhydrophobic surface development, *Soft Matter*: 4, 224–240.

Santhosh, B. et al. 2018 *Shape Memory Materials*. Boca Raton: CRC Press.

Sin, L. T. and Tueen, B. S. 2019, *Polylactic Acid A Practical Guide for the Processing, Manufacturing, and Applications of PLA*. Oxford: William Andrew & Elsevier.

Singh, A. et al. 2005 Electrospinning of Poly[bis(2,2,2-trifluoroethoxy) phosphazene] Superhydrophobic Nanofibers, *Langmuir* 21, 11604–11607.

Smeets, N. M. B. and Hoare, T. 2013 Designing responsive microgels for drug delivery applications, *J. Polym. Sci. A Polym. Chem.*: 51, 3027–3043.

Song, J. et al. 2020 Recent advances on nanofiber fabrications: unconventional state-of-the-art spinning techniques, *Polymers*: 12, 1386–1406.

Srinivasan, A.V., Mc Farland, D.M. 2001 *Smart Structures*. Cambridge: Cambridge University Press.

Sun, S.P. et al, 2009 Alkali etching of a poly(lactide) fiber, *ACS Appl. Mater. Interfaces*: 1, 1572–1578.

Tuljapurkar, S., et al. 2000 A universal pattern of mortality decline in the G7 countries, *Nature*: 405, 789–792.

Venkataraman, M. et al. 2016 Aerogels for thermal insulation in high performance textiles, *Text. Prog.*: 48, 55–118.

Venkataraman, M. et al. 2018 Preparation of Electrosprayed Microporous Membranes, *IOP Conference Series: Materials Science and Engineering*: 460, 012017.

Vigo, T. C., and Frost T. L. 1989 Temperature-adaptable textile fibers and method of preparing same, US Patent 4871615.

Wilson, A. 2011 *The Future of Smart Fabrics, Market and Technology Forecasts to 2021*, Leatherhead: Pira International Ltd.

Yamashita, Y. 2008 Current state of nanofiber produced by electrospinning and prospect for mass production, *J. Text. Eng.*: 54, 199–205.

Yang, Q. 2011 Research on the tensile and surface frictional property of PLA fibers, *Adv. Mater. Res.*: 322, 287–290.

Ziolkowski, A. 2015 *Pseudoelasticity of Shape Memory Alloys-Theory and Experimental Studies*. Amsterdam: Elsevier & Butterworth-Heinemann.

2 Protection against Microbes

State-of-the-Art

Aravin Prince Periyasamy, Mohanapriya Venkataraman and Jiri Militký

Technical University of Liberec, Czech Republic

CONTENTS

2.1 INTRODUCTION

The protection of textiles against the action of microorganisms has a relatively long tradition. The ancient Egyptians (5,000 B.C.) used linen fabrics containing various salts for mummification. In 1867, Joseph Lister impregnated the bandages with carbolic acid and phenol. Around 1900, cellulose acetate-based fibers were found to be resistant to rot and mold. The term antimicrobial fabric and treatment began to be used around 1941. Originally, antimicrobial treatment was used to protect various technical textiles against rot and mold. Examples are tents, tarpaulins, and geotextiles. Later, these modifications began to be used for home textiles (curtains, drapes, carpets, and bathroom mats). Recently, the use of antimicrobial fabrics has significantly expanded to include sports and entertainment clothing fabrics. A separate topic consists of medical textiles, textiles for the food industry, pharmaceutical industry, special filters, protective layers, and masks avoiding transfer of microbes, etc.

One of the classic reasons for using fibers with antimicrobial effects is to control (remove) odors and prevent the survival of microbes on textiles caused by medical problems (infections). In connection with bioterrorism and pandemic caused by air spreading viruses, they are developed also for protective clothing, air filters, and protective masks. It is still a matter of debate to what extent antibacterial textiles can contribute to protection against the spread of diseases. The antimicrobial textiles can help with suppressing the spontaneous spread of microbes responsible for new kinds of diseases. The pandemic spread of COVID 19 from January 2020 till the end of August 2020 was still not under control and the number of infected was about 26 million people over the world. All kinds of tools for avoiding the spread of the responsible virus (coronavirus SARS-CoV-2) are therefore most wanted. A large part of antimicrobial textiles is designed to reduce the odor of socks and stockings, other items classed as underwear, etc. Antimicrobial ability is also often required for textiles used in air filters and for cleaning in households and industry. Many agents used in textile finishes and many dyes also have significant antibacterial effects. Thus, often the antimicrobial effect is associated with another desired effect (antistatic treatment, nonflammable treatment, noncreasing treatment, dyeing, etc.), or microbial protection is part of the protection against the action of chemical substances. Protection against microbes includes protection against the spread of diseases, protection against pathogenic or odor-causing microorganisms (hygienic treatments or textiles), and protection of textiles against microbial degradation caused mainly by fungi and rot. Sometimes insect resistance is also included in this protection, with fungi and bacteria causing the most problems. In very humid conditions, algae can also grow on textiles, which causes problems mainly because they are the food of bacteria and fungi. Fungi (molds) cause stains on textiles, shade changes, and fiber breakage. Bacteria do not cause such significant fiber damage, but especially an unpleasant odor and a mucous, sticky touch. Bacteria and fungi often survive on textiles in symbiosis. A wide range of substances added to the fibers, such as lubricants, antistatic agents, and various types of starch-based auxiliaries, resp. soluble cellulose derivatives can be a food source for bacteria. Even synthetic fibers are not immune to bacteria and fungi (e.g., polyurethane fibers and deposits). Microorganisms act more easily on natural fibers. Wool is more sensitive to bacterial effects and cotton is more sensitive to fungi. This chapter is a survey of common and advanced types of compounds used in protection against different kinds of microorganisms oriented to be used for antimicrobial textile structures. The different kinds of microbes and their peculiarities are discussed. The discussion about the functions of agents acting against coronavirus caused by COVID 19 is included as well.

2.2 MICROBES

Microorganisms (microbes) are single or more multicellular organisms that are unable to form functionally differentiable tissues. Their dimensions are of the order of micrometers. Microorganisms multiply by division. The average generation time (doubling time) of unicellular organisms is about 20 min under optimal conditions. Thus, if *No* divisible bacteria (referred to as colony forming units [CFU]) are initially available, *N*(*t*) bacteria are present at time *t*(*s*). Under optimal conditions (division in 1,200 s), it is valid:

$$N(t) = No\,2^{t/1200} \tag{2.1}$$

Thus, the logarithm of the number of cells is a linear function of growth time. In practical conditions (stationary culture without a supply of nutrients and removal of fumes – metabolites), due to the depletion of nutrients and the increase in the concentration of toxic metabolites, the growth process is slowed down and there is a gradual death. The growth of microorganisms is strongly dependent on the external environment, especially temperature, water content, pH, and the presence of chemicals. The optimal growth temperature is usually 25°C–37°C (for most bacteria, yeasts, and fungi of the so-called mesophilic type). The minimum growth temperature is usually 10°C lower and the maximum growth temperature is about 6°C higher than optimal. A necessary condition for the growth of microorganisms is the presence of free water, ensuring that the cells do not lose intracellular water (making up about 80%–90% of the cell mass). The need for free water is expressed by the so-called water activity

$$a_v = N_v / (N_s + N_v) \tag{2.2}$$

where N_v is the number of moles of water and N_s is the number of moles of dissolved compounds. Most microorganisms grow in the range of $0.6 < a_v < 0.99$; bacteria and algae usually require an upper limit. Yeast grows in the range of $0.88 < a_v < 0.91$, and fungi have smaller requirements for the size of a_v. Most bacteria require a neutral or slightly alkaline environment to grow. Yeast requires an acidic pH (4.8–5.5) and fungi grow best in a neutral environment, even if they grow in the pH range of 1.2–12 as well. Radiation of shorter wavelengths has generally a detrimental effect on microorganisms. The 265 μm radiation is strongly absorbed by nucleic acids. Several microorganisms are needed—from the creation and preservation of the environment, to the production of a wide range of products in the food, chemical, and pharmaceutical industries. On the other hand, there are negative effects of some microorganisms on humans and the environment. Some microorganisms decompose food, textiles, leather, paper, or selected plastics; and microbial corrosion is also known. Some microorganisms are pathogenic and cause the transmission of infection. Fungi and molds are complex slow-growing organisms that can use some bacteria or their decomposition products as food.

2.2.1 Bacteria

Bacteria are the only group of organisms using all known possibilities of obtaining energy. They are usually divided into rod (length 1–3 μm, width 0.5–1.5 μm), spherical (coca), and fibrous (mycelium). The different forms of bacteria are shown in Figure 2.1.

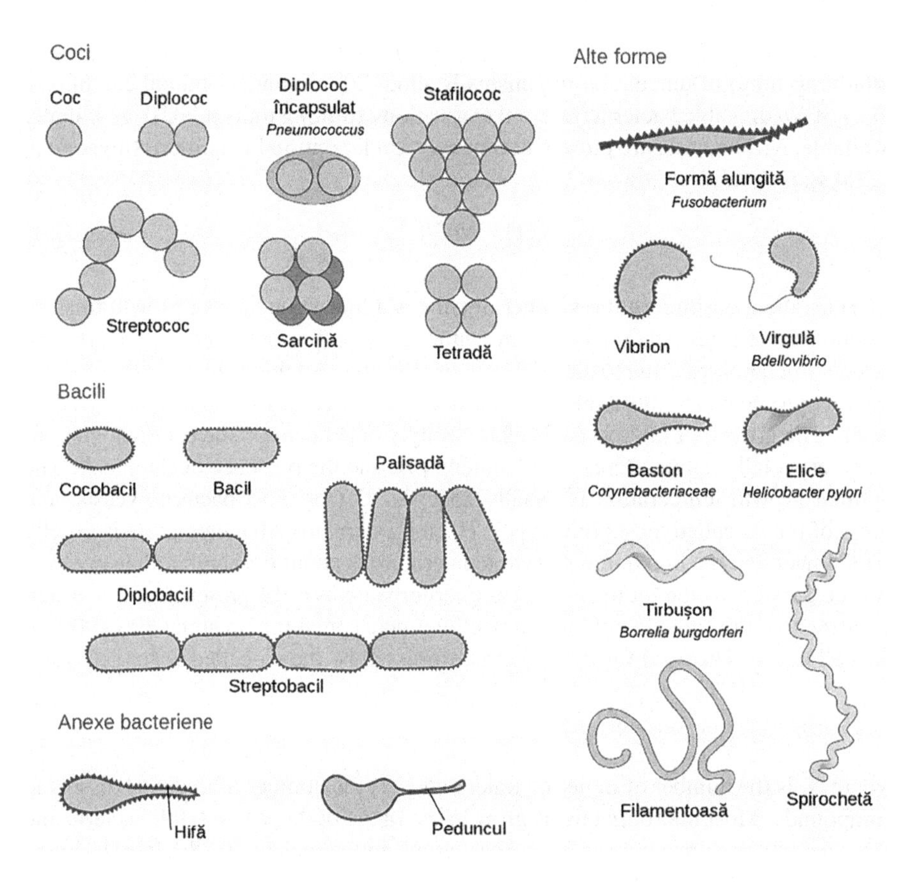

FIGURE 2.1 Different forms of bacteria, (Reprinted under CC 4.0, wikimedia.org/wikipedia/commons/thumb/1/1b/Bacterial_morphology_diagram-ro.svg/2000px-Bacterial_morphology_diagram-ro.svg.png).

Bacteria are one-celled prokaryotic microorganisms (size 0.5–3 μm) composed of cytoplasm covered by a membrane and cell wall (Alberts et al. 2015). A schematic picture of the bacteria cell structure is shown in Figure 2.2.

The cell membrane is a thin bilayer with a hydrophilic surface and hydrophobic core composed of phospholipids and proteins. The main components are a hydrophilic "head" from the phosphate group creating the surface of the bilayer, and two hydrophobic "tails" from fatty acids, joined by an alcohol residue creating the inner part of the bilayer. This membrane is a barrier to protecting the cytoplasm. The cytoplasm contains a gel-like cytosol and organelles (functional units) mainly. Bacteria have a multicomponent cytoskeleton as well playing major roles in cell division, protection, shape determination, and polarity determination. DNA is contained in a plasmid (double-stranded circular molecules) and the chromosome (a continuous piece of DNA containing genes, regulatory elements, and intervening sequences). Organelle ribosome is composed of RNA and proteins. On the surface of cells are flagella, i.e., organelle used by bacteria to move. The most outer layer of bacteria

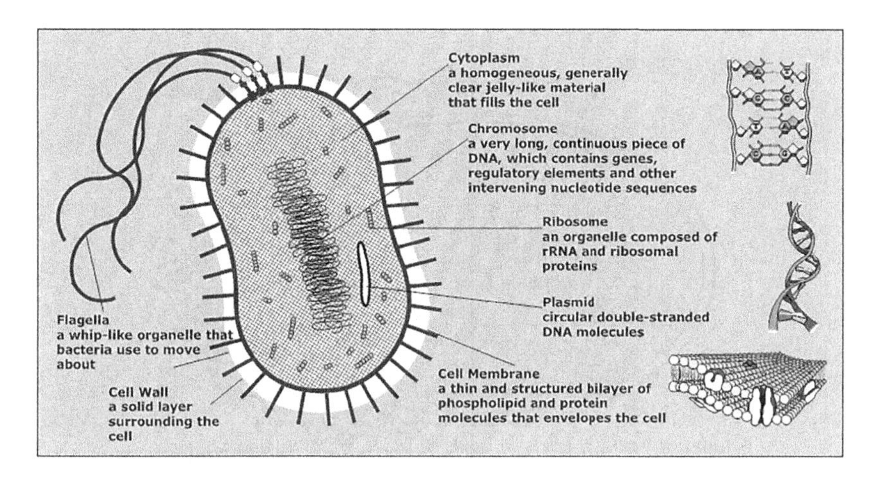

FIGURE 2.2 Structure of bacteria cell.

cells is the solid cell wall from peptidoglycan, which contains polysaccharides chains crosslinked by peptides. Many bacteria have outside of their cell, a rigid S-layer composed from glycoproteins (Alberts et al. 2015). Bacteria are often classified based on the structure of the cell wall. Gram-staining (processing in a solution of crystal violet and iodine followed by rinsing in ethanol) is here used. The so-called Gram-positive bacteria (G^+) are characterized by the violet coloration of the cell wall caused by the fact that the cell walls contain layers of peptidoglycans. Gram-negative bacteria (G^-) are characterized by a pink coloration caused by iodine and an unstained cell wall. These bacteria have cell walls containing glycolipids (complexes of lipids and carbohydrates) and are generally more resistant to antibacterial agents.

Gram-negative bacteria (most common) have a relatively thin cell wall composed of a few layers of peptidoglycan surrounded by a second lipid membrane. Gram-positive bacteria have a thick cell wall containing many layers of peptidoglycan and containing lipopolysaccharides (LPSs) and lipoproteins. The cell walls of Gram-negative bacteria are made of two glycan-rich lipid membranes and a murein layer (negatively charged LPS), while only one layer of lipid membrane (enveloped by a thick murein film) is present in that of the Gram-positive bacteria (Schmidtchen et al. 2014).

Some bacteria as Mycobacteria have special cell wall structures composed of a thick peptidoglycan cell wall like a Gram-positive bacterium, but also a second outer layer of lipids like a Gram-negative bacterium. Due to the cell wall structure, Gram-negative bacteria are relatively resistant and allow only limited diffusion of hydrophobic substances across the surface protected by LPSs. This surface is also resistant to neutral and anionic detergents. Small hydrophilic substances diffuse through the cell wall through water-filled porin channels. The small dimensions of these channels limit the mechanical penetration of larger hydrophilic particles. Because the vast majority of antibacterial agents are hydrophobic in nature, they penetrate the bacterial walls of Gram-negative bacteria relatively poorly. The LPS layer binds cations because it is polyanionic in nature (the negative charge is on toxic lipid A). LPS

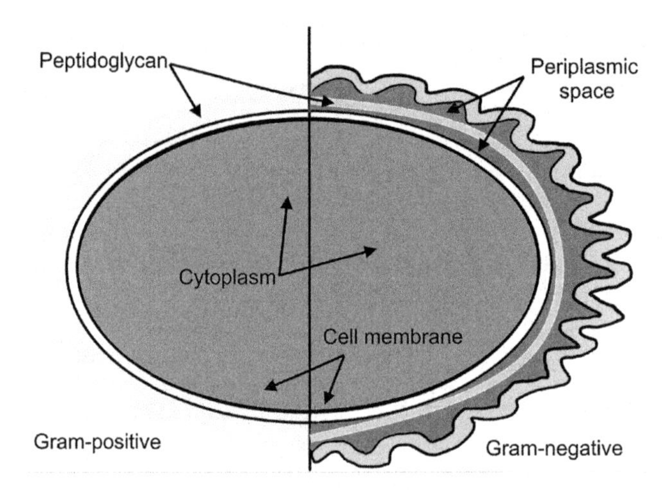

FIGURE 2.3 Schematics of cell walls of Gram-negative and Gram-positive bacteria.

molecules are often electrostatically linked via divalent cations (Mg^{2+} and Ca^{2+}). The surface structure of Gram-negative bacteria is easily destroyed by polycationic substances. Some substances are not antibacterial, but disrupt the primary cell wall of Gram-negative bacteria and allow the penetration of antibacterial substances (Vaara 1992). These substances expand the spectrum of antibiotics' action in medicine.

The cell wall is one of the most important features distinguishing bacterial species. In general, cell walls are porous and are not selectively permeable as cell membranes. The cell walls of Gram-positive bacteria contain up to 20 layers composed mainly of peptidoglycan. These layers are connected by amino acid bridges. The cell walls of Gram-negative bacteria contain mainly lipids and little peptidoglycan. Many Gram-negative bacteria are pathogenic. Pathogenicity is often associated with the presence of a layer of LPSs (endotoxin). Bacteria of this group are responsible for various intestinal diseases. The cell walls of Gram-negative bacteria are significantly weaker and contain only 20% peptidoglycan. These bacteria also have a periplasmic space separating the inner cell membrane from the peptidoglycan layers. There are proteins in this space that destroy dangerous foreign materials penetrating the outer wall. On the surface of the outer wall is a LPS phospholipid bilayer linked to peptidoglycans by lipoproteins. The lipid part of the bilayer contains toxic substances (Lipid A) causing pathogenic effects. Among the basic bacteria found on the human body are:

- *Escherichia Coli* is a Gram-negative moving rod measuring 2 × 0.7 μm forming circular colonies. It colonizes the mucosa of the large intestine and is a major component of the anaerobic physiological intestinal flora. It is the most common cause of primary urinary tract infections. It also colonizes mucous membranes damaged by inflammation or cancer. Enterotoxigenic strains produce a thermolabile or thermostable toxin. Other strains produce verotoxin causing colitis or hemolytic uremic syndrome. It can occur in both underwear and outerwear (trousers). *E. coli* is probably the most used in clinical laboratories. It plays an important role in the digestion of vitamin K production in the large intestine. Pathogenic strains of *E. Coli* cause several diseases, not only diarrhea, but also pneumonia, meningitis, etc. Treatment of *E. Coli* with antibiotic infections can lead to shocks and even death because when a bacterial cell dies, a strong bacterial toxin is released.

- *Staphylococcus Aureus* is a Gram-positive immobile ball (coccus) with a diameter of 1 μm. The colonies are round, smooth, and shiny. The bacteria are osmotically resistant and grow even in soil containing 7.5% NaCl. They form immobile clusters and do not form spores. It occurs in the nose and on the skin. It causes skin inflammations (ulcers), pneumonia, and many other diseases. *S. Aureus* produces several enzymes and toxins that support staphylococcal diseases. It causes several skin infections and purulence. It occurs mainly in underwear. Methicillin-resistant *S. Aurea* (MRSA) has become resistant to most antibiotics. *S. Aureus* is transmitted by air only from 6% to 10% and the rest is by touch transmission.

When walking, a person releases about 5,000 bacteria per minute, and one sneeze can produce up to 1 million bacteria. The dimensions of dust particles are 0.5–500 μm, that of bacteria are 0.2–2 μm, and that of viruses are 0.006–0.03 μm. The most common sources of infections in hospitals are *E. Coli* and *S. Aureus*. Only 5%–10% of *S. Aureus* strains are resistant to penicillin-based agents. The situation is worse for coagulase-negative staphylococci, where more than 50% of the strains are resistant to methicillin (MRSS). There is a real danger that strains will emerge that will be resistant to all available antibiotics (Spera and Farber 1992). There is a paradox that bacteria with increased resistance to drugs and bactericides are increasingly present in hospitals. It has also been shown that hospitals, where Triclosan is widely used, increase the resistance of many bacteria to the drug (Carey and McNamara 2014). It was found (Neely and Maley 2000) that many bacteria can survive on hospital cotton fabrics (medical gowns, bed linen, terry towels, and bathrobes) for 2–3 weeks even in adverse conditions for their growth. The cotton/polyester blend had this period of 1–3 weeks and for pure polyester materials, this period was from 2 weeks to 3 months. Polypropylene was the worst, with bacteria surviving from 8 weeks to 3 months. Microorganisms can survive for more than 90 days on hospital textiles (Slaughter et al. 1996). Bacteria secrete several enzymes as part of their metabolism, which disrupts various substances and converts them into food. These enzymes disrupt textiles and, due to their high toxicity, can be a source of disease if, for example, they enter the bloodstream. Some bacteria can survive in unsuitable conditions by creating highly resistant "spores." Eukaryotic microorganisms (more cell membranes) are significantly more complex and have 10^3–10^4 larger volumes than prokaryotic cells. These include algae, fungi, yeasts, and molds (see Table 2.1).

Under optimal growth conditions, a biofilm is also formed. A biofilm is a matrix composed of microbial populations that are adhered to surfaces and interfaces. Bacteria on the surface in larger quantities secrete slimy, sticky substances, which attach them to the surfaces of materials. Biofilm is usually made up of various types of bacteria, fungi, algae, and organic residues (Kanematsu and Barry 2020). The

TABLE 2.1
Size of Some Objects

Structure	Diameter [nm]
Bacteria *E. Coli*	2,000
Water molecule	0.4
Plasma membrane	6
Liver cell	20,000
Bacteriophage	20
Influenza virus	100
Ion Ag (silver ion)	0.2
SARS-CoV-2 virus	50–200

removal of the biofilm is more complicated because its structure prevents effective action both due to the difficult penetration of the active substance and the preferential oxidation of organic residues.

2.2.2 Viruses

Viruses are submicron agents that replicates inside the living cells and come in various shapes and sizes. Human viruses come in various shapes and sizes with DNA or RNA as their genetic material. Viruses penetrating host cells are rapidly replicated. Viruses are in the form of virions composed of molecules of DNA or RNA responsible for replication, proteins, lipids, and carbohydrates embedded by protecting protein coat and capsid. Especially coronaviruses are enveloped by lipid bilayers similar to cell membranes. Special phospholipids in lipid bilayer can facilitate attachments to host cell. Most of the positive-stranded RNA viruses replicate exclusively in the cytoplasm of the infected cell and intimate contact with intracellular membranes. This enables viral and host factors to concentrate in distinct locations to optimize a new virus particle's formation and evade innate immune responses. The replication of negative-stranded RNA viruses (e.g., Influenza virus) is located in the nucleus. The positive and negative-stranded RNA viruses require therefore a distinct set of lipids in host membranes for replication (Yager and Konan 2019). The virus particle shapes are from simple helical ones to more complex structures with spikes. The Severe Acute Respiratory Syndrome coronavirus (SARS-CoV) is of the type of Betacoronavirus. Responsible for COVID 19 pandemic is single-stranded RNA virus SARS-CoV-2 with an envelope of an approximately spherical shape with protrusions (type S proteins) allowing it to penetrate human cells, where it replicates (see Figure 2.3). Corona of this coronavirus consists primarily of glycoproteins containing oligosaccharides covalently attached to the side groups of the polypeptide chains. Glycoproteins contain oligosaccharide chains (R) covalently bound to amino acid side-chains. Corona composed mainly of lipids can be simply destroyed with alcohol or soap. RNA – ribonucleic acid carrying negative charge codes for amino acid sequences to form proteins. The main part of RNA is helix composed of pentose sugars (ribose), phosphate groups on one side (negatively charged), and nitrogenous bases (derivates of pyrimidine) adenine, guanine, cytosine, and uracil on the other side (see Figure 2.4) (Tinoco et al. 1999). The O–H bond in the ribose of RNA makes the molecule more reactive. RNA is not stable under alkaline conditions, the large grooves in the molecule make it susceptible to enzyme attack. RNA is constantly produced, used, degraded, and recycled. RNA is relatively resistant to UV damage.

The RNA of coronaviruses is terminated at one end by a so-called "cap," which protects the viral RNA from the natural immunity of cells and degradation by cellular enzymes. On the other end, the viral RNA is terminated by a sequence of contiguous adenosine nucleotides (so-called polyadenyl group). These modifications at both ends allow translation and increase the stability of RNA in cells. Viruses must use their host cells for self-reproduction/replication. Receptors for human viruses are mostly glycoproteins present in the plasma membrane (Chen and Liang 2020). Virus replication cycle is composed of the following steps:

a. Adsorption and attachment to the host cell,
b. Penetration or entry through the cell wall and cell membrane,
c. Uncoating,
d. Production of virion components – replication, transcription, and translation,
e. Assembly of naked capsid viruses and nucleocapsids,
f. Release of viral particles.

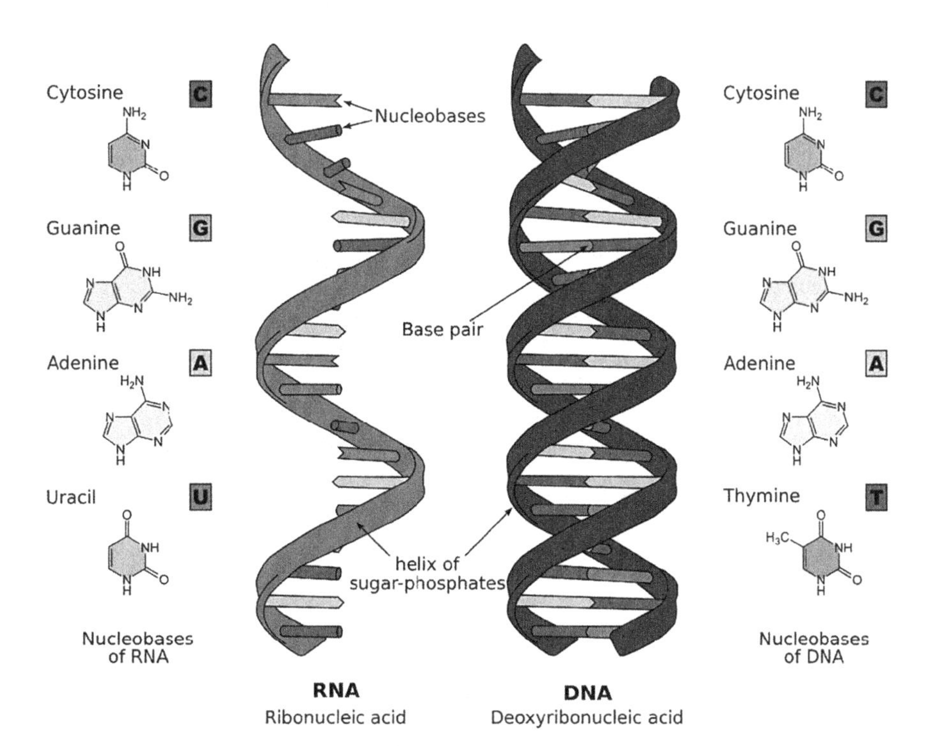

FIGURE 2.4 Composition of RNA and DNA (Image courtesy: https://commons.wikimedia.org/wiki/File:Difference_DNA_RNA-EN.svg).

Antiviral treatment should avoid replication of viruses via their passivation, modification, or break. SARS-CoV-2 has four structural proteins, known as the S (spike), E (envelope), M (membrane), and N (nucleocapsid). N protein holds the RNA genome, and the S, E, and M proteins together create the vial envelope (Figure 2.5). The spike protein (a form of a homotrimer) is responsible for allowing the virus to attach to the membrane of a host cell. S proteins bind to the host receptor known as angiotensin-converting enzyme 2 – ACE2 (Walls et al. 2020).

A key target of the virus includes the surface receptor in human cells, angiotensin-converting enzyme II (ACE2), which is required for efficient uptake in host cells. The C-terminal domain (i.e., a receptor-binding domain of the envelope-embedded spike (*S*) protein) of SARS-CoV-2 binds ACE2. Because binding of the S protein to ACE2 is critical for the first steps of infection, the most common treatment approaches focus on disrupting this process. At present, there are three main strategies to block ACE2 binding (Weiss et al. 2020):

a. Administration of soluble, recombinant ACE2 protein, which acts as a decoy receptor to scavenge the virus and, thus, to prevent uptake into host cells,
b. Vaccination with antibodies that specifically bind to the S protein and interfere with ACE2 interaction,
c. Inhibition of host proteases that process the S protein and are essential for ACE2 binding and subsequent membrane fusion to enable intracellular delivery of the virus.

SARS-CoV-2 is a relatively complex virus with a slow mutation rate. The SARS-CoV-1 virus is highly stable at room temperature and 4°C, but it is inactivated by ultraviolet light at 254 nm, highly alkaline or acidic conditions of pH > 12 or pH < 3, respectively, or by a brief (e.g., 5 min) heat treatment at 65°C. SARS-CoV-2 is expected to be similarly sensitive (Weiss et al. 2020). The stability of SARS-COV-1 (source of SARS disease) and SARS-COV-2 virus in aerosol form (<5 μm) as exhaled from infected persons in air and on various surfaces were investigated (van Doremalen et al. 2020). SARS-CoV-2 remained viable in aerosols throughout the duration of the

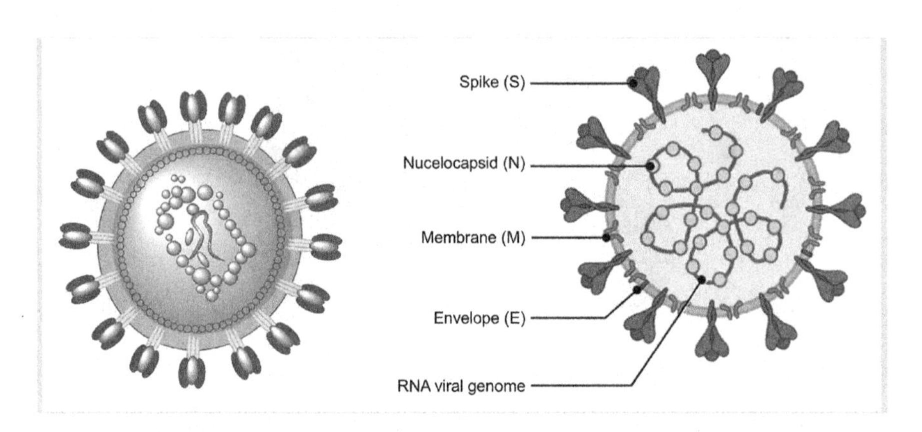

FIGURE 2.5 Structure of SARS-CoV-2 virus.

experiment (3 h), with a low reduction in infectious titer. This reduction was similar to that observed with SARS-CoV-1. SARS-COV-2 was the most stable on plastic and stainless steel and the viable virus was detected up to 72 h after application. On the other hand, after 4 h, no viable virus was found on the copper surface. Copper ions form bonds with glycoproteins (and probably strongly reduce coronavirus replication in human cells).

Several human coronaviruses can be inactivated by classical disinfectants, including bleach, ethanol, povidone–iodine, chloroxylenol, chlorheximide, and benzalkonium chloride. The virus stability on surfaces depends on the composition of the infected material, with inactivation in <3 h on printing and tissue paper, in <2 days on treated wood and cloth, in <4 days on glass and banknotes, and in <7 days on stainless steel and plastic. Conversely, active viruses can remain on the outer layer of a surgical mask even after 7 days (Weiss et al. 2020). The primary coronavirus spread is mainly due to inhalation of exhaled aerosol drops containing viruses produced by infected persons (Tellier et al. 2019). Aerosols resulted from breathing, speaking, or coughing. The size of the cough droplets ranges from 0.6 to 16 μm, but because it is a liquid, the droplets can take any shape to allow penetration through porous hydrophobic materials. The air velocity at inhalation and exhalation is up to 22 m/s; when coughing and sneezing, it is up to 83 m/s. Coronavirus SARS-COV-2 secondary spread is due to deposition on surfaces of aerosol particles with viruses and spreads due to direct contacts or resuspension. So far, the practically unsolved problem is the limitation of the activities of viruses adhering to clothing textiles, especially due to direct infiltration by droplets spread from infected persons. The standard washing conditions at temperatures around 40°C, at pH about or less than 7, used in the majority of clothing textiles now, are not sufficient for killing viruses. Problems related to the virus protection textiles are:

a. Specification of suitable materials and structures obeying efficient mechanisms of antiviral functions (mechanical capturing, inhibition of viral activities),
b. The durability of antiviral effects in maintenance/use cycles and mode of sterilization.

The antiviral textile should provide comfort during use and low impact on the health of wearers associated with the possible release of dangerous particles during long-time accumulation.

2.2.3 Mites

Mites (lat. Acari) belong to the group of spiders and have four pairs of legs. The connection of the chest with the lower body is characteristic. Their main food is parts of the skin (released during its regeneration) and dandruff. Ideal living conditions for mites are especially in beds and bed linen (Colloff 2009). Dermatophagoides mites require 75% RH as optimal humidity. At 50% RH, 87% of the population still survives, and at 30% RH, it is 50% of the population. About 94% of house dust allergy sufferers have a positive skin reaction to an extract from this type of mite. Usually,

0.1 g of house dust contains about 160 mites. The dust also contains the remains of dead mites. The highest concentration of mites is in beds and beddings. The main problem for humans is mite residues containing the allergen DER P1, which causes about 50% of asthmatic problems. Allergic reactions (asthma) can be caused by the mites themselves (170–350 μm), their secretions (10–40 μm), or by their body parts (2–20 μm). Due to the fact that many common bacteria and fungi are part of the food chain of mites, antimicrobial agents also act to reduce the number of mites. For special purposes (bed linen for allergic persons) it is possible to prevent the penetration of mite residues by limiting the porosity of textiles (fabrics with high reach). One model enabling the calculation of the proportion of mite residues passed through fabrics of known porosity is given in (Militký et al. 1998). About 80% of all household mites come from duvet covers, pillows, and mattresses. Dust mites cause skin reactions (dermatitis) and are a source of many allergic reactions (asthma). 10%–30% of the human population suffers from these allergies (Portnoy et al. 2013). It has been found that an important part of the diet of mites is some bacteria, their removal then prevents the survival of mites. Hyfresh (Daiwa) acts as a mite repellent (Maheshwari et al. 2010). Permethrin-impregnated fabrics also show high mite activity.

2.3 INFLUENCE OF MICROBES ON HUMANS

Many bacteria are commonly stored and multiplied on the human body and clothing fabrics. Most of them are not pathogenic and do not affect human health. However, some generate an unpleasant odor. However, there are also bacteria (*S. aureus*, *Klebsiella*, *E. Coli*), resp. fungi (*Trichophyton mentagrophytes*), which can cause skin problems when overgrown or purulence on contact with damaged skin. Some bacteria such as *T. mentagrophytes*, *Candida albicans*, *Penicillium citrinum*, and *S. Aureus* cause the growth of mold on feet, allergies, and the transmission of infections. The skin is the largest area (1.6–2 m^2) of the human body in contact with the environment. It has a thickness of about 1.5–4 mm. It can absorb fat-resistant substances but is impermeable to water. On the surface, there is sweat composed of lactic acid and urea. Lipid-containing sebum is also present on the surface of the skin. Resident bacteria on the skin break down these substances into free fatty acids (the acidic environment of the skin), resulting in a naturally bacteriostatic effect on several microorganisms. On the surface of the skin, the skin fat is composed of free fatty acids, bound fatty acids, triglycerides, and other components (squalene and cholesterol). The surface layer of the skin has a pH of 4–7. Human sweat has a pH of 5.5–6.7. Healthy human skin has a pH of 5, a temperature below 37°C, and is relatively dry. Contains lysozymes. There are around 100 bacteria/cm^2 in dry areas of the skin and up to 107 bacteria/cm^2 in wet skin. Gram-positive cocci and sticks appear on the skin. The main component is *S. Epidermis* bacteria, which can form a biofilm. These bacteria can be pathogenic under certain conditions if they get into body tissues, for example, during surgery. Human sweat contains up to 99% water, has a pH of 3.8–6.4, and does not smell. Unpleasant odors are caused by bacteria *S. Aureus*, *Corynebacterium minutissimum, and Tripophylic diphtheroid*, which break down sweat and subcutaneous fat into a variety of substances including propionic and valeric acid, free higher fatty acids, trimethylamine, ammonia, mercaptans, and aldehydes. The ammonia concentration of 1.5 ppm is a level where one already smells an unpleasant odor.

Some fatty glands also secrete fat, which forms an emulsion with sweat, dirt, and dust, adheres to the fibers, is difficult to remove, and serves as food for bacteria. Because a neutral pH is optimal for the growth of bacteria, their multiplication is suppressed in the first stages after the elimination of sweat and fat. Gradually, however, the bacteria break down these substances, which results in an increase in pH and the promotion of their multiplication.

Sweat and skin fat are food for bacteria and fungi, but on the other hand, acidic pH limits their living conditions. Healthy skin is therefore in balance and there is no overgrowth of bacteria or their mass death. Disruption of this balance (e.g., with an antimicrobial agent) can result in serious problems and skin diseases. For these reasons, the use of antimicrobials for products that are in long-term contact with the skin is not recommended. The odor should be an indication of the need to change the fabric or its washing. The chemical composition and moisture of the skin determine what bacteria and in what amounts will be present on the skin. The surface of the skin (epidermis) is not very suitable for the growth of bacteria, because it is often dry and has a low pH. Most of the microbes are around the sweat glands and hair follicles, where the environment is wetter and nutrients are present. These are mainly urine, amino acids, salts, and lipids excreted by the skin. The microbial flora on healthy human skin is described in (Korting et al. 1988). Microorganisms are an integral part of human skin. There is about 100–1,000 microorganisms/cm^2 on clean skin, which do not cause health or odor problems.

Corneybacterium Diptheriae, *S. Aureus*, *Micrococcus Luteus*, *S. Epidermis*, *and Pityrosporum ovale* are the most common on the skin. *S. Albus* (white staphylococcus) is also a frequented bacterium on the skin surface, which is a white spherical cluster-forming bacterium. The predominant part is aerobic Gram-positive cultures (catalase-positive coca). Most of these bacteria are beneficial to healthy skin because they prevent the colonization of pathogenic bacteria and regulate the presence of other microorganisms. When the skin is broken, these bacteria can penetrate the skin and cause various diseases. The skin itself acts as an antibacterial barrier. The upper layer of the epidermis, called the stratum corneum, is composed of flat cells (corneocytes) interconnected to form a stratum corneum of keratin mixed with skin lipids. The connecting material (lipids) and the corneocytes themselves form the first protective layers of the skin (there are about 15 of them). 107 particles are released from healthy skin daily, of which about 10% contain viable bacteria. The stratum corneum layer is renewed approximately once a week (Evans and Stevens 1976). Usually, various strains of staphylococci survive in it, which are removed from the skin surface, for example, by conventional antimicrobial agents. The number of colonies (CFU) remaining after hand disinfection is 10^3–10^4, after disinfection by washing 10^5, and after sterilization by the order of 10^6. Fatty acids on the surface of the skin have fungicidal and bactericidal effects. In a favorable environment (wet and in the presence of nutrients), larger bacterial colonies (around 105 CFU) survive two to eight times longer than small colonies (around 102 CFU) (Hendley and Ashe 2003). Survival time ranges from 5 to 50 days depending on the type of materials. Also, several nonsoiling treatments that increase hydrophobicity and reduce dirt-trapping have secondary antibacterial effects. Penetration of nanoparticles through the skin can be either through the hair follicles and sweat glands or through the superficial stratum corneum, which is not continuous but consists of fragments with crevices around 50 nm (Baroli 2010). Skin fragments have a wide distribution of characteristic size with a median around 20 μm and 10% below 10 μm. Thus, these particles can pass freely through dense fabrics with a pore size of 10–15 μm. However, densely finished fabrics are impermeable to corneocytes (keratinized skin cells), which in the hydrated state are larger than 30 μm (Mackintosh 1978). Fabrics can also generally irritate the skin mechanically (many people have increased sensitivity to wool) or cause allergic reactions. These usually cause type IV allergic reactions with a delay of 24–72 h from contact with the fabric. They are manifested by redness and blisters accompanied by increased itching. Clothing fabrics that are in contact with the skin usually have enough moisture and heat around 37°C, which supports the growth of bacteria. Thus, underwear and socks are most prone to odors, as there is a problem with removing moisture by evaporation. In many cases, therefore, the presence of bacteria is not a problem. However, larger populations of staphylococci cause at least an unpleasant odor, especially in a humid environment. In the case of pathogenic bacteria, transmission or spread of the infection can occur. Viral infections caused by coronaviruses of SARS types were pandemic. Covid-19 caused by SARS-CoV-2 is a highly infectious disease. The virus can be transmitted from person to person, it is spread by droplets during sneezing, coughing, or body contact. The incubation period is approximately between 1 and 14 days and even during it, the disease is infectious. Transmission of the virus to other persons was confirmed in the period 1–3 days before the onset of symptoms of the disease. The median incubation period is approximately 5 days. The disease has symptoms and course very similar to more severe influenza or other similar influenza diseases. It is very difficult at first glance to distinguish coronavirus infection from influenza. The first symptom of COVID 19 is fever (approximately 90% of cases), severe fatigue, and shortness of breath. Later, dry irritating cough or muscle and joint pain are added. The vital signs of admitted patients are usually stable. Loss of smell (hence the taste) called anosmia (in 10%–30% of cases) is also possible without other symptoms. More severe cases can lead to pneumonia (pneumonia), acute inflammation of the heart muscle, organ failure, and death.

2.4 EFFECTS OF MICROORGANISMS ON TEXTILES

It is known that natural macromolecular substances (cellulose and polypeptides) are attacked by degraded microorganisms. Also, under certain conditions, several synthetic polymers are microbiologically degraded (e.g., locally colored) or upset. Surface degradation usually occurs, which often contains easily biodegradable substances (antistatic agents, plasticizers, hardeners, etc.). The destruction of polymer chains is caused by metabolites resp. extracellular enzymes produced by microorganisms. The result is discoloration, odor generation, and surface itching in the first phase, and loss of strength resp. weight after prolonged exposure. Polymeric materials are attacked mainly by fungi of the Aspergillus Niger (see Figure 2.6).

FIGURE 2.6 Head of Aspergillus niger (Image courtesy: https://en.wikipedia.org/wiki/Aspergillus_niger).

It has been found that biodegradation is favorably affected by the hydrophilicity of polymers and increases with decreasing molecular weight (Doi and Fukuda 1994). Biodegradation is also affected by the supramolecular structure of fibers. More ordered and crystalline structures resist the action of microorganisms better than disordered amorphous regions (Eldsaeter et al. 2000). Antimicrobial degradation is not always undesirable. On the contrary, at the end of the product's life, it is advantageous if the degradation can be carried out in a natural way. Also, in the field of medical textiles (surgical sutures, bandages, tissue replacements, etc.) controlled biodegradation is often required. Thus, microorganisms can disrupt textiles both by degradation and by changing the shade of the dyes and creating an unpleasant odor. However, the odor is also caused by excrements of the human body, cosmetics on textiles, food, and beverage residues, resp. dirt and dust. The reason for these differences is the fact that water is available on synthetic materials at all times. At Lyocell, moisture diffuses inward and then the fiber surface remains (relatively smooth) with a minimum of moisture insufficient for bacteria to grow. After 15 times washing, the number of *S. Aureus* bacteria in untreated cotton fabrics increases about 30% (Barnes and Warden 1971). Microbial cultures on textiles cannot be removed by washing at lower temperatures (40°C–60°C) and boiling is required, which is not always possible. The odor is caused by waste products of bacterial metabolism, which are of low molecular weight and volatile. The perceptible odor is caused by approximately 10^4–10^5 bacteria. Some odors are sensitive to low pH around 2–3 (acetals).

Cellulose materials degrade due to enzymatic action on water and carbon dioxide. Cellular fungi are active in the pH range 4–6.5 and cellular bacteria in the pH range 7–9. Their action first results in enzymatically catalyzed disruption of ether bridges between glucopyranose rings in the presence of water. Thus, the insoluble polymer is gradually converted to soluble sugars, which are metabolized within the microbial cells. Microorganisms damage cotton by gradually breaking the primary and secondary walls It was found that their moisture content correlates the most with the biodegradability of cellulose fibers (Park et al. 2004). In textile technology, anti-rot agents are added to sizing baths. In compost and sewage, cellulose materials degrade very easily. After 6 weeks, the viscose fibers are completely degraded and the cotton is 80% degraded. Under the same conditions, the degradation of polyester and polypropylene fibers is negligible. Fibers made of natural proteins usually resist fungi much better than cellulose. Keratin fibers are degraded by proteolytic enzymes (collagenosis, chymotrypsin) (Aray et al. 2004). However, regenerated protein materials, which have shorter chains and are poorly organized, are degraded significantly more. Synthetic fibers generally cast mold and bacteria very well. Some bacterial strains can use parts of short-chain polyamides as a source of carbon and nitrogen. One strain of *C. aurantiacum* readily degrades oligomers of PA6.

Phanerockaete chrysosporium reduces the molecular weight of polyamide 6 by 50% after 3 months of treatment (Klun et al. 2003). *Bjerkandera adusta* fungi can completely decompose polyamide 6 fibers within 40 days. Microbes are also able to damage the surface of polyester fibers. Antifungal agents are usually based on organic copper compounds (copper naphthenate). Especially materials containing wool fibers are also attacked by insects, especially moths. At temperatures above 60°C, for example, insects die at all stages of life. Even a temperature of 40°C is enough to kill all the eggs of the clothing pier (*Tineola bisselliella*) and kill 80% of its larvae. Low temperatures have a similar effect. Permethrin is mainly used as a relatively universal agent against moths and insects. Permethrin (3-phenoxiphenyl) methyl 3-(2,2-dichloroethenyl)-2,2-dimethylcyclopropane carboxylate is a biodegradable synthetic pyrethroid with limited washability. It is a colorless crystalline powder with very low solubility in water (0.2 mg/L at 20°C) and a melting point of 35°C. It acts as a contact and food poison. It has low toxicity to animal cells but in contact with the skin or may cause allergic reactions if inhaled. It is effective against microorganisms and rots even in very low concentrations. Permethrin is used in the form of aqueous suspensions (Dodd et al. 1981).

2.5 ANTIMICROBIAL ACTION

Antimicrobial agents are characterized by a negative effect on bacterial growth or virus replication. They are divided into microbiostatic (bacteriostatic/virostatic), which stop reproduction, and microbiocidal (bacteriocidal/virucidal), which kill microorganisms. Antibacterial agents are:

- Bacteriostatic – inhibiting the growth of bacteria, causing their death over time;
- Bactericidal – more or less selectively killing bacteria.

Interestingly, at very low concentrations, most antimicrobials have a stimulating effect (supports growth rate) because they accelerate microbial metabolism. *Virustatic* (often nontoxic) substances are functioning outside the cell inhibiting penetration of viruses into cells. The reversible binding to viruses prevents their use as a drug, because, upon dilution, the inhibition is lost. Current virustatic materials (heparin, polyanions, etc.) are focused on virus-cell interactions that are common to many viruses. *Virucidal* substances cause irreversible viral deactivation and their effect is retained even if dilution occurs after the initial interaction with the virus (Cagno et al. 2020).

There are plenty of virucidal substances from simple detergents to strong acids, some polymers and some nanoparticles. Virucidal activity is in some cases caused by the releasing of ions (e.g., Cu^{2+}). These substances chemically damage the virus, but they are not affecting the host cells. In the ideal case, the drug should have all the positive properties of virustatic drugs such as broad-spectrum efficacy and low toxicity and a virucidal mechanism. According to the mechanism of action are antibacterial substances divided into the following groups (Vodrážka 1996):

- Cell wall damaging agents (certain antibiotics),
- Substances disrupting the structure of the cell (cytoplasmic) membrane (formaldehyde, strong oxidizing and reducing agents, fatty solvents, etc.),
- Substances acting on microbial enzymes inside the cell (oxidizing agents, metal ions),
- DNA-reactive substances (certain antibiotics).

In an aqueous medium, the active substances can either disrupt the cell walls (e.g., by oxidation) or penetrate through the cell wall to the cell membrane or inside the cell. There are several approaches in which microbes are killed by contact without the need to release the active substance. An example is the grafting of antimicrobial polymers (e.g., poly(*N* alkyl vinyl pyridine salts). Also, a surface coating containing photocatalytic TiO_2 (Agrios and Pichat 2005) contacts the destruction of microbes, because light-induced hydroxyl radicals have a very small area of active action of the order of nanometer (Ohko et al. 2001). In terms of action, antibacterial agents are divided into the following:

- Gradually soluble in water (elution type),
- Firmly attached to the fiber (nonelution type).

Antimicrobial agents of the elution type function are based on the controlled release of the active substance. Their deactivation occurs due to the depletion of the active substance. Due to their release into the environment (whether in the air or in the presence of moisture), they can destroy the flora on healthy human skin and cause problems in biological (enzymatic) processing of textiles, resp. in biological wastewater treatment. Examples of gradually soluble substances in water are metal salts, where the antibacterial effect is induced by metal ions. In the case of substances gradually soluble in water, metal ions penetrate the cells and bind to the –SH group of enzymes, which causes a decrease in activity and death, resp. suppression of the growth of microorganisms. The required concentrations of metal ions are around one-millionth of a percent (ppm).

Nonelution-type antimicrobial agents are firmly bound to the fibers and act in direct contact with bacteria. Their deactivation occurs either by mechanical clogging of active sites or by abrasion of the fabric surface. These agents include low molecular weight organic antimicrobial agents such as tributyltin oxide (CH_3-CH_2-CH_2-CH_2-)$_3$-Sn-O-Sn-(CH_2-CH_2-CH_2-CH_3)$_3$, dichlorophen, and 3-iodopropynyl butyl carbamate. These compositions are relatively effective, but due to their limited stability to textiles and negative effects on the environment, their use is very limited. The use of low molecular weight quaternary ammonium salts for textile purposes is limited by their high solubility in water. Other examples of firmly attached substances are the quaternary ammonium salts chitin and resin-bound chitosan. For quaternary ammonium salts and chitin in contact with microorganisms, the metabolism of enzymes is altered, resulting in cell wall disruption and cell destruction. Due to the firm attachment on the surface of the fibers, there is no penetration into the cells, so that some bacteria are resistant to these agents.

When testing antimicrobial activity, a sample is placed on a petri dish containing a bacterial strain in a nutrient medium. The growth of microorganisms is evident in the sample without antimicrobial treatment. In eluents where the active substance is eluted, an inhibition zone appears. Over time, the microorganism may adapt and the antimicrobial activity decreases over time. Noneluting agents do not have an inhibition zone. Antibacterial or antifungal treatments can cause severe allergic reactions if fabrics are not rinsed sufficiently before first use (Matthies 2003). Dyestuffs can also be contact allergens. Dispersion Blue 35, 106, and 124, for example, have significant allergenic effects. The activation energy of thermal destruction of bacteria is sufficiently high 240–400 J/mol, which allows short-term sterilization, for example, food without their deterioration. Even if complete sterilization does not occur, a reduction in the number of colonies of bacteria and viruses is sufficient to prevent infection, because many pathogenic bacteria cause infections in large quantities. For example, approximately 10^5 bacteria are needed for the disease caused by *S. Aureus*. One of the potentially worst problems of the future is the emergence of bacterial strains resistant not only to antimicrobials but also to drugs. There are several recommendations (e.g., the American Medical Association) to limit the use of antimicrobials only in necessary cases. Many biocidal substances have toxic and ecotoxic behaviors that are potentially dangerous to human health. This is the case, for example, with organic tin-containing compounds that accumulate in cells and are ecotoxic. Some quaternary ammonium compounds cause allergies and are toxic in the aquatic environment. There is Directive 98/8/EC of February 16, 1998, on biocidal products on the market. Textiles are in the second main group. This directive is gradually being supplemented by lists of permitted substances. According to several regulations in different countries, the use of different products with a biocidal effect is restricted or prevented. This applies in particular to substances based on chlorinated phenols (including their salts and esters).

2.6 ANTIMICROBIAL AGENTS

Many antimicrobial agents used in textiles are well-known substances widely used in cosmetics and the food industry. This is a certain advantage when obtaining hygienic and ecological certificates. Another advantage is the need for very low concentrations (max. around 2%–5%), especially for organic compounds. If antimicrobial additives are used in polymers, the requirements are:

- Very low solubility in water of alkalis and acids,
- Chemical resistance to acids, alkalis, oxidizing agents, and possibly UV, resp. light radiation,
- Sufficient thermal stability.

For fibers with antimicrobial functions, it is required that antimicrobial agents do not adversely affect the fiber preparation process (spinning) or the properties of the resulting fibers (especially mechanical). Ingredients of the elution type must be able to migrate to the surface of the fibers. When selecting a suitable antimicrobial additive for synthetic fibers, a particle size below 2 μm, stability and nonvolatility up to 300°C, the ability to disperse uniformly, and compatibility with the fiber-forming polymer is required. Ultrafine powders (typical size below 1 μm) and nanopowders (typical size below 0.1 μm) are commonly used. The price of nanopowder is about 1.3 times higher than the price of ultrafine powders. With the help of mechanical milling, finenesses of 0.7–1 μm can be achieved. Nanopowders are prepared by nanomilling or by chemical means. Their main benefit is the denser covering of space (small interparticle distances) and thus a greater probability of contact with microorganisms (Hett 2005). Another advantage of nanoparticles is the compatibility of dimensions with the dimensions of polymeric chains, which results in improved interactions and an increase in the number of interconnections.

2.6.1 General Aspects

It is advantageous if the antimicrobial agents act in combination, i.e., they affect different functions of the bacteria and their cells. This can reduce the possibility of resistant strains of bacteria. Antimicrobial agents are generally divided into two categories according to their action on a particular group of microbes. Bactericidal agents destroy bacteria and bacteriostatic agents prevent the growth of bacteria. Similarly, fungicides are divided into fungicidal and fungistatic agents, algae growth agents are algicidal, resp. angiostatic. Mites effective against mites are acaricidal or acaristatic. Germicides destroy special bacteria and fungi caused by higher humidity and temperatures during the storage of textiles. The basic requirements for antibacterial fabrics are:

- Prevention of the transmission of the disease by pathogenic microorganisms,
- Compatibility with skin (skin cells) and skin microorganisms,
- Reduction of microbial contamination,
- Stopping microbial metabolism to prevent odors,
- Protection of textiles against stains, changes in shade, and deterioration of useful properties.

The ideal requirement is to remove the cell walls of microorganisms and possibly toxic products of their decomposition selectivity only for pathogenic microorganisms resp. harmful organisms.

In terms of the use of antimicrobial agents, there are two main possibilities. The first is to use suitably treated fibers containing active agents either alone or (more economically advantageous) in mixtures with conventional fibers. Concerning textile processing, it is advantageous if there are as few antimicrobial fibers as possible in the mixture, which ensures the desired effect. The second possibility is to use the antimicrobial agent as a part of the finishing either as a separate treatment or its part of the standard finishing (noncreasing, nonflammable). These agents can be delivered to the dyeing bath as well. There are several possibilities for attaching an antimicrobial agent to fibers. The basic ones are:

- Use of special polymers, copolymers, or modifiers;
- Mixing of polymer melts or solutions of special and common fiber-forming polymers;
- hybrid spinning, where the active polymer component or the component with the antimicrobial agent forms the outer layer (sheath) and the classic fiber-forming polymer forms the core;
- Addition of the active substance as an additive to the polymer melt (solution) before spinning, in the form of particles or capsules containing the active substance inside;
- Preparation of fibers containing sites to which the active agent subsequently binds;
- Addition of active substance to the precipitation bath during wet spinning;
- Binding of the active agent to the finished fibers by dyeing and printing techniques (bath extraction, thermal insulation, screen printing);
- Fixing the active agent to the surface of the fibers using resin treatments (coating);
- Fixation of the active agent on the surface of the fiber using functional modifications (dextrins, sol–gel procedures);
- Fixation of the active agent on the surface of the finished fibers by grafting and surface copolymerization.

The purpose is for the antimicrobial agent of the elution type to be available on the surface of the fibers via a diffusion mechanism (thermally activated transport, first-order kinetics) or only by leaching/elution (nonthermally activated transport, zero-order kinetics). For nonelution type compositions, resp. photocatalysts must be the active agent in contact with microbial organisms. In the case of eluents, the rate of diffusion or leaching is decisive for the binding of active ions or on the transfer of the capsules with the active substance. In some cases, diffusion is possible only in the presence of moisture and some ions (Na^+). For photocatalytic agents, light containing UV radiation part (wavelength below 400 μm) is required. In terms of action, antimicrobial agents can be divided into:

- Exhausting in time (most of the elution type). Here, the antibacterial effect is lost over time. It is advantageous if these compositions can be used to "revive" the effect during washing, resp. cleaning (e.g., N-halamines).

- Not exhausted in time (most of the nonelution type). Here, the antibacterial effect is lost over time because of dead bacteria, resp. cell wall residues mechanically limit the action of the active substance. Washing removes these residues and can reuse the fabric. With photocatalysis agents, dead bacteria and cell wall debris are automatically destroyed.

Passive, antimicrobial textiles do not contain active substances. The suitable rough surface structure (lotus effect) and the choice of fibers make the conditions for the growth and life of bacteria difficult. Due to the "lotus effect," bacteria are poorly maintained on the surface of textiles and it is difficult to form colonies (Ramaratnam et al. 2008). The relatively smooth fibers, which quickly remove surface moisture, actually dehydrate the environment occupied by microorganisms. Active antimicrobial textiles contain antimicrobial agents acting on the cell surface or penetrating inside. Cell wall, metabolism, or replication of microbes are usually disrupted. The main field of application of antimicrobial agents is for sanitary textiles and treatments that prevent the transmission and spread of pathogenic bacteria. Other fields of application are deodorant fabrics and treatments that prevent odors due to microbial degradation or treatments preventing degradation and decomposition of fibers by microbes. According to the type of binding of the active substance to the substrate, the antimicrobial effects are divided into permanent and limited to maintenance (washing, cleaning). The needs and limitations of antimicrobial textiles are specified in the article (Hober 2004). Antibacterial fabric products must not be toxic, irritating, allergic, mutagenic, or carcinogenic. They must not penetrate the skin or adversely affect normal skin flora. It must not be dangerous for the environment or form toxic substances during normal maintenance operations and exposure to sunlight, resp. UV sources.

Protection of humans against the biological attack of microbes comprise an aesthetic point of view (suppression of odors caused by bacteria), hygienic point of view (suppression of skin diseases caused mainly by fungi), and medical point of view (protection against pathogenic and parasitic microorganisms especially in hospitals and public spaces and protection against microbes responsible for illnesses). Important is the protection of textiles against biological degradation caused mainly by fungi and rot-producing fungi. There are three basic mechanisms by which antimicrobial agents protect textiles and their wearers:

a. Controlled release of the active substance without regeneration of the active substance,
b. Controlled release of the active substance with the regeneration of the active substance,
c. Barrier or blocking the function of the active substance.

An example of controlled release without regeneration is Letilan fiber. It is a polyvinyl alcohol fiber (PVA) that is converted to an acetal by reaction with 5-nitro-furyl acrolein in the presence of acids as catalysts (Volf et al. 1963). Antimicrobial activity is caused by the gradual release of active nitro-furyl acrolein in the presence of moisture. Another example of controlled release is the Permox process, in which zinc and peroxide are gradually released (Bajaj (2001)). The effectiveness of a controlled release process depends on the distribution of the active substance coefficient, surface absorption, vapor pressure, and water solubility. An example of controlled release with regeneration is halamines, where the regeneration of the active substance is carried out, for example, by adding hypochlorite-type bleaching agents to washing baths (Hui and Debiemme-Chouvy 2013). This group can also include cases where active substances are formed from moisture and atmospheric oxygen by photo-oxidation under irradiation in the UV region (photo-oxidation by use of TiO_2). Between the controlled release and the regeneration of the active substance lie microencapsulation processes, in which the active substance in microcapsules diffuses to the surface of the textile, where it is released either due to the presence of moisture or due to degradation by radiation. Barrier and blocking functions are realized either using an inert film that is impermeable to bacteria or utilizing a layer that kills bacteria by direct contact. The formation of an inert layer impermeable to bacteria usually requires a larger deposit than the active layer killing bacteria. An example of an active layer killing bacteria is an organosilicon polymer containing quaternary ammonium salts. Hydrolysis and condensation of 3-(trimethoxyl silyl) propyl dimethyl octadecyl ammonium chloride on the surface of a fabric containing reactive (OH) groups produces a polysiloxane. This technology has several practical disadvantages associated with the release of methanol and the incompatibility of alkoxysilanes with aqueous emulsions used in refining, etc. (Bajaj 2001). Known antimicrobial agents include several substances, only some of which are suitable for antibacterial fabrics. In many cases, these substances can be added as additives to polymer melts before spinning. They can also be applied in the finishing phase.

2.6.2 Groups of Antimicrobial Agents

The basic groups of antimicrobial agents suitable for textiles are:

- **Coagulants**, especially alcohols, cause irreversible denaturation of proteins. Highly reactive substances such as halogens, peroxides, and isothiazones contain free electrons. They react with all organic structures and cause oxidation of thiols (SH) in amino acids. Even at low concentrations, they are dangerous because they promote mutations and dimerizations. The application is mainly for disinfection.
- **Oxidizing agents** such as aldehydes, halogens, and chemical warfare agents attack the cell membrane, enter the cytoplasm, and act on enzymes in microorganisms. These salts bind to synthetic fibers, for example, through association with dye molecules. Direct durable bonds can be achieved on natural fibers. Glyoxal, which is used as a crosslinking agent in noncreasing treatments, has significant bactericidal effects at concentrations of 1%–2%. It was found that glyoxal-treated cotton has significantly better antibacterial properties than chitosan-treated cotton (Kittinaovarat and Kantuptim 2005). The problem is the low efficiency for some Gram-negative bacteria and viruses.
- **Polycationic absorbents** are represented by quaternary ammonium salts, biguanides, amines, and glucoprotamines. They bind to cell membranes and

disrupt LPS structures, resulting in the rupture of cell walls. It has been shown that hydrophobic polycations need to be bound to the surface of the fibers to achieve antibacterial activity. For short chains, it is difficult (if not possible) to penetrate the bacterial wall. The minimum length of polymer chains, expressed as average molecular weight, is around several thousand.

- **Metals and metal complexes** containing silver, copper, zinc, oxides, metal sulfides, metal-containing ceramics, cadmium, and mercury cause inhibition of active enzyme centers and thus prevent metabolism.
- **Antibiotics** such as tetracycline, gentamicin sulfate, and Garamycin can act on bacteria by several mechanisms. Their applications are for drugs mainly.
- **Isothiazolinones** are stable over a wide pH range and act mainly as antibacterial agents (Silva et al. 2020).
- **Pyrethroids** act at pH 6–8 and have a significant effect on fungi. Natural pyrethroids are obtained by extraction from chrysanthemum flowers (chrysanthemum cineraria folium). A representative of synthetic pyrethroids is pyrethrin.
- **Aldehydes** Glutaraldehyde $CHO\text{-}(CH_2)_3\text{-}CHO$ works best in an acidic environment and the rate of elimination of microbes is high. Formaldehyde generating agents work best in the alkaline range and the elimination of microbes is fast.
- **Haloorganic compounds** work best in the pH range 4–8. The elimination of microbes is fast, but their long-term effect is limited. The effects depend on pH. The most popular antimicrobial agents included diphenyl ethers, especially Triclosan, which have been used in soaps, toothpastes, deodorants, mouthwashes, and various skin creams for many years. Triclosan is very sparingly soluble in water and its molecules gradually diffuse to the surface of the fiber, where they form an inhibitory environment. Triclosan is banned by EU regulation and it is no longer used.

A highly effective biocidal agent is formaldehyde, which is released from conventional resin treatments (noncreasing). However, its use is practically impossible by the regulations of most countries defining its permissible quantity, resp. prohibiting its presence on certain types of clothing textiles (baby clothes). For antimicrobial modification of fibers, it is possible to easily use, for example, the addition of 0.5%–2% of organic nitro compounds. Many polymers can be modified in the prespinning phase with substances containing carboxyl or sulfo groups (e.g., modified polyesters (Militký et al. 1990)). After spinning, the cationic antimicrobial agents can then be bound to these groups by ionic bonds. A popular additive of the nonelution type is "octadecyl aminodimethyl trimethoxy silylpropylammonium chloride." This composition is applied to the fibers either by exhaustion from the bath or by means of padding. After drying, condensation follows to form a siloxane polymer on the surface of the fibers. Another frequently used composition of this type is polyhexamethylene biguanide, which is applied in the same way as compositions. A process for the formation of chloramines on cellulose fiber is also described (Chen et al. 2019). First, methylol 5,5-dimethydiantoin is covalently bound to the OH groups of the cellulose, which is converted to chloramine after treatment with hypochlorite. The problem with using an antimicrobial treatment of this type is the risk of yellowing during ironing and the relatively significant loss of strength caused by the formation of oxicellulose. A new class of SAM (sustainable active microbicidal) polymers is made from biologically inactive monomers (Goldschmidt and Streitberger 2018). Thus, only the polymer, which does not release any active substances, has an antimicrobial effect. SAM polymers have a carbon backbone with densely distributed amino-containing side functional groups. Polymers are insoluble in water but soluble in many organic solvents (especially polar). They can be prepared from commercially available monomers. The typical molecular weight is around 50,000–500,000 and the glass transition temperature is around 40°C. Their thermal stability is up to 180°C. Many metal ions Ag^+, Cu^{2+}, Cu^+, and Zn^{2+} have significant antimicrobial effects. Metals are bonded to textile materials via carboxyl or hydroxyl groups. They are currently used not only as particles but also in the form of complexes. The silver and copper particles or ions are probably most frequently used especially for antibacterial and antiviral textiles.

2.6.3 Silver

The silver (Ag) particles can be supplied directly into fibers as fillers to the polymers. In the case of nanoparticles, it is possible to use techniques known as dyeing with disperse dyes. These are the following basic mechanisms of action of silver (Ag^+ ions) on bacteria:

a. Interference with electron transport in bacteria,
b. Binding to bacterial DNA after entry of Ag^+ ions into cells and loss of replication ability,
c. Interaction with thiol groups of proteins (SH) after entry of Ag^+ ions into cells,
d. Interaction with cell walls without penetrating inside and the formation of complexes on their surface preventing dehydrooxigenation processes.

The action of Ag^+ ions, which are supposed to deactivate cysteine groups, is expressed by relation;

$$R - SH + Ag^+ \rightarrow R - SAg + H^+$$

Silver ions penetrate microbial cells and affect nucleic acids, resp. they cause poisoning of enzymes on the surface of the cell wall. The amount of silver ions is limited by the diffusion rate and charge level. The contact time required to destroy microbes is in hours. Silver ions are deactivated by chemical reactions with oxidants, sulfur, and chlorine. The advantage of silver is its nontoxicity to human and animal cells, although long-term contact can cause greying of skin. The use of silver as an antimicrobial agent has several practical limitations due to inappropriate particle size and inappropriate treatment with inappropriate concentration. In the region below 10 nm, nanoparticles have a significant increase in surface area, which has a positive effect on antibacterial activity. The main silver disadvantage is that it is active only against a limited number of bacteria, loses its effectiveness by oxidation (e.g., after washing), and is relatively expensive. There are two main types of silver on the market suitable for textile applications: powder (20 nm) and colloidal solution (30 nm). Colloidal silver is an aqueous solution containing 70–150 ppm of silver particles with sizes from 11.6 nm. Although particle agglomeration occurs, the size of the agglomerates does not exceed 50 nm. Colloidal silver using ethanol instead of water allows the use of silver particles with a size of 2–5 nm. It turns out that for this particle size, only 10 ppm of silver is sufficient to completely inhibit bacterial growth. The antibacterial activity of aqueous colloidal silver solutions depends on their concentration and particle size. For particles of 11.6 nm, concentrations of at least 50 ppm must be used to completely remove bacteria, and for particles of 5 nm, a concentration of 10 ppm is sufficient (concentrations are calculated from the weight of the textile substrate). Elemental silver, which is applied to surfaces either by chemical vapor deposition (Szabo and Winefordner 1997), by platting (Blair 2002), using low-temperature magnetron deposition (Dowling et al. 2001), or by ion beam assisted deposition (Woodyard et al. 1996). These procedures have both advantages and disadvantages. In the article (Ho et al. 2004), a combined method of antimicrobial protection is proposed, consisting of the application of a polymer network that can effectively capture silver nanoparticles. Polyethylene glycol is applied to its surface, which reduces the deposition of bacteria (repels bacteria). The resulting coating thus prevents the attachment of bacteria (repellent to bacteria), acts in a contact manner, and still allows the release of the active substance (elution type). A 0.005 molar solution of $AgNO_3$ was used to prepare silver nanoparticles, which were stirred at reflux with 1% sodium citrate and heated until a pale-yellow solution was obtained. After cooling, a dispersion of silver nanoparticles was obtained (Kamat et al. 1998). To deposit silver nanoparticles on the surface of materials, surface activations can be used to formaldehyde R-CHO groups that are capable of reacting with Tolens reagent to form metallic silver (Figure.2.7). The reagent containing $Ag^{+}(NH_3)_2$ ions is prepared by mixing an aqueous solution of silver nitrate with ammonium hydroxide $(NH_4)OH$ in an alkaline medium (Jing et al. 2004).

The properties of silver nanoparticles concerning their antimicrobial action are described in the article (Williams et al. 1998). In sewing threads coated with nanosilver, the growth of culture in the müeller-hinton agar was inhibited. The growth of culture was determined by turbidity and also read at 600 nm (Venkataraman et al. 2014). Under optimal conditions, the growth rate constant of bacteria *S. Epidermis* in a bath containing sufficient food was 0.78 h^{-1} (first-order kinetics was considered). In the presence of polyester fabric, the rate constant of growth 0.72 h^{-1} was slightly lower. In the presence of silver-coated polyester fabric, the growth kinetics was significantly lower with a growth rate constant of 0.15 h^{-1} due to the release of silver ions from the surface (Klueh et al. 2000). Direct binding of silver clusters (by treatment in $AgNO_3$ solution) to polyester fabrics activated by oxygen plasma or by

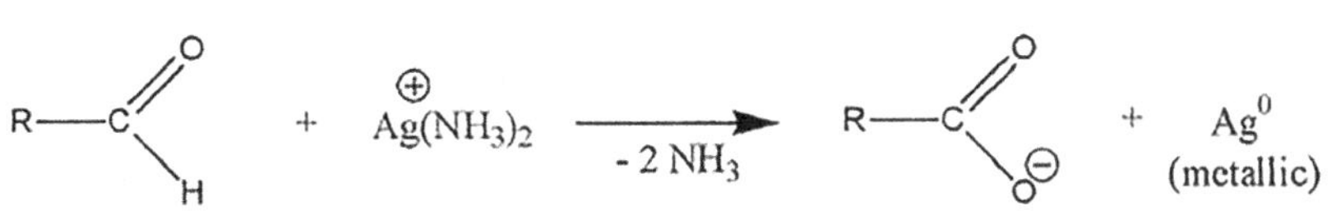

$$\mathrm{R{-}CHO} + \mathrm{Ag(NH_3)_2^{\oplus}} \xrightarrow[-\,2\,\mathrm{NH_3}]{} \mathrm{R{-}COO^{\ominus}} + \underset{\text{(metallic)}}{\mathrm{Ag^{0}}}$$

FIGURE 2.7 Tolens reaction for preparation of nanosilver.

irradiation with a UV lamp in a vacuum was performed (Yuranova et al. 2003). The maximum silver bacterial activity was obtained by applying silver corresponding to a concentration of 1.5 g $AgNO_3$ per liter. The binding of silver particles on cotton with subsequent sonification is described as well (Yuranova et al. 2006). The treatment of cotton, polyester, and blended fabrics with 20 and 50 ppm colloidal nanosilver in ethanol was applied (Lee, H. J. et al. 2003). It has been shown that fabrics treated in this way have good resistance to washing and retain long-term antibacterial activity. The sonochemical process of impregnating silver nanoparticles on nylon, polyester, and cotton fabrics was developed (Perelshtein et al. 2008). The coating was stable on the fabric for at least 20 washing cycles in 40°C water. The intensity of the release of Ag^+ ions from polyamides increases with increasing silver concentration and decreases with increasing particle size. Using polyamide monofilaments containing elemental silver particles, it has been shown that their antimicrobial activity did not manifest itself until a 1-week deposit in an aqueous medium (Kumar and Muenstedt 2005). Silver can also be applied using supercritical CO_2, which penetrates the fibers faster than liquids and causes swelling, especially of nonpolar polymers. In some cases, the support to which the silver is bound promotes Ag^+ diffusion and shortens the initial time for which the silver is inactive. This time is also significantly shortened by the use of supercritical CO_2. However, the release of silver ions from $AgNO_3$ or other salts can cause changes in the concentration of calcium and zinc on the skin surface. Silver complexes in zeolites are also used. Zeolites are aluminosilicates characterized by a 3D structure and a hole in the middle with a diameter of about 0.2–1 nm. Access to the main cavity of zeolite X is controlled by a 12-oxygen ring with a crystallographic diameter of 0.74 nm. The void volume is 53%. Ag^+ ions are formed here in the presence of moisture. Ion formation is also the result of the dissociation of complexes in zeolites. The controlled release of silver ions depends to a large extent on the complex used. AgION technology uses an antibacterial agent, where the silver ion carrier is zeolite (silver is bound in a complex). Due to the presence of sodium ions present in moisture (e.g., sweat), the ion-exchange mechanism results in the controlled release of Ag^+ ions and thus the inhibition of bacteria. The problem is that improper washing can rapidly reduce antimicrobial activity. Bioactive glass particles containing 60% SiO_2, 2% Ag_2O, 34% CaO, and 4% P_2O_5 were used for wound healing dressings (Blaker et al. 2004). The combination of silver complexes with poly-hexamethyl biguanide is preferred. The cationic polymeric biguanide effectively enhances the action of Ag^+ (Parquet et al. 2001). Zinc, which affects intracellular transport processes, is also markedly antibacterial. It can be combined with silver and zeolites. Silver–zirconium salts were used especially for dressing materials because they promote healing and have a relatively strong effect against a wide range of bacteria.

2.6.4 Copper

Copper (Cu)-based antibacterial agents are effective against a wide range of bacteria, fungi, viruses, and mites. They are significantly more stable to oxidation during washing and significantly more cost-effective. The antimicrobial activity of Cu is known since ancient times, and surfaces containing a significant amount of Cu have demonstrated their efficacy to inactivate viruses. Copper ions play a dual role in the healing of burn injuries, i.e., prevent the wound from infection and help in the formation of bone matrix. Copper reversibly denatures DNA competing with hydrogen bonding within the macromolecule (Figure 2.8).

Copper is a redox-active metal and can donate and accept electrons to shift between reduced (Cu^{+}) and oxidized (Cu^{2+}) states. Copper is one of the most potential materials with a broad spectrum of antibacterial, antiviral, antifungal, and antimite properties. Cu particles are used for their catalytic and bactericidal properties in the dark. Cu ions are biocidal by binding to specific sites in the DNA. These ions enter the bacterial cell wall and disrupt the cytoplasm. Cu also produces reactive oxygen species (ROS) (highly oxidative radicals), leading to the damage of iron–sulfur enzymes (Solioz 2018).

The proposed inactivation mechanisms include both toxicities toward virions of Cu ions released from the Cu-containing surface and attack of viral proteins and lipids by ROS generated from Cu reacting with exogenous hydrogen or molecular oxygen (Grass et al. 2011). The redox reaction of copper ions in the cytoplasm by reducing environment in cells (see Figure 2.9) are producing so-called ROS (Sunada et al. 2012);

$$Cu^{2+} + O_2^{\bullet -} = Cu^{+} + O_2$$

$$Cu^{+} + H_2O_2 = Cu^{2+} + OH^{-} + {}^{\bullet}OH$$

$$H_2O_2 + O_2^{\bullet -} = O_2 + OH^{-} + {}^{\bullet}OH$$

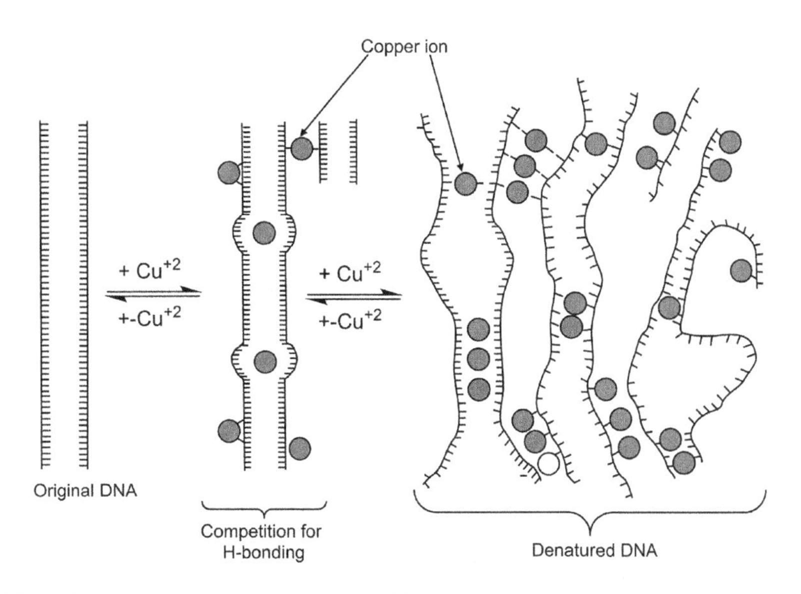

FIGURE 2.8 Reversible denaturation of DNA by copper ions.

Cu^{2+} is firstly reduced by superoxide ($O_2^{\bullet-}$), and after is re-oxidized by hydrogen peroxide (H_2O_2) to produce hydroxyl radicals ($^{\bullet}OH$). These redox reactions require $O_2^{\bullet-}$or H_2O_2 for producing reactive hydroxyl radicals. Bacteria often produce $O_2^{\bullet-}$ and H_2O_2 during their metabolic processes in the presence of oxygen. Viruses are not metabolically active, and therefore it is unlikely that ROS were produced in this way. Hydroxyl radicals can react nonspecifically with lipids, proteins, and nucleic acid. Cu^{+} can also lead to thiol depletion not only in the GSH pool, but also in proteins and free amino acids. Under anaerobic conditions, glutathione-copper complexes (GS–Cu–SG) act as copper-donors for metalloenzymes (Figure 2.9). The dominant toxicity mechanism in bacteria is the displacement of iron from iron–sulfur cluster proteins by Cu^{+} (Solioz 2018). Cu ions are preferentially bound to the N terminal amino group, the side chain of lysine, histidine, and arginine, and the deprotonated C-terminal carboxyl group. Cu ions are strongly attached to the SH or SO_3H side groups in proteins via an intramolecular proton transfer. In solution, side chains of basic amino acids (arginine, lysine, and histidine) are protonated, which reduces the Cu^{+} and Cu^{2+} binding energies of N-donor ligands, whereas in the gas phase the Cu^{+} and Cu^{2+} have strong preferences for binding to arginine, lysine, and histidine (Wu et al. 2010).

Several solid-state cuprous compounds, including cuprous oxide (Cu_2O), sulfide (Cu_2S), iodide (CuI), and chloride (CuCl), have highly efficient antiviral activities, whereas those of solid-state silver and cupric compounds as (CuO). Cu_2O is adsorbing and denaturing more proteins than CuO, which suggests the difference in the inhibitory activity. Infectious activity is stopped due to direct contact with the solid-state surface of cuprous compounds, but not due to the generations of ROS or copper ions (Sunada et al. 2012). Deposition of cuprous oxide Cu_2O on viscose fabric was realized in an aqueous solution of Cu^{2+}-sodium citrate complex alkalized with Na_2CO_3 at elevated temperatures of 50°C–60°C in a water bath. At lower temperatures, the reaction is much slower but may result in better surface distribution and smaller particles of the precipitate with better adhesion to the fiber surface. Results are shown in Figure 2.10.

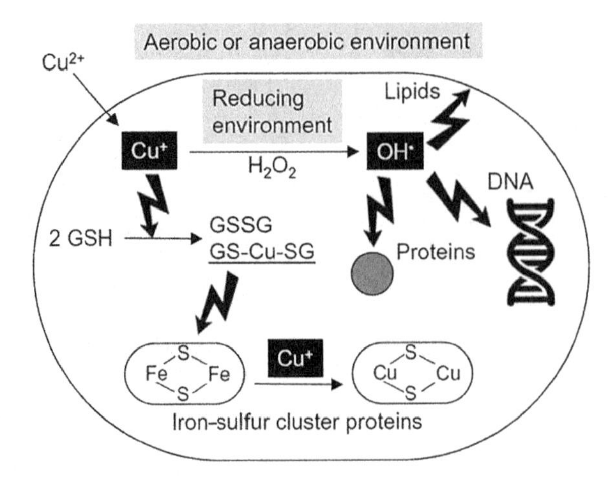

FIGURE 2.9 Copper toxicity (Solioz 2018).

FIGURE 2.10 Deposition of Cu_2O on viscose fabric surface (different magnification).

Cupron Inc. patented the technology of antimicrobial treatment based on copper, resp. copper oxide CuO, which can be used as an additive in polymer melts and solutions or as an additive to tempering baths (Borkow and Gabbay 2004). After only 2 h, materials containing CuO reduce *E. Coli* bacteria by 99.9%. Blended fabrics where at least 20% of the fibers containing Cupron kill mites (50% of the mites die within 12 days). Cu^{2+} ions also have a beneficial effect on wound healing. This is used, for example, in socks suitable for diabetics, where 10% of Cupron-containing fibers are enough to heal chronic wounds. Copper generally has a low risk of harmful effects on the skin (Hostynek and Maibach 2003). Cupron-based fabrics passed both the guinea pig test (chemical product allergy test) and the rabbit test (ISO-10993-10). The advantage of Cupron is that it can be used for its versatile antibacterial, antifungal, and racial effect combined with improved wound healing from bed linen, socks, and clothing textiles to special technical and soldier textiles. Each tissue cell has to maintain the concentration of each metal ion including copper within a narrow and specific range to avoid a detrimental alteration of the metal ion homeostasis (Kornblatt et al. 2016).

2.6.5 Other Metals

Preferably, ZnO is often used as an antimicrobial agent because it is stable to oxidation (unlike copper and silver) and does not cause discoloration. It also works as a UV absorber (El-Naggar et al. 2003). Method of creating a permanent attachment of ZnO on cotton and blended fabrics (cotton–polyester) using a crosslinking agent and heat or radiation curing was developed (Zohdy et al. 2003). Titanium dioxide TiO_2 (titania) is one versatile material creating the pairs of free electrons and holes in the conduction and valence band region under the light of a specific wavelength (<390 nm). During photocatalytic action on illuminated TiO_2 particles, the light with energy greater than the bandgap of the TiO_2 excites electrons in the valence band to the conduction band, with remaining holes in the valence band. The electrons and holes then migrate to the surface of the TiO_2 particles and here react with oxygen and adsorbed hydroxyl group to produce superoxide ($^{\cdot}O_2-$) and hydroxyl radicals ($^{\cdot}OH$) (called ROS) which possess strong oxidation destroying a wide range of organic pollutants, bacteria, and viruses. Simultaneously, electrons and holes are recombined on the way to the surface, which is the major reason for the reduction of the photocatalytic efficiency. Bacteria have natural enzymatic antioxidant defense systems that inhibit the effects of ROS radicals. When those systems are exceeded, a set of redox reactions can lead to the death of cell by the alteration of different essential structures (cell wall, cell membrane, DNA, etc.) and metabolism routes (Kiwi and Rtimi 2018). DNA is particularly sensitive to oxidative damage because oxygen radicals, specially $^{\cdot}OH$ may attack the sugar–phosphate helix and cause saccharide fragmentation. The particle size, crystal structure, and morphology play important roles in photocatalytic activity of TiO_2, which are mainly dependent on synthesis methods and reaction conditions including titanium salt, pH value, reaction temperature, time, additives, etc. Huge efforts have been made to create surfaces that not only repel water but also decompose organic contaminants including microorganisms. However, the preparation of superhydrophobic, photo catalytically active surfaces is complicated due to loss of superhydrophobicity upon irradiation with light or deterioration of photocatalytic activity by the creation of superhydrophobic surfaces. The efficiency of TiO_2 is generally limited due to the use of limited light energy and due to the recombination of photogenerated charge carriers, which is typically faster than the rate production of reactive oxidation species. The facile synthesis route for the preparation of TiO_2 nanoparticles with enhanced self-cleaning capability was developed (Khan et al. (2019). The principle is the in situ growth of 3D-shaped TiO_2 nanoflower particles on the surface of PES fabrics (Figure 2.11) by the hydrothermal method followed by silanization using sol–gel technology.

Interestingly, the biocidal activity of TiO_2 depends on the type of microorganism. Some bacterial phages are inactivated mainly by hydroxyl radicals throughout, but *E. Coli* is also inactivated on the surface (López et al. 2020). Fe(III)–Ti(IV) binary oxides are the dispersion of single metal oxides, Fe_2O_3 and TiO_2. They have faster bacterial inactivation kinetics. In colloidal formulations up to 10%, Fe_2O_3 can be added and will disperse well in the TiO_2 lattice. The addition of higher percentages of FeOx leads to Fe-phase segregation. This is a major limitation for use. The use of ferrates (Fe(VI)), effective over the entire pH range, is relatively new. Active

FIGURE 2.11 TiO_2 Nanoflowers grown on polyester surface.

oxidizing substances based on ferrates have various antibacterial effects (Anquandah et al. 2011). Ferrates are very effective for the removal of antibacterial agents from water and for disinfection purposes.

2.7 ANTIMICROBIAL TEXTILES

Antimicrobial textiles can protect against different kinds of microorganisms. Basic kind of products with hygienic and medical effects include:

- Socks – usually knitted polyamide, polyester, or cotton.
- Wipes – usually nonwoven fabrics based on chemical cellulose fibers (viscose).
- Protective clothing – a range of fibers according to the type of protection and mostly multilayer structures containing fabrics, knitwear, nonwoven fabrics, or membranes.
- Uniforms – standard fabrics made of cotton, polyester, polypropylene, and blends.
- Bed linen – cotton, polyester, and blend fabrics as standard.
- Surgical gowns, caps, and masks – woven or nonwoven materials with the possibility of using multiple layers (e.g., membranes) of cotton, polyester, polypropylene, and viscose.

Known antimicrobial agents include several substances (see Section 2.6), only some of which are suitable for antibacterial fabrics. Depending on the main desired effect, the antimicrobial fabrics are divided into three main categories:

- Hygienic and medical, where the purpose is to remove pathogenic organisms.
- Deodorant, where the purpose is to remove odors.
- Anti-rot, where the purpose is to protect textiles against microorganisms, especially when exposed to moisture for a long time.

It is important to distinguish whether antibacterial protection applies to clothing textiles that come into contact with humans frequently (home and furniture textiles) or other technical textiles. For technical textiles, the antimicrobial ability is required in connection with the effects of the weather. Ropes, tents, tarpaulins, blinds, sieve tents, etc., require protection against rot and mold. Household textiles such as carpets, curtains, mattresses, and upholsteries are also preferably resistant to bacteria, fungi, and the like mites. A special problem is the protection of historical textiles in museums. Textiles and clothing that come into contact with pathogenic bacteria and viruses must not only protect against their action but also prevent their transmission, resp. spread. The use of antimicrobial fabrics for socks and underwear and sportswear is required both to reduce unpleasant odors and to protect against mold. Antimicrobial treatments are also relatively important for household textiles, which are cleaned or rarely washed. Examples are furniture textiles (upholstery), textile wallpapers, draperies, and mattress covers. For these textiles, it is especially important to prevent odors due to bacterial action and to prevent the accumulation of pathogenic microorganisms. For antimicrobial fibers used in wound healing (bandages, tampons, gauze), in addition to antibacterial effects, it is also required to prevent the formation of pus and speed up the healing process. Many bacteria are commonly stored and multiplied on the human body and clothing fabrics. Most of them are not pathogenic and do not affect human health. However, some generate an unpleasant odor. However, there are also bacteria (*S. aureus*, *Klebsiella*, and *E. Coli*), resp. fungi (*T. mentagrophytes*), which can cause skin problems when overgrown or purulence on contact with damaged skin. Microorganisms can also break down textiles by both degrading and changing the shade to create an unpleasant odor. However, the odor is also caused by excrements of the human body, cosmetics on textiles, food, and beverage residues, resp. dirt and dust. Bacteria such as *S. aureus*, *C. minutissimum*, and *Tipophylic diphtheroid* are almost always present on the skin and cause unpleasant odors. Their components are acetic acid, propionic acid, valeric acid, and free higher fatty acids, trimethylamine, ammonia, mercaptans, and aldehydes. Effective antimicrobials for textiles include metals and metal compounds (silver, copper, zinc, oxides, metal sulfites, metal-containing ceramics), quaternary ammonium salts, *N*-phenylamides, animal polysaccharides (chitin, chitosan), fatty acid esters, and phenolic compounds (chloroxylenol). In many cases, these substances can be added as additives to polymer melts before spinning. They can also be applied in the finishing of textiles, for example, by surface coating. It is still a matter of debate whether antimicrobial agents of the elution or nonelution type are more preferred.

Elution-type agents are advantageous in that they reach the bacteria either by the action of moisture on the textiles or by the active substances flowing into the environment. However, in the case of clothing textiles, this often leads to the destruction of bacteria on the skin (or to the action on human cells), which can be a source of many problems as a result.

Nonelution-type agents react in direct contact with microorganisms (or in the immediate vicinity). They are usually slower, but less dangerous to the human skin. The aim is to use covalent bonds for the binding of antimicrobial agents of both types to the fibers.

Probably the oldest antimicrobials used against rot are the salts of mercury ($HgCl_2$) and silver ($AgNO_3$). Phenol and its chlorinated compounds were used for antiseptic purposes. Due to their toxicity, these antimicrobial treatment agents are no longer used. Quaternary ammonium salts, hexachlorophene, and salicylanilide have been used for antimicrobial treatment of heavy fabrics based on cellulose fibers. These agents are generally less effective against rot, but they prevent the spread of pathogenic bacteria and the formation of odors as a result action of bacteria. One of the most common antimicrobials was Triclosan, also used in mouthwashes, toothpastes, and deodorants. The Danish Environmental Protection Agency has published an article (Rastogi et al. 2003) examining the antimicrobial treatment agents used and their concentrations based on data from manufacturers and standard antimicrobials on the Danish market. Triclosan was found to be present in concentrations of 0.0007% to 0.0195% (i.e., 7–195 ppm) in some products. This is significantly less than the permitted dose of 3% in cosmetics. Triclosan was also found in materials labeled as Sanitized (which, according to the manufacturer Kathon 893, i.e., a mixture of chloro-2-(2,4-dichloro phenoxy)-phenol and 2-N-octyl-isothiazolin). This indicates that in many cases the information on antimicrobials is incorrect. Triclosan is according to the harmonized classification and labeling (CLP00) approved by the European Union, very toxic to aquatic life with long-lasting effects, causes serious eye irritation and skin irritation (European Chemical Agency [ECHA]). It is a common notion that textile experts do not need to know what the nature of an antimicrobial substance is and what mechanism it acts on microbes, resp. mites. However, this often leads to a state where inappropriate adjustment, resp. textile maintenance considerably weakens the antimicrobial effect and often expensive materials are practically ineffective and can have adverse effects on human health. Besides, it is a natural effort of antimicrobial fiber manufacturers, resp. published information that positively influences marketing. Various restrictions, resp. the different types of potential problems can often be estimated from the knowledge of the active substances and their effects (including information on the persistence of the effect over time and not after repeated washing under mild conditions at very short intervals). Unfortunately, information and details on the behavior of various antimicrobial fibers and treatments are scattered throughout specialized journals and often difficult to access for textile professionals. Many antimicrobials used in textiles are well-known substances widely used in cosmetics and the food industry. This is a certain advantage when obtaining hygienic and ecological certificates. Another advantage is the need for very low concentrations (max. around 2%–5%), especially for organic compounds. If antimicrobial additives are used in polymers, the requirement is:

- Very low solubility in water, alkalis, and acids;
- Chemical resistance to acids, alkalis, oxidizing agents, and possibly UV, resp. light radiation;
- Sufficient thermal stability.

Furthermore, it is required that the additives do not adversely affect the fiber preparation process (spinning) or the properties of the resulting fibers (especially mechanical). Ingredients of the elution type must be able to migrate to the surface of the fibers. When selecting a suitable antimicrobial additive for synthetic fibers, a particle size below 2 μm, stability and nonvolatility up to 300°C, the ability to disperse uniformly, and compatibility with the fiber-forming polymer are required. Ultrafine powders (typical size below 1 μm) and nanopowders (typical size below 0.1 μm) are commonly used. It is advantageous if the antimicrobial agents act in combination, i.e., they affect different functions of the bacteria and their cells. This can reduce the possibility of resistant strains of bacteria. Recently, several specialized companies supply antimicrobial agents of various types under the same name, but they differ significantly in the composition of the active substance (differences are expressed by different numbers). For controlled delivery of active substances, it is also possible to use microcapsules, where the active substance is surrounded by a protective layer (barrier) controlling the rate of its penetration into the environment. For the production of microcapsules, solvent evaporation, precipitation crosslinking polymerization, and phase separation (coacervation) techniques are mostly used.

Another possibility is the use of cyclic molecules (dextrins), where the active substance is trapped in cavities of a hydrophobic nature. Sometimes microspheres are also used where the active substance is dispersed in a suitably selected polymeric carrier. The concentration of most antimicrobials on textiles ranges from 0.2% to 1.5%. Antimicrobial agents of the elution type are deactivated due to depletion of the active substance. Due to their release into the environment (whether in the air or in the presence of moisture), they can destroy the flora on healthy human skin and cause biological problems. Nonelution-type antimicrobial agents are deactivated either by mechanical clogging of the active sites or by abrasion of the fabric surface. The adhesion of bacteria to the surface of polymers depends on the type of polymer and the type of bacterium. However, adhesion is not significantly affected by surface biodegradation. The free energy of bacterial adhesion is negative, indicating that adhesion is thermodynamically advantageous (Yahyaei et al. 2019). The most negative is the free adhesion energy for *P. Aeruginosa* and the least negative is the free adhesion energy for *S. Epidermis*. *E. Coli* bacteria have the greatest negative free adhesion energy for polyethylene (Barton et al. 1996). Due to the so-called lotus effect, the deposition of microbes is respected, resp. their removal is achieved, for example, due to the presence of water drops. The "lotus" effect, i.e., self-cleaning, occurs when there are nanoscale irregularities on the hydrophobic surface (Guittard and Darmanin 2018). It has been found that a certain ratio between the height and width of the unevenness must be observed. The distances between the irregularities can range from a few mm to 10 μm (Guittard and Darmanin 2018). The lotus effect can be achieved, for example, by treating the surface of the fibers with silver nanoparticles and then covering the surface with a layer of polyglycidyl methacrylate (Ramaratnam et al. 2008). The treatment is performed after dyeing and is suitable for all textile materials. The lotus effect is preferably used to create surfaces that are better cleaned and maintained (Sethi et al. 2019). It is also a natural antibacterial treatment that limits the attachment of bacteria to the surface of textiles (Arfin et al. 2019). Mechanically, bacteria can be prevented from settling on the surface of textiles either by hydrogel-forming deposits, for example, by grafting polyethylene glycol or by creating an ultrahydrophobic surface with a contact angle greater than 150° (Kumar and Nanda 2019). In both cases, the adhesion of the microbes is severely limited.

For medical gowns, a mechanical barrier to the penetration of viruses by means of a nonporous membrane can be used, which, however, does not prevent the transport of moisture by a diffusion mechanism. The physical barrier to the penetration of microorganisms is to form a film on the surface of the fabric with a pore size smaller than the size of the microorganism. For bacteria, the lower pore limit is around 0.5 µm. Another possibility is to use a membrane that is not porous but allows moisture to be removed by a diffusion mechanism or a microporous membrane with sufficiently small pores. Clothing fabrics for surgical gowns and surgical upholstery fabrics must meet three essential requirements (Leonas and Jinkins 1997):

- Impermeability to liquids,
- Absorption capacity,
- Permeability to air and water vapor.

Especially for surgical gowns, in addition to antibacterial properties, protection against pathogens contained in patients' blood (e.g., HIV) is also required. Coats must therefore prevent blood from penetrating. Combined repellent and antimicrobial treatments can be achieved with quaternary ammonium salts containing perfluoroalkyl groups and diallyl groups. These preparations can be used either as an additive in polymers or as a final antimicrobial repellent treatment (Morais et al. 2016).

2.8 CONCLUSION

From the above overview, it is clear that there are several options for protection against microorganisms and depends on the requirements or restrictions on their use. In the future, we can expect the selection of methods to optimize with regard to efficiency, lasting effect, and price.

ACKNOWLEDGEMENT

This work was supported by the Ministry of Education, Youth and Sports of the Czech Republic and the European Union – European Structural and Investment Funds in the frames of Operational Programme Research, Development and Education under project Hybrid Materials for Hierarchical Structures [HyHi, Reg. No. CZ.02.1.01/0.0/0.0/16_019/0000843].

REFERENCES

Agrios, A.G. and Pichat, P. 2005 State of the art and perspectives on materials and applications of photocatalysis over TiO_2, *J. Appl. Electrochem.* 35: 655–663.

Alberts, B. et al. 2015 *Molecular Biology of the Cell.* New York: Garland Science, Taylor & Francis Group.

Anquandah G. et al. 2011 Oxidation of trimethoprim by ferrate (VI): Kinetics, products, and antibacterial activity, *Environ. Sci. Technol.* 45: 10575–10581.

Aray, T. et al. 2004 Biodegradation of bombyx mori silk fibroin fibers and films, *J. Appl. Polym. Sci.* 91: 2383–2390.

Arfin, T. et al. 2019 Biological adhesion behavior of superhydrophobic polymer coating. In *Superhydrophobic Polymer Coatings*, eds. Samal, S. K. et al., 161–177. Amsterdam: Elsevier B.V.

Bajaj, P. 2001 Finishing of textile materials *J. Appl. Polym. Sci.* 83: 631–659.

Barnes, C. and Warden, J. 1971 Microbial degradation: fiber damage from staphylococcus aureus, *Text. Chem. Color.* 3: 52–56.

Baroli, B. 2010 Penetration of nanoparticles and nanomaterials in the skin: fiction or reality? *J. Pharm. Sci.*, 99: 21–50.

Barton A. J. et al. 1996 Bacterial adhesion to orthopedic implant polymers, *Biomed. Mater. Res.* 30: 403–410.

Blair, A. 2002 Silver plating, *Met. Finish.* 100: 284–290.

Blaker, J. J. et al. 2004 Development and characterization of silver-doped bioactive glass-coated sutures for tissue engineering and wound healing applications, *Biomaterials* 25: 1319–1329.

Borkow, G. and Gabbay, J. 2004 Putting copper into action: copper-impregnated products with potent biocidal activities, *FASEB J.* 18: 1728–1747.

Cagno, V. et al. 2020 Broad-spectrum non-toxic antiviral nanoparticles with a virucidal inhibition mechanism, *Nat. Mater.* 17, 195–203.

Carey, D. E. and McNamara, P. J. 2014 The impact of triclosan on the spread of antibiotic resistance in the environment, *Front. Microbiol.* 5: 1–11.

Chen, L. and Liang, J. 2020 An overview of functional nanoparticles as novel emerging antiviral therapeutic agents, *Mater. Sci. Eng. C*, 112: 1–15.

Chen, Y. et al. 2019 Construction of pyridinium/N-chloramine polysiloxane on cellulose for synergistic biocidal application, *Cellulose*, 26: 5033–5049.

Colloff, M. J. 2009 *Dust Mites*. Collingwood: CSIRO Australia.

Dodd, G. D. et al. 1981 The use of permethrin in new insect. Proofing agents for wool. *J. Soc. Dyers Col.* 97: 125–127.

Doi, Y. and Fukuda, K. 1994 *Biodegradable Plastic and Polymers*. Amsterdam: Elsevier.

Dowling, D.P. et al 2001 Deposition of anti-bacterial silver coatings on polymeric substrates, *Thin Solid Films*.

Eldsaeter, C. et al. 2000 The biodegradation of amorphous and crystalline regions in film-blown poly(e-caprolactone), *Polymer* 41: 1297–1304.

El-Naggar, A. M. et al. 2003 Antimicrobial protection of cotton and cotton/polyester fabrics by radiation and thermal treatments. I. Effect of ZnO formulation on the mechanical and dyeing properties, *J. Appl. Polym. Sci.* 88: 1129–1137.

Evans A. and Stevens R. J. 1976 Differential quantitation of surface and subsurface bacteria of normal skin by the combined use of the cotton swab and the scrub methods, *J. Clin. Microbiol.* 3: 576–581.

Goldschmidt, A. and Streitberger, H. 2018 *BASF Handbook Basics of Coating Technology*. Hannover: Vincentz Network.

Grass, G. et al. 2011 Metallic copper as an antimicrobial surface, *Appl. Environ. Microbiol.* 77: 1541–1547.

Guittard, F. and Darmanin, T. 2018 *Bioinspired Superhydrophobic Surfaces Advances and Applications with Metallic and Inorganic Materials*. Singapore: Pan Stanford Publishing Pte. Ltd.

Hendley J. O. and Ashe K. M. 2003 Eradication of resident bacteria of normal human skin by antimicrobial ointment, *Antimicrob. Agents Chemother.* 47: 1988–1990.

Hett A. 2005 *Nanotechnology, Small Matter, Many Unknowns*. Zurich: Swiss Reinsurance Co

Ho, C. H. et al 2004 Nanoseparated polymeric networks with multiple antimicrobial properties, *Adv. Mater.* 16: 957–961.

Hober, D. 2004 Antimicrobial textiles – curse or benefit, Poc. 7th Dresder Textiltagung, June.

Hui, F. and Debiemme-Chouvy, C. 2013 Antimicrobial N-Halamine polymers and coatings: A review of their synthesis, characterization, and applications, *Biomacromolecules*, 14: 585–601.

Jing, H. et al. 2004 Plasma-enhanced deposition of silver nanoparticles onto polymer and metal surfaces for the generation of antimicrobial characteristics, *J. Appl. Polym. Sci.* 93: 1411–1422.

Kamat, P. V. et al. 1998 Picosecond dynamics of silver nanoclusters. Photoejection of electrons and fragmentation, *J. Phys. Chem. B* 102: 3123–3128.

Kanematsu, H. and Barry, D. M. 2020 *Formation and Control of Biofilm in Various Environments*. Singapore: Springer.

Khan, M. Z. et al. 2019 Self-cleaning properties of polyester fabrics coated with flower-like TiO_2 particles and trimethoxy(octadecyl)silane, *J. Ind. Text.* 50: 543–565.

Kittinaovarat, S. and Kantuptim, P. 2005 Comparative antibacterial properties of glyoxal and glyoxal and chitosan treated cotton fabrics, *AATCC Rev.* 5: 22–24.

Kiwi, J. and Rtimi, S. 2018 Mechanisms of the antibacterial effects of TiO_2 -FeO_x under solar or visible light: Schottky barriers versus surface plasmon resonance, *Coatings* 8: 391–404.

Klueh, V. et al. 2000 Efficacy of silver-coated fabric to prevent bacterial colonization and subsequent device-based biofilm formation, *Appl. Biomater.* 53: 621–631.

Klun, U. et al. 2003 Polyamide-6 fibre degradation by a lignolytic fungus, *Polym. Degrad. Stab.* 79: 99–104.

Kornblatt, A. P. et al. 2016 The neglected role of copper ions in wound healing, *J. Inorg. Biochem.* 161: 1–8.

Korting, H. C. et al. 1988 Microbial flora and odor of the healthy human skin, *Hautarzt* 39: 564–568.

Kumar, R. and Muenstedt, H. 2005 Silver ion release from antimicrobial polyamide/silver composites, *Biomaterials* 26: 2081–2088.

Kumar, A. and Nanda, D. 2019 Methods and fabrication techniques of superhydrophobic surfaces. In *Superhydrophobic Polymer Coatings*, eds. Samal, S. K. et al., 43–75. Amsterdam: Elsevier B.V.

Leonas, K. K. and Jinkins, R. S. 1997 The relationship of selected fabric characteristics and the barrier effectiveness of surgical gown fabrics, *Am. J. Infect. Control* 25: 16–23.

López, C. et al. 2020 Antimicrobial Effect of Titanium Dioxide Nanoparticles. In ed. Antimicrobial Resistance, 10.5772/intechopen.90891, Available from: https://www.intechopen.com/online-first/antimicrobial-effect-of-titanium-dioxide-nanoparticles.

Mackintosh, C. A. 1978 The dimensions of skin fragments dispersed into the air during aktivity, *J. Hyg. Camb.* 81: 471–479.

Maheshwari, D. K. et al 2010 *Industrial exploatoion of Microorganisms*. p. 136 New Delhi: International Publ. House.

Matthies, W. 2003 Irritant dermatitis to detergents in Textiles, *Curr. Probl. Dermatos.* 31: 123–128.

Militký, J. et al. 1990 *Modified PES fibres* Amsterdam: Elsevier

Militký, J. et al. 1998 Penetration of mites through textile layer In *Medical Textiles´96*, ed. Anand. S., 56–59. Cambridge: Woodhead Publ.

Morais D. S. et al. 2016 Antimicrobial approaches for textiles: From research to market, *Mater.* 9, 498–519.

Neely, A. N. and Maley, M. P. 2000 Survival of enterococci and staphylococci on hospital fabrics and plastic, *J. Clin. Microbiol.* 38: 724–726.

Ohko, Y. et al. 2001 Self-sterilizing and self-cleaning of silicone catheters coated with TiO_2 photocatalyst thin films: A preclinical work, *J. Biomed. Mater. Res.* 58: 97–101.

Park, C. H. et al. 2004 Biodegradability of cellulose fabrics, *J. Appl. Polym. Sci.* 94: 248–253.

Perelshtein, I. et al. 2008 Sonochemical coating of silver nanoparticles on textile fabrics (nylon, polyester and cotton) and their antibacterial activity, *Nanotechnology*, 19: 245705–245711.

Portnoy, J. J. D. et al. 2013. Environmental assessment and exposure control of dust mites: a practice parameter. *Ann. Allergy Asthma Immunol.* 111: 465–507.

Ramaratnam, K. et al. 2008 Ultrahydrophobic textiles: Lotus approach, *AATCC Rev.* 8: 42–48.

Rastogi, S. C. et al. 2003 Survey of Chemical Substances in Consumer Products Antibacterial Components in Clothing Articles, Survey No 24: Danish Ministry of Enviroment

Schmidtchen, A. et al. 2014 Effect of hydrophobic modifications in antimicrobial peptides, *Adv. Colloid Interf. Sci.* 205: 265–274.

Sethi, S. K. et al. 2019 Fundamentals of superhydrophobic surfaces. In *Superhydrophobic Polymer Coatings*, eds. Samal, S. K. et al., 3–29. Amsterdam: Elsevier B.V.

Silva, V. et al. 2020 Isothiazolinone biocides: Chemistry, biological, and toxicity profiles, *Molecules* 25: 991–1013.

Slaughter, S. et al. 1996 A comparison of the effect of universal use of gloves and gowns with that of glove use alone on acquisition of vancomycin-resistant enterococci in a medical intensive care unit, *Ann. Intern. Med.* 125: 448–456.

Solioz M. 2018 *Copper and Bacteria*, Cham: Springer Nature.

Spera, R. V. and Farber, B. G. 1992 Multiply-resistant Enterococcus faecium, *J. Am. Med. Assoc.* 268: 2563–2567.

Sunada, K. et al. 2012 Highly efficient antiviral and antibacterial activities of solid-state cuprous compounds, *J. Hazard. Mater.* 235–236: 265–270.

Szabo, N. J., and Winefordner, J. D. 1997 Surface-Enhanced Raman scattering from an etched polymer substrate, *Anal. Chem.* 69: 2418–2425.

Tellier, R. et al. 2019 Recognition of aerosol transmission of infectious agents: a commentary, *BMC Infect. Dis.* 19: 101–110.

Tinoco, I. et al. 1999 How RNA folds, *J. Mol. Biol.* 293: 271–281.

Vaara, M. 1992. Agents that increase the permeability of the outer membrane, *Microbiol. Rev.* 56: 395–411.

van Doremalen, N. et al. 2020 Aerosol and surface stability of SARS-CoV-2 as compared with SARS-CoV-1, *N. Engl. J. Med.* 382: 1564–1567.

Venkataraman, M. et al. 2014. Application of silver nanoparticles to industrial sewing threads: Effects on physico-functional properties & seam efficiency, *Fibers and Polymers* 15: 510–518.

Vodrážka, Z. 1996 *Biochemie*, Praha: Academia.

Volf, L. A. et al. 1963 Biologically active polyvinylalkohol Leitan fiber, *Chimičekie volokna* 6: 16–18.

Walls, A. C. et al. 2020 Structure, function, and antigenicity of the SARS-CoV-2 spike glycoprotein, *Cell* 180 (2020): 281–292.

Weiss, C. et al. 2020 Toward nanotechnology-enabled approaches against the COVID-19 pandemic, *ACS Nano* 14, 6383–6406.

Williams, D. E. et al. 1998 Is free halogen necessary for disinfection? *Appl. Environ. Microbiol.* 54: 2583–2585.

Woodyard, L. L. et al. 1996 A comparison of the effects of several silver-treated intravenous catheters on the survival of staphylococci in suspension and their adhesion to the catheter surface, *J. Control. Release* 40: 23–30.

Wu, Z. et al. 2010 Amino acid influence on copper binding to peptides, *J. Am. Soc. Mass Spectrom.* 21: 522–533.

Yager, E. J., and Konan, K. V. 2019 Sphingolipids as potential therapeutic targets against enveloped human RNA viruses, *Viruses* 11, 912–926.

Yahyaei, H. et al. 2019 Superhydrophobic coatings for medical applications. In *Superhydrophobic Polymer Coatings*, eds. Samal, S. K. et al., 321–338. Amsterdam: Elsevier B.V.

Yuranova, T. et al. 2003 Antibacterial textiles prepared by RF-plasma and vacuum-UV mediated deposition of silver, *J. Photochem. Photobiol. A* 161: 27–34.

Yuranova, T. et al. 2006 Performance and characterization of Ag–cotton and Ag/TiO_2 loaded textiles during the abatement of E. Coli, *J. Photochem. Photobiol. A*, 181: 363–369.

Zohdy, M. M. et al 2003 Microbial detection, surface morphology, and thermal stability of cotton and cotton/polyester fabrics treated with antimicrobial formulations by a radiation method, *J. Appl. Polym. Sci.* 89: 2604–2610.

3 Virology of SARS-CoV-2

Dan Wang, Dana Kremenakova and Jiri Militký
Technical University of Liberec, Czech Republic

CONTENTS

3.1 INTRODUCTION

Human coronaviruses are part of the family Coronaviridae, order Nidovirales, and subfamily Coronavirinae. The subfamily members can be divided into four categories – alpha, beta, gamma, and delta coronaviruses based on genome structure and phylogenetic studies. Most mammals are infected by alpha and beta coronaviruses, while birds and a small number of mammals are usually infected by gamma and delta coronaviruses. Gastroenteritis often occurs in animals infected with α-coronavirus and β-coronavirus, but humans can cause respiratory distress because of these two viruses. Four human coronaviruses can cause mild upper respiratory tract infections that are HCoV-229E, HKU1, HCoV-NL63, and HCoV-OC43. There are two other types of coronaviruses that are related to the human severe respiratory syndrome and are more pathogenic, including severe acute respiratory syndrome coronavirus (SARS-CoV) and Middle East respiratory syndrome coronavirus (MERS-CoV) (Gurung et al. 2020). From early December 2019, the outbreak of unknown epidemic pneumonia in the Wuhan, a city in the Hubei Province of China, is identified to be caused by a novel coronavirus, named SARS-CoV-2 by the International Committee on Taxonomy of Viruses (ICTV) and the disease was named as Coronavirus Disease 2019 (COVID-19) by the World Health Organization (Liu et al. 2020a, 2020b).

3.2 STRUCTURE OF SARS-COV-2

SARS-CoV-2 is a kind of the positive-sense single-stranded RNA and the enveloped virus, and its RNA sequence is nearly 30,000 bases in length. The coronavirus particles are spherical in shape with a diameter of about 120 nm (Akram and Mannan 2020). SARS-CoV-2 has four main structural proteins, including the S (spike) protein, E (envelope) protein, M (membrane) protein, and N (nucleocapsid) protein like other coronaviruses (see Figure 3.1).

The S protein of the coronavirus is a kind of large multifunctional class I viral transmembrane protein (Belouzard et al. 2012). The coronal appearance of the virus particle is attributed to the trimer located on the surface of the virus particle. Besides, it allows infectious virus particles to enter cells through interaction with various host cell receptors because the S proteins are capable of inducing host immune response (Beniac et al. 2006). Furthermore, it plays an important role in tissue tropism and the determination of host range. S proteins can be divided into two domains, S1 and S2. S1 functions as the receptor-binding domain (RBD), while S2 acts as a membrane fusion subunit (Li 2016). The E protein of the coronavirus is the smallest structural protein in the virion, but the toxicity of the virus is related to it. The E protein also plays an important role in the pathogenesis, assembly, and release of viruses (Nieto-Torres et al. 2014). The amino acid sequences of E proteins of different coronaviruses have some common structures, but they are also quite different (Godet et al. 1992). Compared with SARS-CoV, SARS-CoV-2 E protein shows a similar amino acid composition without any substitution (Wu et al. 2020a, 2020b). Since the E protein does not seem to be essential for the replication of SARS-CoV, when the gene encoding the E protein is deleted, the amplification speed of the virus will slow down (DeDiego et al. 2007). M protein is the most abundant structural protein in coronavirus particles. The M protein generally maintains structural similarity, but its amino acid sequence is also different in different coronaviruses (Arndt et al. 2010). It is also responsible for maintaining the shape of the virion due to a small protein with three transmembrane domains in the structure (Neuman et al. 2011). It has a short

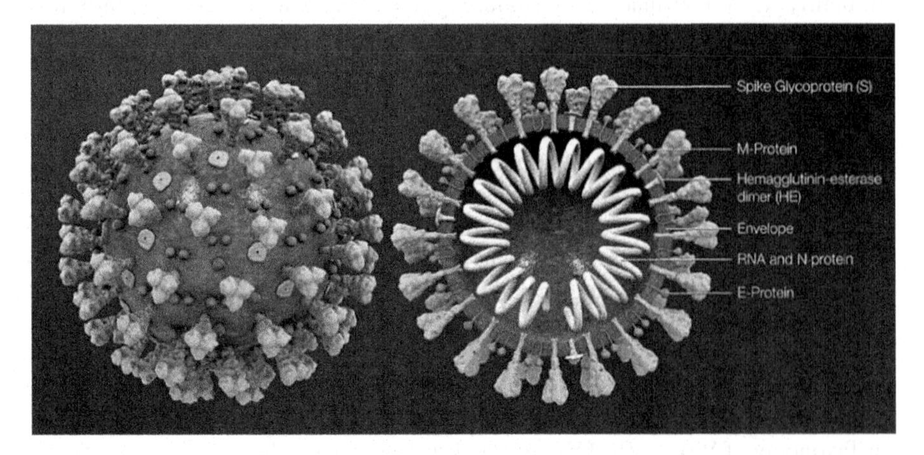

FIGURE 3.1 Plane structure of SARS-CoV-2, (Reprinted from [https://www.scientificanimations.com 2020], with kind permission of scientific animations).

N-terminal glycosylated domain outside the virion and has a much longer C-terminal domain (CTD) inside the virion that extends 6–8 nm into the viral particle (Nal et al. 2005). Most M proteins are co-translationally inserted into the ER membrane without a signal sequence. The viral scaffold is preserved through the interactions between M proteins. Some researchers have found that the M protein exists as a dimer in the virion, and may promote membrane bending and binding to the nucleocapsid by using two different conformations (Neuman et al. 2011). The N protein is the only structural protein present in the nucleocapsid. It consists of two independent domains, an N-terminal domain (NTD) and a CTD. Both can bind RNA in vitro, but the binding mechanism is different (Mcbride et al. 2014). The N protein has two RNA substrates that have already been identified, the transcriptional regulatory sequence (TRS) and the genomic packaging signal (Stohlman et al. 1988). The high content of N protein and the high hydrophilicity of N protein is believed to be beneficial to the effective immunity after the coronavirus infects cells. The N protein of coronavirus is often used as a diagnostic test marker.

3.3 CHARACTERISTICS OF SARS-COV-2

This part will introduce some new studies and discoveries about SARS-CoV-2 from different countries' scientists. These detailed researches of SARS-CoV-2 will broaden our understanding of diseases and also promote us to find a way to treat the disease as soon as possible.

3.3.1 Main Protease

SARS-CoV-2 is similar to SARS and MERS in protein composition. The genome is encoded by structural proteins, nonstructural proteins, and accessory proteins. Structural proteins include S proteins containing receptor binding domains, while nonstructural proteins include 3-chymotrypsin-like protease (3CLpro), papain-like protease (PLpro), helicase, and RNA-dependent RNA polymerase (RdRp).

The main protease (Mpro, also known as 3CLpro) in the new coronavirus is an important potential drug target, which is essential for inhibiting virus replication. The researchers found that the three-dimensional structure of the main protease of the new coronavirus is highly similar to the three-dimensional structure of the main protease of SARS-CoV, with a sequence identity of 96%. This enzyme works at no less than 11 cleavage sites on the large protein 1ab, which is essential for processing polyproteins converted from viral RNA (see Figure 3.2). Therefore, inhibiting the activity of this enzyme will Prevent virus replication. Since no human proteases with a similar cleavage specificity are known, inhibitors are unlikely to be toxic (Anand et al. 2003; Zhang et al. 2020).

3.3.2 Hemagglutinin-Esterase Lectin

SARS-CoV-2 is a member of the *Sarbecovirus* subgenus (Betacoronavirus Lineage β), just like the SARS-related coronavirus strain involved in the SARS outbreak in 2003. Therefore, the researchers conducted a series of research on OC43 and HKU1 from the Betacoronavirus Lineage α (see Figure 3.3).

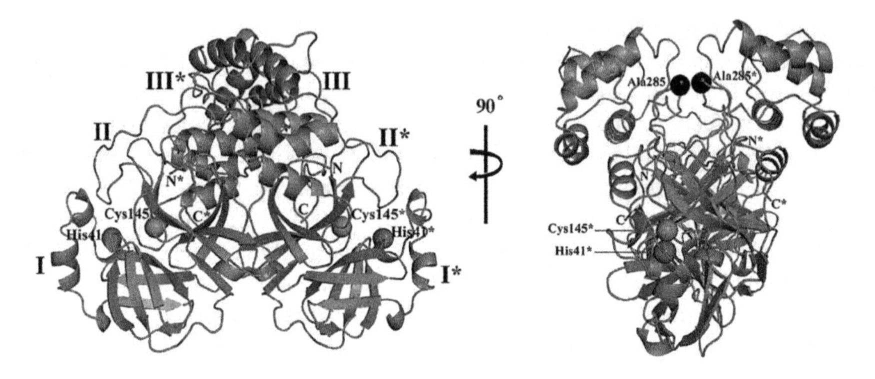

FIGURE 3.2 Three-dimensional structure of SARS-CoV-2 Mpro, in two different views, (Reprinted from Zhang et al. (2020), with kind permission of The American Association for the Advancement of Science).

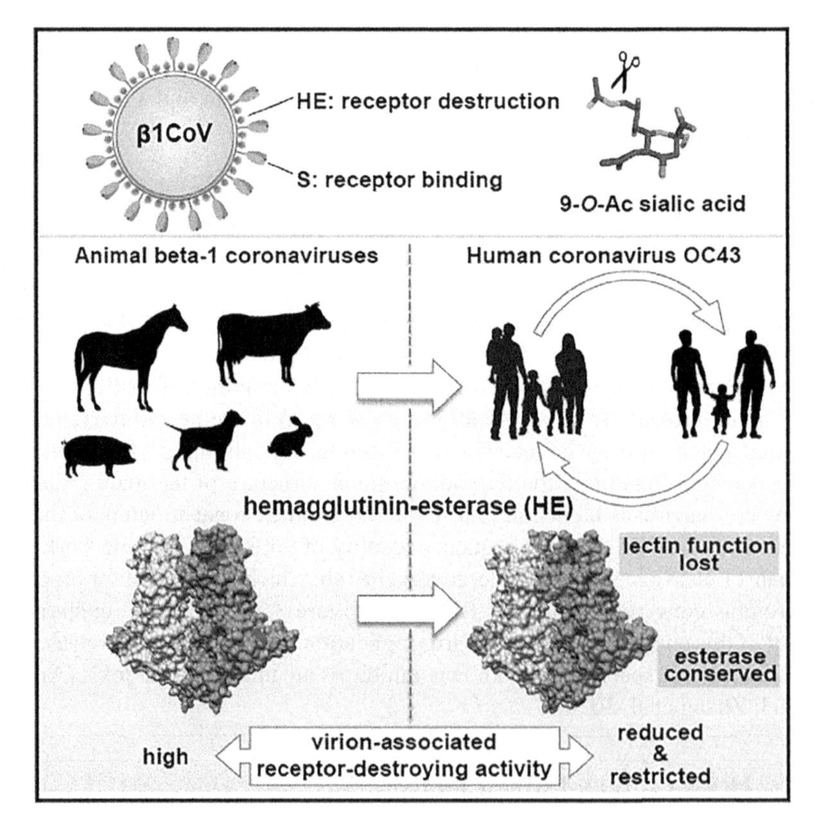

FIGURE 3.3 Hemagglutinin-esterase lectin (Reprinted from Bakkers et al. (2017), with kind permission of Elsevier Publications).

The recent appearance of human β1-coronavirus (β1CoV) OC43 was spread through zoonotic diseases. Similar to other animal β1-coronaviruses, 9-O-acetylated sialic acid is the receptor determinant of this coronavirus. The hemagglutinin esterase protein HE that binds/fusions synaptic protein S and receptor binding/destroying receptor usually control the binding of the β1CoV receptor. Researchers have found that when OC43 begins to intervene in the human body, the HE-mediated receptor binding is initially resisted, but it is lost due to the gradual accumulation of mutations in the HE lectin domain. Such a result will reduce and change the virion-related receptor damaging activity against the multivalent glycoconjugate, and eventually make some clustered receptor populations no longer cleaved. In another respiratory human coronavirus, HKU1, researchers also observed the loss of function of HE lectin. Therefore, this seems to be an adaptation to the sialoglycome of the human respiratory tract and replication in the human airway. This study can show that the host tropism benefits from the kinetics of virion–glycan interaction. Researchers also observed other zoonotic human respiratory viruses such as influenza A virus. Different from OC43, hemagglutinin esterase protein plays a receptor binding role in influenza C, but in coronavirus, it is a receptor destroying enzyme. So, as mentioned before, the binding affinity between the virus and the human body can be enhanced due to the lack of lectin action, this is also considered to be an evolutionary mechanism that successfully replicates in respiratory epithelial cells and adapts to human respiratory sialoglycoprotein (Bakkers et al. 2017).

3.3.3 Angiotensin-Converting Enzyme 2 (ACE2)

Some researchers analyzed the full-length genome sequence obtained from patients and found that 2019-nCoV and SARS-CoV have 7 conservative nonstructural protein amino acid sequences with a similarity of 94.6%. Computer homology modeling shows that 2019-nCoV and SARS-CoV have similar receptor-binding regions. Then the nCoV-2019 virus was isolated from the bronchoalveolar lavage fluid of severely ill patients, which can be neutralized by the serum of several patients. Therefore, researchers have confirmed that this new CoV uses the same cells as SARS-CoV to enter the receptor angiotensin-converting enzyme 2 (ACE2) (Zhou et al. 2020). This similarity to SARS-CoV is crucial because ACE2 is a functional SARS-CoV receptor both in vitro and in vivo to enter the host cell and subsequent virus replication is necessary (Li et al. 2003; Kuba et al., 2005). In vitro experiments have confirmed that cells expressing ACE2 will be infected, while cells that do not express ACE2 will not be infected (Zhou et al. 2020). Other researchers have used cryo-electron microscopy technology to speculate from the structural level that the binding strength of 2019-nCoV and ACE2 is 10–20 times that of SARS-CoV. This explains why 2019-nCoV is stronger than SARS-CoV (Wrapp et al. 2020). Based on single-cell sequencing data, it was found that the expression levels of ACE2 in human organs from high to low were ileum, heart, kidney, bladder, esophagus, lung, and trachea. Taking the ACE2 expression of alveolar type 2 cells as 1%, the positive expression rates of ACE2 in ileal epithelial cells and cardiomyocytes were 30.0% and 7.5%, respectively (Zou et al. 2020). However, clinically, 2019-nCoV infection is mainly caused by lung damage. This may be because the virus is mainly transmitted through the respiratory

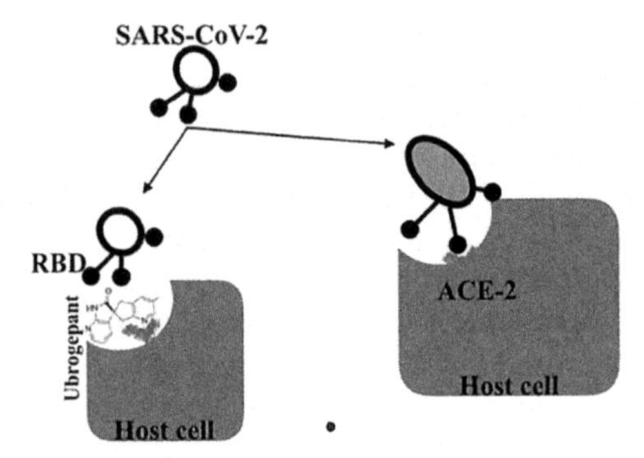

FIGURE 3.4 Function of ACE2 and RBD (Reprinted from Omotuyi et al. (2020), with kind permission of Creative Commons Attribution License).

tract, and the lung tissue is more exposed and the exposed area is larger. Besides, previous studies on the coronavirus have found that animals There are persistent intestinal infections but no symptoms. Although infection with 2019-nCoV rarely shows gastrointestinal symptoms, virus particles have been found in feces. Whether there is a similar mechanism remains to be further studied (Ren et al. 2003). In short, it should be noted that in addition to lung tissue, a variety of human tissues and organs that express ACE2 may be attacked by 2019-nCoV. In the mouse model experiment, when the researchers injected the SARS-CoV peak into the mice, it would aggravate the lung damage of the mice. Crucially, this damage will be attenuated by blocking the renin-angiotensin pathway, which means that overexpression of human ACE2 will increase the severity of the disease (Yang et al. 2007; Imai et al. 2005). Therefore, for the pathogenesis of SARS-CoV, ACE2 is not only an entry receptor for the virus, but also prevents lung damage. SARS-CoV is very lethal because the virus can relieve the regulation of lung protection pathways (see Figure 3.4) (Kuba et al. 2005; Imai et al. 2005). In short, the SARS-CoV-2 spike protein directly binds to the ACE2 receptor on the host cell surface to promote virus entry and replication.

3.3.4 Receptor Binding Domain (RBD)

The coronavirus' S protein split into two functional units, including S1 and S2. S1 promotes viral infection through the binding to host receptors. It contains the NTD and the C-terminal RBD domain that directly interacts with host receptors (Li 2012). Like SARS-CoV, the S protein of SARS-CoV-2 binds to their common receptor ACE2 through the RBD to mediate the virus into host cells. During the infection process, the S protein is cleaved by the host protease (such as TMPRSS2) into the N-terminal S1 subunit and the C-terminal S2 subunit, and changes from the prefusion state to the postfusion state. S1 and S2 are composed of an extracellular domain (ECD) and a single transmembrane helix, which mediate receptor binding and membrane fusion, respectively. S1 consists of an NTD and an RBD, which are essential for determining tissue tropism and host range. The RBD of the SARS-CoV-2 spike protein is immunodominant, and it is also the target of 90% neutralizing antibodies present in the SARS-CoV-2 immune serum. The researchers used monoclonal antibodies to structurally determine the RBD antigen profile, and serologically quantified the serum antibodies specific for different RBD epitopes, thereby identifying the two main receptor binding motif antigen sites. These research results explain the immunodominance of the receptor-binding motif and will guide the design of COVID-19 vaccines and drug development (Piccoli et al. 2020). In the existing knowledge, the RBD mutation of the virus is responsible for the higher affinity of SARS-CoV-2 and hACE2 (Ou et al. 2020).

3.3.5 Transmembrane Protease, Serine 2 (TMPRSS2)

It is known that SARS and other coronavirus S proteins enter target cells depending on their binding to hACE2 receptors and the activation of S proteins by cellular proteases, the binding of SARS S protein to its receptor ACE2 is triggered by the cellular serine protease TMPRSS2 (Bertram et al. 2012). In 2019, some studies have shown that TMPRSS2 can activate the synaptic proteins of highly pathogenic human coronaviruses, such as the coronavirus SARS-CoV that causes severe acute respiratory syndrome and the coronavirus MERS-CoV that causes Middle East respiratory syndrome, in vitro, TMPRSS2 activation can induce the virus to fuse with the cell membrane. Experiments show that the lack of TMPRSS2 attenuates the spread of the virus in the airways of mice and the severity of lung pathology (Iwata-Yoshikawa et al. 2019). As early as 2011, the German Institute of Virology evaluated whether the S protein of SARS-CoV was proteolyzed by TMPRSS2. Analysis of the results of the WB experiment showed that the SARS S protein was cleaved into several fragments by TMPRSS2, and the cis-cut resulted in the release of SARS S fragments. In the cell supernatant, the antibody-mediated neutralization reaction is inhibited, and the SARS S protein is activated by trans-lysis to fuse with the target cell. It shows that TMPRSS2 may promote the spread and disease of the virus by reducing the recognition of the virus by the neutralizing antibody and activating the fusion between the cell and the virus (Glowacka et al. 2011).

Besides, 2019-nCoV has a unique four amino acid insertion (681-PRRA-684) in the spike protein or nucleotide positions. This insertion is unique to 2019-nCoV. When compared with other CoV family members, they found that similar insertions have been identified and are located in the structural boundary between the S1 and S2 domains of the spike protein. Therefore, mutations or indels that change the S1–S2 subunits will significantly affect virus infection (Wu et al. 2020a, 2020b).

3.4 LIFE CYCLE OF SARS-COV-2

3.4.1 Attachment and Entry

Coronavirus replication is initiated by the binding of S protein to cell surface receptors. The S protein is composed of two functional subunits, S1 (spheroid) for receptor binding, and S2 (stem) for membrane fusion. The location of the RBD in the S1 region of the coronavirus S protein varies from virus to virus. Some of them have RBD at the N-terminus of S1, while others have RBD at the C-terminus of S1

(Kubo et al. 1994, Cheng et al. 2004). The specific interaction between S1 and the homologous receptor triggers the conformational change of the S2 subunit, which leads to the fusion of the viral envelope and the cell membrane and the release of the nucleocapsid into the Cytoplasm. Receptor binding largely determines the host range and tissue tropism of the coronavirus. Some HCoV has adopted cell surface enzymes as receptors. For example, SARS-CoV and HCoV-NL63 use ACE2 as receptors. MHV enters through CEACAM1. The recently identified MERS-CoV binds to dipeptidyl peptidase 4 Into human cells (Fehr and Perlman 2015). Usually, the S protein of coronavirus is further cleaved into two subunits S1 and S2 by the host's protease. The cleavage of S1/S2 is mediated by one or more host proteases. As an example, the activation of the S protein of SARS-CoV requires sequential cleavage by the endosomal cysteine protease cathepsin L and another trypsin-like serine protease. Two furin cleavage sites belong to the S protein of MERS-CoV. As an example, interferon-induced transmembrane protein (IFITM) exhibits a broad-spectrum antiviral function against a range of RNA viruses. IFITM restricts the entry of SARS-CoV, MERS-CoV, HCoV-229E, and HCoV-NL63. But on the contrary, IFITM2 or IFITM3 promotes HCV-OC43 infection. Studies have shown that several organic compound residues in IFITM control the restriction of HCoV entry and enhance the activity (Zhao et al. 2018). S protein cleavage occurs at two sites within the S2 part of the protein. The primary cleavage is vital for separating the fusion domain of RBD and S protein, and therefore the second cleavage is important for exposing the fusion peptide (Belouzard et al. 2009). Cleavage at S2 exposes the fusion peptide, which is inserted into the membrane, then two heptapeptide repeats are joined at S2 to make an antiparallel six-helix bundle (Bosch et al. 2003). The formation of this bundle allows the blending of the viral and cell membranes, leading to the fusion of the viral genome and its release into the cytoplasm (Fehr and Perlman 2015).

3.4.2 Replicase Protein Expression

After the coronavirus enters the cell, the genomic RNA acts as a transcript to perform cap-dependent translation of ORF1a to produce the polyprotein pp1a. There is an unstable sequence (slippery sequence) and an RNA pseudoknot near the end of ORF1a, which allows 25–30% of ribosomes to perform a-1 frameshift, thereby continuing to translate on ORF1b, resulting in a longer polyprotein pp1ab. Automatic proteolysis of pp1a and pp1ab produces 15–16 kinds of nonstructural proteins (nsps) with various functions. Among them, nsp12 has RdRP activity, while nsp3 and nsp5 have PLPro and master protease Mpro activities, respectively. Nsps 3, 4, and 6 also induce cell membrane rearrangement to form double-membrane vesicles or spherules, which are also the RTC assembly and anchor sites of the coronavirus replication and transcription complex (Baranov et al. 2005; Brierley et al. 1989). Next, many of the nsps combine into the replicase – transcriptase complex to form an environment that suits RNA synthesis, and finally are liable for replication of the RNA and transcription of the subgenomic RNAs. The nsps also have the other enzyme domains and functions that play an important part in RNA replication. As an example, nsp12 encodes the RNA-dependent RNA polymerase domain. After research, it is found that some nsps have another function besides replication function, like blocking innate immune reaction (Snijder et al. 2003). Additionally to RNA secondary structure, programmed ribosomal frameshifting (PRF) may also be regulated by virus or host factors. As an example, within the porcine reproductive and respiratory syndrome virus (PRRSV), nsp1β interacts with the PRF signal through a putative RNA binding motif and activates PRF. The host RNA binding protein ANXA2 can bind to the pseudoknot structure in the infectious bronchitis virus IBV genome. Host factors appear to be involved within the formation of double-membrane vesicles DMV and therefore the assembly of the viral replication transcription complex RTC (Napthine et al. 2016).

3.4.3 Replication and Transcription

The first step of replication is to unzip the two strands. There is a helicase in the cell that breaks the hydrogen bond in a certain area of the DNA strand to unzip the double helix and maintain a temporary unwinding state. A replication bubble is formed where it is unwound, and the unwound nucleotides will be exposed, which is conducive to the subsequent binding of free nucleotides. The replication is carried out at both ends of the replication bubble at the same time, forming a replication fork. In the second step, both strands of DNA are used as templates for replication, and free nucleotides are combined through the principle of base complementary pairing to form hydrogen bonds so that free nucleotides are fixed at complementary positions. Finally, the condensation reaction takes place, combining two adjacent nucleotides through five-carbon sugar and phosphate groups to form a new chain. A newly synthesized strand will spiral with an old strand to form a new DNA molecule. The transcription process only requires RNA polymerase, which recognizes the promoter of a gene on the DNA. A certain gene fragment on DNA first opens the double helix, and then there are two strands, called coding strand/sense strand and template strand/antisense strand, which will be used as templates to transcribe single-stranded mRNA in the form of complementary base pairing. The mRNA floats through the nuclear pore to the ribosome in the cytoplasm. The ribosome is equivalent to a barcode reader, which reads the code on the mRNA and guides the synthesis of protein. Then another kind of RNA, tRNA is involved in the protein translation process (Clegg 2015).

Perhaps the newest research of coronavirus replication is how the leader and body of the TRS segments merge during the production of subgenomic RNAs. In the beginning, it was thought that will happen during the synthesis of the positive-strand. Later, it is generally believed that it occurs during the discontinuous extension of the negative-strand RNA (Sawicki et al. 2007). Finally, coronaviruses also can use homologous and nonhomologous recombination. The recombination ability of these viruses is related to the chain switching ability of RdRp. Recombination may play an important role in virus evolution and is the basis for targeted RNA recombination. RNA recombination is a reverse genetics tool used to engineer viral recombinants at the 3'end of the genome (Keck et al. 1988; Lai et al. 1985).

3.4.4 Assembly and Release

After the virion is transported to the perinuclear region, the coronavirus RNA exudes from the vesicles and virus particles and enters the nucleus for reverse transcription and replication. The DNA replicon is then transcribed into RNA and enters the Golgi apparatus/ER/microtubule organization center from the nucleus. That is, after replication and synthesis of subgenomic RNA, the viral structural proteins S, E, and M are translated and inserted into the endoplasmic reticulum (ER) (Krijnse-Locker et al. 1994; Tooze et al. 1984). Initially, the nucleocapsid (N) protein binds to the RNA copy and binds to the vesicle membrane, and then the N and E proteins further mature, which is the basic virus-like assembly required for particles (virus-like particles [VLP]). The M protein can direct most of the protein–protein interactions required for the assembly of coronaviruses. But when only M protein acts, it is not enough to form VLPs. However, when the M protein and the E protein are expressed together, a VLP will be formed, which indicates that the two proteins work together to produce a coronavirus envelope (Bos et al. 1996). The role of the N protein is to enhance the formation of VLPs. When the N protein that is wrapped by the viral genome germinates into the ERGIC membrane containing the viral structural protein, then a mature virion is formed (De Haan and Rottier 2005). If Spike (S) protein is co-expressed, it will be incorporated into virus particles. Under the synergistic action of a variety of cytoskeletons and membrane regulatory proteins, the assembly is assisted by concentrated packaging components. The ability of S protein to transport to ERGIC and interact with M protein is critical for its incorporation into viral particles (Siu et al. 2008). According to the findings of the scientists, it is likely that M protein interaction provides the impetus for envelope maturation. Because E protein is only present in a small amount in virus particles, M protein is relatively abundant. It is not clear how E protein assists the M protein to assemble virions. Some studies have shown that E protein plays a role in inducing membrane curvature, but other studies have shown that E protein prevents M protein aggregation (Fehr and Perlman, 2015). The M protein also binds to the nucleocapsid, and this interaction promotes the completion of virion assembly. These interactions have been mapped to the C-terminus of the endodomain of M with CTD 3 of the N-protein (Hurst et al. 2005). However, there is currently no study to explain how the nucleocapsid interacts with the viral particle RNA to transport to ERGIC and also interacts with the M protein to bind to the viral envelope. There is also no answer to the question of how the N protein selectively packages the forward full-length genomes of many different RNA species produced during infection (Kuo and Masters 2013). The virion is transported to the cell surface within the vesicle and released by exocytosis. The genetic fusion of the coronavirus nucleocapsid or spike protein and GFP makes it possible to trace noninfectious virus particles through fluorescence microscopy. Some studies used this system to observe the discharge of SARS-CoV and located vesicles fused into multiparticle clumps.

This transport is sensitive to Brefeldin A, indicating that the secretory pathway is being utilized. Other studies have found that nocodazole can effectively inhibit the transport of virions to the semipermeable membrane, indicating that microtubules are a very important part of the virus release outlet. However, it is not yet known that the

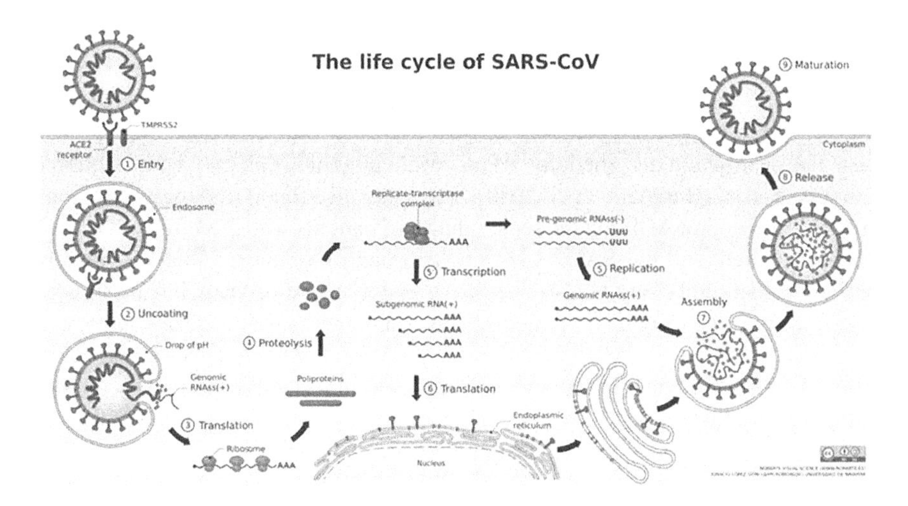

FIGURE 3.5 Coronavirus SARS-CoV-2 replication process, (Reprinted from Vega Asensio (2020), with kind permission of Creative Commons Attribution-Share Alike 4.0 International).

virions use traditional methods to ship large cargoes from the Golgi or the viruses have changed a unique way for their own exports already. There is also a special situation where the S protein that has not been assembled into a virion migrates to the cell surface, allowing the mix between infected and uninfected cells (see Figure 3.5). This can cause the virus to spread within the infected organism without being detected or neutralized by virus-specific antibodies (Fehr and Perlman 2015; Siu et al. 2008).

3.5 PATHOGENESIS AND IMMUNE RESPONSE OF SARS-COV-2

Viral infectious diseases are caused by the interaction between the virus and the body's immune response. Among the coronaviruses known to infect humans, COVID-19, SARS-CoV, and MERS-CoV can all cause severe respiratory diseases. ACE2 is the cell receptor of SARS-CoV, which is widely expressed in body tissues. In addition to respiratory epithelial cells and intestinal mucosa, SARS-CoV also infects renal tubular epithelium and brain neurons, causing damage to corresponding tissues and organs. Besides, the virus can also directly infect circulating lymphocytes, monocytes, macrophages, and lymphoid tissues such as the spleen and lymph nodes, resulting in a significantly weakened immune defense ability, thereby worsening the condition (Gu et al. 2005). Judging from the current existing studies on COVID-19, COVID-19 and SARS-CoV are similar, mainly causing respiratory diseases, severe cases can progress to acute respiratory distress syndrome, and the viral load in respiratory specimens the severity of lung disease was positively correlated (Liu et al. 2020a, 2020b). At present, several pathological anatomical studies on COVID-19 death cases have found that COVID-19 mainly causes deep airway and alveolar damage. This not only provides valuable information for clinical treatment but also suggests that the virus infects the respiratory tract and replicates in it as COVID-19, one of the important factors of pathogenesis (Xu et al. 2020).

2019-nCoV and SARS-CoV have the same cell receptor ACE2, so in addition to the lungs, 2019-nCoV may also infect other tissues. Individual COVID-19 patients have detected the virus in stool specimens, and the intestinal mucosal epithelial cells express ACE2, suggesting the possibility of the virus infecting intestinal epithelial cells. The destruction and shedding of mucosal epithelial cells may be the source of the virus in stool (Burgueño et al. 2020). The study also found that the expression of 2019-nCoV receptor ACE2 in the renal tubules of patients who died from COVID-19 was upregulated, and the SARS-CoV nucleoprotein immunostaining was positive, suggesting that the kidney is also one of the target organs of virus infection, which may occur in some COVID-19 patients. This is a major cause of kidney damage (Su et al. 2020). Besides, the 2019-nCoV nucleic acid and antigen detection on various organ tissue pathological specimens are helpful to understand the target cell spectrum of virus infection. After the respiratory virus invades the body, it infects respiratory epithelial cells and macrophages; induces an innate immune response; produces cytokines, chemokines, and chemotaxis; and recruits immune cells to the infected foci to play an antiviral effect together. Appropriate innate immunity will also initiate an adaptive immune response, which will produce antigen-specific T cell immunity and antibody immunity after 1–2 weeks to eliminate pathogens. An appropriate inflammatory response exerts an antiviral effect, while an excessive inflammatory storm can cause immune pathological damage and aggravate the disease. Excessive inflammation is related to lung inflammation and extensive lung damage, which may lead to a poor prognosis of the disease (Zhang et al. 2004). The abnormally excessive immune response leads to long-term lung damage and fibrosis, dysfunction, and reduced quality of life in surviving patients (Ngai et al. 2010). Similar to SARS-CoV infection patients, COVID-19 infection causes acute respiratory distress syndrome in severely ill patients with COVID-19. Some studies have shown that the levels of serum inflammatory factors in patients are also higher than normal controls. Some of the serum concentrations of severely ill patients are higher than those of noncritically ill patients, suggesting that the cytokine storm is related to disease severity and may be an important cause of organ damage (Huang et al. 2020). Therefore, the innate immune response is too weak and the inflammatory response is too strong, both may be involved in the pathogenesis of COVID-19. In addition to inflammatory factors, innate immunity also includes innate immune cells, such as macrophages, natural killer cells, monocytes, dendritic cells, and neutrophils. It has been reported in the literature that the case of critically ill children with natural killer cells decreased significantly. In COVID-19 patients, the number and function of these immune cells, as well as their immune protection and immune pathological mechanisms in the disease still need further research (Chen et al. 2020).

After viral infection induces an innate immune response, the body will further initiate an adaptive immune response to control the infection. In response, the virus will evolve a set of strategies to escape the body's immune response, such as directly infecting immune cells to induce apoptosis and inhibit cell function, to promote virus replication and spread, leading to disease progression. SARS-CoV histopathological examination revealed that there are a large number of SARS-CoV particles and genome sequences in circulating lymphocytes, spleen and lymph nodes, and other lymphoid tissues, suggesting that the virus can directly infect lymphocytes, and the lymphoid tissues and germinal centers. Lymphocytes, especially T lymphocytes, induced apoptosis by virus infection or destroyed by other immune cells may be one of the reasons for the continuous decrease of lymphocytes in SARS-CoV patients (Gu et al. 2005). Similar to SARS, some current papers on COVID-19 point out that most adult patients have different degrees of a significant or progressive decrease in lymphocytes, and the absolute value and percentage of lymphocytes are highly correlated with acute lung injury, which can be used as a measure of disease severity (Liu et al. 2020a, 2020b). Among children, the number of lymphocytes in most mild to moderate cases is normal, but T cells in critically ill children are also significantly reduced (Chen et al. 2020). The lymphocyte count can reflect the patient's adaptive immune response to a certain extent. Combined with COVID-19 cases, critically ill patients usually progress within 7–10 days after the onset of disease, which indirectly reflects the impaired adaptive immune response function involved in the immune pathogenic mechanism of disease progression. The normal number of lymphocytes in children may also be one of the reasons why most children are in mild to moderate cases (Ruggiero et al. 2020). Through cellular immune response studies, it was found that all recovered patients from COVID-19 produced the spike protein-specific helper T cells, and virus-specific killer T cells were detected in 70% of the subjects. Besides, COVID-19 specific T cells have also been detected in some normal individuals who have never been infected. It is speculated that this may be a cross T cell response caused by these individuals who have been infected with other coronaviruses before. This is important for understanding the body. Immune protection mechanisms and immunotherapy are of great significance (Grifoni et al. 2020).

3.6 CONCLUSION

In the past few decades, the emergence of many different coronaviruses that cause a larger kind of human and veterinary diseases has occurred. It is likely that these viruses will continue to emerge and evolve and cause both human and veterinary outbreaks because of their ability to recombine, mutate, and infect multiple species and cell types (Fehr and Perlman 2015). In this report, some research results have showed how viral factors could manipulate the host cell to expedite its replication cycle and pathogenesis. For years, HCoVs are believed to be mild respiratory pathogens that will not have a big effect on the human population. However, it had been the emergence of SARS-CoV, MERS-CoV, and SARS-CoV-2 that thrust these human viruses into the spotlight of the research field. That is likely that more emerging HCoVs might surface to threaten the worldwide public health, as seen from the high mortality rates within the past two outbreaks: SARS-CoV (10%) and MERS-CoV (35%). Therefore, a study of the pathogenesis of all HCoVs would gain more insights into the event of antiviral therapeutics and vaccines.

ACKNOWLEDGEMENT

This work was supported by the Ministry of Education, Youth and Sports of the Czech Republic and the European Union – European Structural and Investment Funds in the Frames of Operational Programme Research, Development and

Education – project Hybrid Materials for Hierarchical Structures (HyHi, Reg. No. CZ.02.1.01/0.0/0.0/16_019/0000843) and student grant competition No. 21408 at Technical University of Liberec.

REFERENCES

Akram A, Mannan N. Molecular structure, pathogenesis and virology of SARS-CoV-2: A review. *Bangladesh Journal of Infectious Diseases*, 2020: S36–S40.

Anand K, Ziebuhr J, Wadhwani P, et al. Coronavirus main proteinase (3CLpro) structure: basis for design of anti-SARS drugs. *Science*, 2003, 300(5626): 1763–1767.

Arndt A L, Larson B J, Hogue B G. A conserved domain in the coronavirus membrane protein tail is important for virus assembly. *Journal of Virology*, 2010, 84(21): 11418–11428.

Bakkers M J G, Lang Y, Feitsma L J, et al. Betacoronavirus adaptation to humans involved progressive loss of hemagglutinin-esterase lectin activity. *Cell Host & Microbe*, 2017, 21(3): 356–366.

Baranov P V, Henderson C M, Anderson C B, et al. Programmed ribosomal frameshifting in decoding the SARS-CoV genome. *Virology*, 2005, 332(2): 498–510.

Belouzard S, Chu V C, Whittaker G R Activation of the SARS coronavirus spike protein via sequential proteolytic cleavage at two distinct sites. *Proceedings of the National Academy of Sciences*, 2009, 106(14): 5871–5876.

Belouzard S, Millet J K, Licitra B N, et al. Mechanisms of coronavirus cell entry mediated by the viral spike protein. *Viruses*, 2012, 4(6): 1011–1033.

Beniac D R, Andonov A, Grudeski E, et al. Architecture of the SARS coronavirus prefusion spike. *Nature Structural & Molecular B*, 2006, 13(8): 751–752.

Bertram S, Heurich A, Lavender H, et al. Influenza and SARS-coronavirus activating proteases TMPRSS2 and HAT are expressed at multiple sites in human respiratory and gastrointestinal tracts. *PloS one*, 2012, 7(4): e35876.

Bos E C W, Luytjes W, Van Der Meulen H, et al. The production of recombinant infectious DI-particles of a murine coronavirus in the absence of helper virus. *Virology*, 1996, 218(1): 52–60.

Bosch B J, Van der Zee R, De Haan C A M, et al. The coronavirus spike protein is a class I virus fusion protein: structural and functional characterization of the fusion core complex. *Journal of Virology*, 2003, 77(16): 8801–8811.

Brierley I, Digard P, Inglis S C. Characterization of an efficient coronavirus ribosomal frame-shifting signal: requirement for an RNA pseudoknot. *Cell*, 1989, 57(4): 537–547.

Burgueño J F, Reich A, Hazime H, et al. Expression of SARS-CoV-2 Entry Molecules ACE2 and TMPRSS2 in the Gut of Patients With IBD. *Inflammatory Bowel Diseases*, 2020, 26(6): 797–808.

Chen Feng, Liu Zhisheng, Zhang Furong. China's first case of severe novel coronavirus pneumonia in children. Chinese *Journal of Pediatrics*, 2020, 58(3): 179.

Cheng P K C, Wong D A, Tong L K L, et al. Viral shedding patterns of coronavirus in patients with probable severe acute respiratory syndrome. *The Lancet*, 2004, 363(9422): 1699–1700.

Clegg C J. Biology for the IB Diploma Second Edition. *Hodder Education*, 2015.

De Haan C A M, Rottier P J M. Molecular interactions in the assembly of coronaviruses. *Advances in Virus Research*, 2005, 64: 165–230.

DeDiego M L, Álvarez E, Almazán F, et al. A severe acute respiratory syndrome coronavirus that lacks the E gene is attenuated in vitro and in vivo. *Journal of virology*, 2007, 81(4): 1701–1713.

Fehr A R, Perlman S. *Coronaviruses: an overview of their replication and pathogenesis// Coronaviruses*. Humana Press, New York, NY, 2015: 1–23.

Glowacka I, Bertram S, Müller M A, et al. Evidence that TMPRSS2 activates the severe acute respiratory syndrome coronavirus spike protein for membrane fusion and reduces viral control by the humoral immune response. *Journal of Virology*, 2011, 85(9): 4122–4134.

Godet M, L'Haridon R, Vautherot J F, et al. TGEV corona virus ORF4 encodes a membrane protein that is incorporated into virions. *Virology*, 1992, 188(2): 666–675.

Grifoni A, Weiskopf D, Ramirez S I, et al. Targets of T cell responses to SARS-CoV-2 coronavirus in humans with COVID-19 disease and unexposed individuals. *Cell*, 2020.

Gu J, Gong E, Zhang B, et al. Multiple organ infection and the pathogenesis of SARS. *Journal of Experimental Medicine*, 2005, 202(3): 415–424.

Gurung A B, Ali M A, Lee J, et al. Structure-based virtual screening of phytochemicals and repurposing of FDA approved antiviral drugs unravels lead molecules as potential inhibitors of coronavirus 3C-like protease enzyme. *Journal of King Saud University-Science*, 2020, 32 (6): 2845–2853.

Huang C, Wang Y, Li X, et al. Clinical features of patients infected with 2019 novel coronavirus in Wuhan, China. *The Lancet*, 2020, 395(10223): 497–506.

Hurst K R, Kuo L, Koetzner C A, et al. A major determinant for membrane protein interaction localizes to the carboxy-terminal domain of the mouse coronavirus nucleocapsid protein. *Journal of Virology*, 2005, 79(21): 13285–13297.

Imai Y, Kuba K, Rao S, et al. Angiotensin-converting enzyme 2 protects from severe acute lung failure. *Nature*, 2005, 436(7047): 112–116.

Iwata-Yoshikawa N, Okamura T, Shimizu Y, et al. TMPRSS2 contributes to virus spread and immunopathology in the airways of murine models after coronavirus infection. *Journal of Virology*, 2019, 93(6).

Keck J G, Matsushima G K, Makino S, et al. In vivo RNA-RNA recombination of coronavirus in mouse brain. *Journal of Virology*, 1988, 62(5): 1810–1813.

Krijnse-Locker J, Ericsson M, Rottier P J, et al. Characterization of the budding compartment of mouse hepatitis virus: evidence that transport from the RER to the Golgi complex requires only one vesicular transport step. *The Journal of Cell Biology*, 1994, 124(1): 55–70.

Kuba K, Imai Y, Rao S, et al. A crucial role of angiotensin converting enzyme 2 (ACE2) in SARS coronavirus–induced lung injury. *Nature Medicine*, 2005, 11(8): 875–879.

Kubo H, Yamada Y K, Taguchi F. Localization of neutralizing epitopes and the receptor-binding site within the amino-terminal 330 amino acids of the murine coronavirus spike protein. *Journal of Virology*, 1994, 68(9): 5403–5410.

Kuo L, Masters P S. Functional analysis of the murine coronavirus genomic RNA packaging signal. *Journal of Virology*, 2013, 87(9): 5182–5192.

Lai M M, Baric R S, Makino S, et al. Recombination between nonsegmented RNA genomes of murine coronaviruses. *Journal of Virology*, 1985, 56(2): 449–456.

Li F. Evidence for a common evolutionary origin of coronavirus spike protein receptor-binding subunits. *Journal of Virology*, 2012, 86(5): 2856–2858.

Li F. Structure, function, and evolution of coronavirus spike proteins. *Annual Review of Virology*, 2016, 3: 237–261.

Li W, Moore M J, Vasilieva N, et al. Angiotensin-converting enzyme 2 is a functional receptor for the SARS coronavirus. *Nature*, 2003, 426(6965): 450–454

Liu C, Yang Y, Gao Y, et al. Viral architecture of SARS-CoV-2 with post-fusion spike revealed by Cryo-EM. Biorxiv, 2020a.

Liu Y, Yang Y, Zhang C, et al. Clinical and biochemical indexes from 2019-nCoV infected patients linked to viral loads and lung injury. *Science China Life Sciences*, 2020b, 63(3): 364–374.

McBride R, Van Zyl M, Fielding B C. The coronavirus nucleocapsid is a multifunctional protein. *Viruses*, 2014, 6(8): 2991–3018.

Nal B, Chan C, Kien F, et al. Differential maturation and subcellular localization of severe acute respiratory syndrome coronavirus surface proteins S, M and E. *Journal of General Virology*, 2005, 86(5): 1423–1434.

Napthine S, Treffers E E, Bell S, et al. A novel role for poly (C) binding proteins in programmed ribosomal frameshifting. *Nucleic Acids Research*, 2016, 44(12): 5491–5503.

Neuman B W, Kiss G, Kunding A H, et al. A structural analysis of M protein in coronavirus assembly and morphology. *Journal of Structural Biology*, 2011, 174(1): 11–22.

Ngai J C, Ko F W, Ng S S, et al. The long-term impact of severe acute respiratory syndrome on pulmonary function, exercise capacity and health status. *Respirology*, 2010, 15(3): 543–550.

Nieto-Torres J L, DeDiego M L, Verdiá-Báguena C, et al. Severe acute respiratory syndrome coronavirus envelope protein ion channel activity promotes virus fitness and pathogenesis. *PLoS Pathog*, 2014, 10(5): e1004077.

Omotuyi O, Nash O, Ajiboye B, et al. The disruption of SARS-CoV-2 RBD/ACE-2 complex by Ubrogepant Is mediated by interface hydration. Preprints, 2020, 2020030466.

Ou J, Zhou Z, Dai R, et al. Emergence of RBD mutations in circulating SARS-CoV-2 strains enhancing the structural stability and human ACE2 receptor affinity of the spike protein. bioRxiv, 2020.

Piccoli L, Park Y J, Tortorici M A, et al. Mapping neutralizing and immunodominant Sites on the SARS-CoV-2 spike receptor-binding domain by structure-guided high resolution serology. *Cell*, 2020, 183 (4): 1024–1041.

Ruggiero A, Attinà G, Chiaretti A. Additional hypotheses about why COVID-19 is milder in children than adults. *Acta Paediatrica*, 2020, 109: 1690.

Sawicki S G, Sawicki D L, Siddell S G. A contemporary view of coronavirus transcription. *Journal of Virology*, 2007, 81(1): 20–29.

Siu Y L, Teoh K T, Lo J, et al. The M, E, and N structural proteins of the severe acute respiratory syndrome coronavirus are required for efficient assembly, trafficking, and release of virus-like particles. *Journal of Virology*, 2008, 82(22): 11318–11330.

Snijder E J, Bredenbeek P J, Dobbe J C, et al. Unique and conserved features of genome and proteome of SARS-coronavirus, an early split-off from the coronavirus group 2 lineage. *Journal of Molecular Biology*, 2003, 331(5): 991–1004.

Stohlman S A, Baric R S, Nelson G N, et al. Specific interaction between coronavirus leader RNA and nucleocapsid protein. *Journal of Virology*, 1988, 62(11): 4288–4295.

Su H, Yang M, Wan C, et al. Renal histopathological analysis of 26 postmortem findings of patients with COVID-19 in China. *Kidney International*, 2020, 98(1): 219–227.

Tooze J, Tooze S, Warren G. Replication of coronavirus MHV-A59 in sac-cells: determination of the first site of budding of progeny virions. *European Journal of Cell Biology*, 1984, 33(2): 281–293.

Vega Asensio, 27 March 2020, https://commons.wikimedia.org/wiki/File:SARS-CoV-2_cycle.png

Wrapp D, Wang N, Corbett K S, et al. Cryo-EM structure of the 2019-nCoV spike in the prefusion conformation. *Science*, 2020, 367(6483): 1260–1263.

Wu A, Niu P, Wang L, et al. Mutations, Recombination and Insertion in the Evolution of 2019-nCoV. bioRxiv, 2020a.

Wu A, Peng Y, Huang B, et al. Genome composition and divergence of the novel coronavirus (2019-nCoV) originating in China. *Cell Host & Microbe*, 2020b, 27(3): 325–328.

Xu Z, Shi L, Wang Y, et al. Pathological findings of COVID-19 associated with acute respiratory distress syndrome. *The Lancet Respiratory Medicine*, 2020, 8(4): 420–422.

Yang X, Deng W, Tong Z, et al. Mice transgenic for human angiotensin-converting enzyme 2 provide a model for SARS coronavirus infection. *Comparative Medicine*, 2007.

Ren Yi, Ding Huiguo, Wu Qingfa, et al. Determination of SARS CoV RNA in the stool and mouthwash of SARS patients by RT-PCR and its clinical significance. *Chinese Journal of Microbiology and Immunology*, 2003, 23(12): 930–932.

Zhang Y, Li J, Zhan Y, et al. Analysis of serum cytokines in patients with severe acute respiratory syndrome. *Infection and Immunity*, 2004, 72(8): 4410–4415.

Zhang L, Lin D, Sun X, et al. Crystal structure of SARS-CoV-2 main protease provides a basis for design of improved α-ketoamide inhibitors. *Science*, 2020, 368(6489): 409–412.

Zhao X, Sehgal M, Hou Z, et al. Identification of residues controlling restriction versus enhancing activities of IFITM proteins on entry of human coronaviruses. *Journal of Virology*, 2018, 92(6).

Zhou P, Yang X L, Wang X G, et al. A pneumonia outbreak associated with a new coronavirus of probable bat origin. *Nature*, 2020, 579(7798): 270–273.

Zou X, Chen K, Zou J, et al. Single-cell RNA-seq data analysis on the receptor ACE2 expression reveals the potential risk of different human organs vulnerable to 2019-nCoV infection. *Frontiers of Medicine*, 2020: 1–8. https://www.scientificanimations.com/wiki-images/.

Section II

Disinfection Mechanism

4 Characterization, Indication, and Passivation of COVID-19

Sajid Faheem
Technical University of Liberec, Czech Republic
Nazia Nahid
Government College University, Faisalabad, Pakistan
Muhammad Shah Nawaz ul Rehman
University of Agriculture Faisalabad, Pakistan
Jiri Militký and Jakub Wiener
Technical University of Liberec, Czech Republic

CONTENTS

4.1 INTRODUCTION

Viruses as infectious agents are continuously emerging and causing serious threats to the health of humans and animals. Viruses have been known to infect human beings since prehistoric times. Several epidemics due to virus infection are recorded causing devastating losses to humans. The recent outbreaks related to severe respiratory illness that have severely impacted the economy and the health systems of many are caused by the subcellular entity, the viruses from a group known as Coronavirus. Coronaviruses form a vast group of viruses that causes deadly diseases in birds and mammals ranging from mild respiratory illness to lethal respiratory infections. These viruses also infect other body organs including the liver, kidney, brain, and gastrointestinal tract of both animals and humans. Recently, Severe Acute Respiratory Syndrome Coronavirus-2 (SARS-CoV-2) is responsible for a pneumonia-like respiratory disease called Coronavirus disease-19 (COVID-19). The infection from this virus was first reported in December 2019 in Wuhan city, Hubei province in China. High fever, difficulty in breathing, dry cough, and typical pneumonia characterize COVID-19. The disease symptoms are very similar to the SARS outbreak caused by the SARS-CoV during 2013, hence the name 2019-novel coronavirus (2019-nCoV) was used initially for this unique virus. Later, the International Committee on Taxonomy of Viruses (ICTV) named it SARS-CoV-2. In comparison to other members of respiratory disease-causing coronaviruses, SARS-CoV was first reported in China in 2002, and Middle East respiratory syndrome coronavirus (MERS-CoV) was first recorded in Saudi Arabia in 2012. The SARS-CoV-2 is more deadly. Most likely originated in animals SARS-CoV-2 has spread to 216 different countries and territories all over the world. It is a very contagious virus and a novel strain that was not known to infect humans previously. Coronaviruses causing infections in humans have zoonotic origins, spread to humans from animals including bats, dromedary camels (*Camelus dromedarius*), and civet cat (*Paguma larvata*). These animals act as reservoirs for the coronaviruses from where the virus was transmitted to humans through an intermediate unknown animal host. SARS-CoV-2 is considered as the third pathogenic coronavirus causing high mortality. The incubation period is from 2 to 14 days with general symptoms of coughing, sneezing, runny nose, headache, fever, fatigue, sore throat, and breathing difficulties. In rare cases, this virus can cause pneumonia or bronchitis. The deaths in the case of SARS-CoV-2 infection are mainly caused by the complications associated with the disease. The complications of the disease are sepsis, acute respiratory distress syndrome (ARDS) induced by the massive release of cytokine, and kidney failure. These frequent outbreaks of coronaviruses, the potential of these coronaviruses to mutate and the identification of several coronaviruses from bats represent a probable threat for future pandemics. Regarding COVID-19, the World Health Organization (WHO) announced the outbreak as a Public Health Emergency of International Concern (PHEIC) on January 30, 2020. It was declared as a pandemic on March 11, 2020, just after 3 months of its emergence on December 31, 2019. The world faced a global recession due to the economic disruption during this pandemic and the control measures taken to overcome the disease. Globally the governments are involved in establishing the measures to combat the disease, curtailing the devastating effects of pandemic, and improving the health facilities. Scientists and researchers from every field are working day and night to understand the disease that can help in developing any antiviral drug and vaccine.

4.2 GENERAL CLASSIFICATION OF VIRUSES

Biologists have employed several methods for the classification of viruses based on the morphology, genome composition, and mode of replication. Previously, viruses were named randomly based on the host organism from the type of disease, name of scientists who discovered the virus, and from where they were first isolated. Lwoff, Horne, and Tournier proposed a comprehensive scheme for the classification of viruses in 1962 (Lwoff, Horne, and Tournier 1962) and (Lwoff and Tournier 1966). At that time, an enormous amount of information about viruses was available related to the virion morphology, biology (serology and virus properties), nature, and size of virion due to advancement in molecular biology techniques and the advent of electron microscopy. Lwoff and colleagues suggested the use of classical Linnaean hierarchical classification comprising of phylum, class, order, family, genus, and species. The characteristics of viruses used for the classification include:

- Morphology of virion: size, shape, symmetry (helical or icosahedral), and presence or absence of an envelope.
- Virus genome: nature of nucleic acid (RNA or DNA genome), genome size, single- or double-stranded genome, linear or circular genome, positive-sense or negative-sense, segmented genome or non-segmented, and presence of characteristic features including repetitive elements, isomerization, 5′-terminal cap, 5′-terminal covalently linked protein, and 3′-terminal poly-A tail).
- Genome organization: gene number, gene order, and different open reading frames.

About 50 years ago in 1966, the International Committee on Nomenclature of Viruses (ICNV) was established for the creation of a universal taxonomic scheme for all viruses. The first report on virus classification was published in 1971. In 1974, ICNV was renamed as ICTV and it authorized the universal taxonomy and nomenclature of viruses. It was established to describe, name, and classify every virus known to infect any organism. Generally, the classification of viruses is based on morphology, chemical composition, and mode of replication of that particular virus. The most commonly used system of virus classification is the Baltimore classification (Baltimore 1971). A Nobel laureate biologist David Baltimore suggested this scheme in the 1970s. According to this system now viruses can be classified in one of the seven groups based on the type of virus genome (DNA or RNA) and the way the messenger RNA (mRNA) is produced during the replication cycle of the virus (see Figure 4.1).

Viral genome expression studies for the production of viral protein provided a significant landmark in the field of virology. DNA genome contains the genes that encode for a specific protein. mRNA is produced from the information encoded in

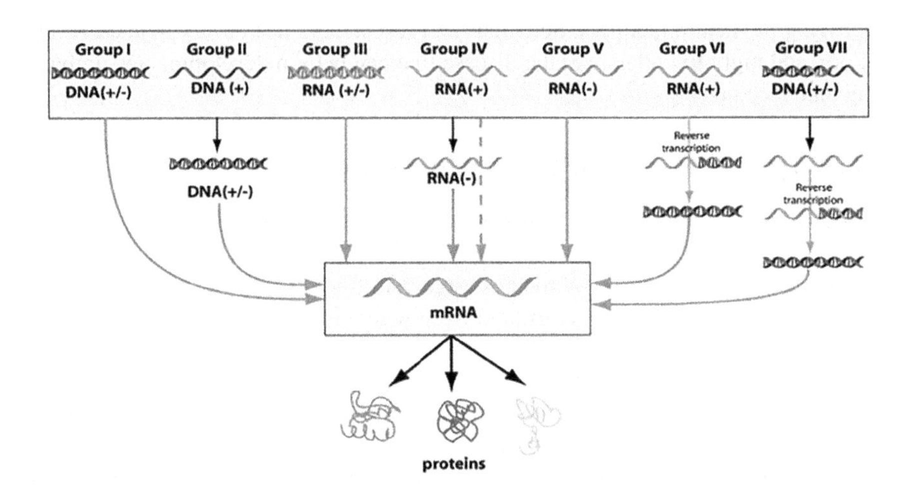

FIGURE 4.1 Baltimore classifications of viruses (https://microbenotes.com/classification-of-virus/).

the form of genes on DNA. mRNA acts as an intermediate for the synthesis of proteins. The central dogma of molecular biology explained by Francis Crick states (Watson and Crick 1958):

$$\text{DNA} \rightarrow \text{RNA} \rightarrow \text{Protein}$$

Viruses also need to synthesize mRNA for the production of proteins. Viruses recruit the host protein synthesis machinery and host proteins for perpetuation. No viral genome is completely independent to produce proteins. In Baltimore classification, mRNA occupies a central position and illustration of the pathways involved in the synthesis of mRNA from DNA or RNA genomes are highlighted. This design emphasizes the mandatory link between the genome of the virus and the mRNA, and it is widely used alongside the taxonomy of viruses, but it is the authority of the ICTV to specifically name and further classify the virus. Baltimore system of classification originally grouped virus genomes in six classes. Later on, with the discovery of viruses with a gapped DNA genome (hepadnaviruses, hepatitis B virus), a seventh group was introduced (see Table 4.1).

4.3 HISTORY OF CORONAVIRUSES

Coronaviruses have been known to be associated with acute respiratory diseases of birds since the 1930s (Sturman and Holmes 1983). An outbreak of respiratory illness had occurred in chickens due to an avian infectious bronchitis virus (IBV) now known to widespread in the whole world. Schalk and Hawn in 1931 described this disease in detail for the first time (Schalk 1931). In 1947 and 1951, two more viral infections known as JHM were discovered from laboratory mice, causing brain disease (encephalomyelitis) and mouse hepatitis virus (MHV). Human coronaviruses B814 and 229E

TABLE 4.1
Baltimore Classification of Viruses*

Group	Virus Genome	mRNA Production Pathway	Examples	Disease
I	Double-stranded DNA (dsDNA) viruses	mRNA transcription is direct from the DNA genome	Herpesvirus Adenovirus	Chickenpox Common cold
II	Single-stranded DNA (ssDNA) viruses	ssDNA is converted into dsDNA to act as a template for mRNA synthesis	Canine parvovirus	Gastrointestinal illness in puppies
III	Double-stranded RNA (dsRNA) viruses	mRNA is synthesized directly from the genome	Rotavirus	Gastroenteritis in human and animals
IV	Positive-sense single-stranded RNA (+ssRNA) viruses	+ssRNA genome acts as the mRNA	Coronavirus Hepatitis A virus Poliovirus	Covid-19 Hepatitis A Polio
V	Negative-sense single-stranded RNA (−ssRNA) viruses	mRNA is synthesized from −ssRNA genome	Rabies virus	Rabies
VI	Single-stranded RNA viruses with a DNA intermediate [reverse transcription (ssRNA-RT)] in their life cycle	DNA produced by reverse transcription is incorporated in the host genome that further act as a template for mRNA synthesis	Human immunodeficiency virus (HIV)	Immunodeficiency disease
VII	Double-stranded DNA viruses with an RNA intermediate in their life cycle	The gaps in dsDNA are filled, RNA is produced that may directly act as mRNA or act as a template to make mRNA	Hepatitis B virus	Hepatitis B

* Adapted and reproduced from (OpenStax College, Biology. OpenStax CNX. http://cnx.org/contents/185cbf87-c72e-48f5-b51e-f14f21b5eabd@11.2)

associated with common colds were discovered in the 1960s (Tyrrell and Bynoe 1961; Hamre and Procknow 1966). The transmission electron microscopy revealed that B814 and 229E were having the same morphology and it was identical to the IBV, JHM, and transmissible gastroenteritis virus of swine (Almeida and Tyrrell 1967). As seen through the electron microscope these virus particles were 80–150 nm, pleomorphic, enveloped with club-like surface projections. These surface projections when viewed through an electron microscope appeared like a crown (Latin word "corona"), and the name "coronavirus" was proposed for these newly discovered viruses. Later on, it was accepted as a genus of the newly established monogeneric family "*Coronaviridae*" (Tyrrell et al. 1975). Previously, four coronaviruses (HKU1, NL63, 229E, and OC43) were known to infect humans that generally cause mild respiratory disease.

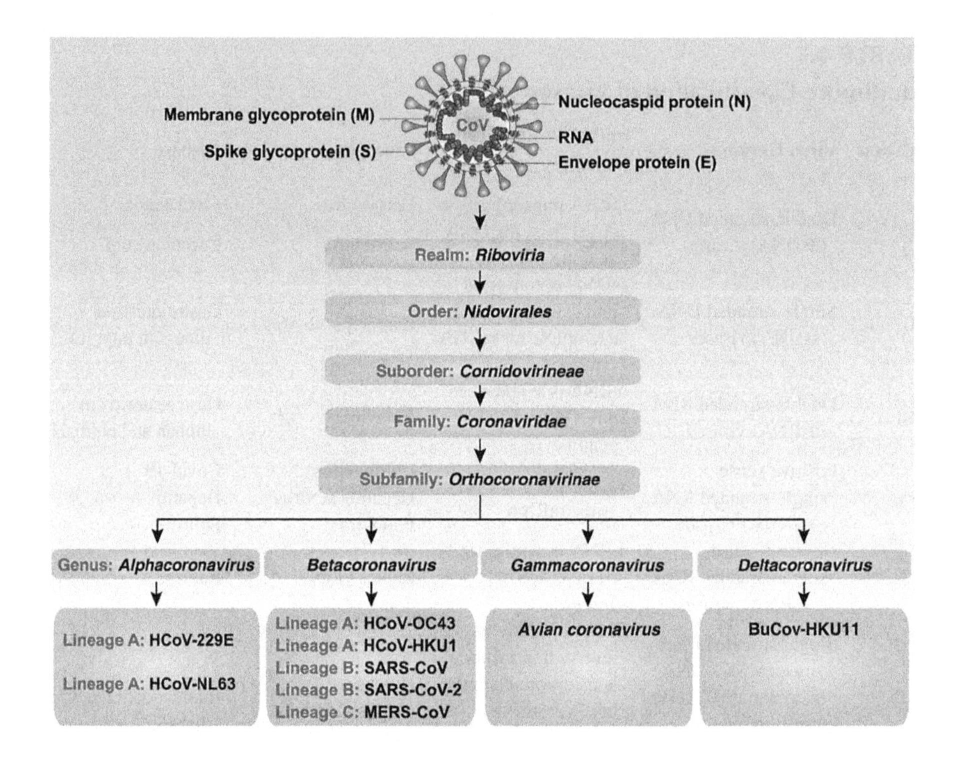

FIGURE 4.2 Classification scheme of coronaviruses (Reprinted under the terms of the Creative Commons Attribution license from Plos Pathogens, Plos Publications (Tang, Comish, and Kang 2020)).

4.4 CLASSIFICATION OF CORONAVIRUSES

The family *Coronaviridae* is comprised of two subfamilies, five genera, 26 subgenera, and 46 species. These are enveloped by positive-sense single-stranded RNA viruses of mammals (coronaviruses and toroviruses), birds (coronaviruses), and fish (bafiniviruses).

The subfamily *Coronovirinae* (also known as *Orthocoronavirinae*) is further classified into four genera, namely, *Alphacoronavirus*, *Betacoronavirus*, *Gammacoronavirus*, and *Deltacoronavirus* (see Figure 4.2). The alphacoronaviruses infect camel, cat, pig, bat, and human; beta coronaviruses infect horse, mouse, cow, bat, and human; delta coronaviruses infect sparrow and bulbul; and gamma coronavirus infect chicken and whale (Chen, Liu, and Guo 2020).

4.5 TAXONOMY OF SARS-COV-2

SARS-Cov-2, causing the present-day COVID-19 is the member of the order *Nidovirales*, family *Coronaviridae*, subfamily *Coronovirinae*, and genus *Betacoronavirus*. It is considered a novel human-infecting Betacoronavirus (Lu et al. 2020). The WHO named this virus initially "2019 novel coronavirus" (2019-nCoV)

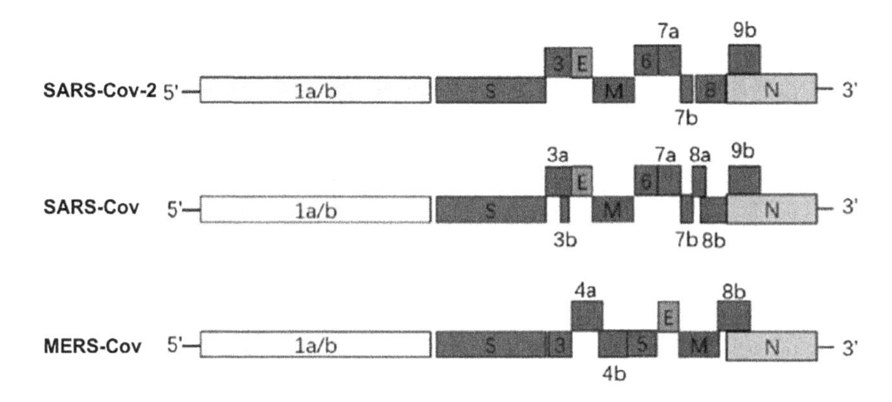

FIGURE 4.3 Genome comparison of three important human coronaviruses (Reprinted under the terms of the Creative Commons Attribution license from PeerJ-Life and Environment, PeerJ Publications (Hu et al. 2020)).

on February 11, 2020, about 2 months after its first detection as a virus causing novel coronavirus symptoms. On March 2, 2020, ICTV officially named it as SARS-CoV-2. It shares the genus with reparatory illness-causing SARS-CoV and MERS-CoV that also caused epidemics in 2002–2003 and 2012–2015, respectively. Genome sequence analysis revealed that SARS-CoV2 is distantly correlated to SARS-CoV (~79%) and MERS-CoV (~50%) (Wu et al. 2020). The genome comparison of these three beta coronaviruses further revealed that the arrangement of important proteins, nucleocapsid (N), envelope (E), and membrane protein (M) is also different (see Figure 4.3).

Seven different coronaviruses are reported to infect humans, among these two species belong to alphacoronaviruses and five species are from the betacoronaviruses group. A list of coronaviruses infecting human beings is presented in Table 4.2.

4.6 TRANSMISSION AND PERSISTENCE OF SARS-COV-2

The disease outbreak in Wuhan was found associated with the seafood market. The patients were identified as the workers of this live animal market or the visitors. These seafood wholesale markets are considered as a potential source of SARS-CoV-2 and hence the CoVID-19 spread. The studies showed that infection was transmitted from animal to human. The zoonotic agents as suggested by the studies are bats, pigs, pangolins, and snakes (Guan et al. 2003; Lau et al. 2005; Zhang, Wu, and Zhang 2020). The increased number of infected patients even after the closure of the seafood market suggested the transmission of disease through person-to-person contact. A person with close contact of about 6 ft (1.8 m) to an infected person can get the infection (Chan et al. 2020). The incubation period of COVID-19 is on an average 5–7 days but can be up to 14 days, which in some cases is extended up to 24 days (Bai et al. 2020). This is the period of the first exposure of a person to a virus till the onset of the symptoms and is also known as the "presymptomatic" period. An infected person can spread the disease during this presymptomatic period. One

TABLE 4.2
List of Coronaviruses Causing Human Diseases*

Genus	Virus	Natural Host	Symptoms
Alpha coronaviruses	Human coronavirus 229E (HCoV-229E)	Bats	Mild upper respiratory tract infection
	Human coronavirus NL63 (HCoV-NL63, New Haven coronavirus)	Bats	
Beta coronaviruses	Human coronavirus OC43 (HCoV-OC43)	Rodents	
	Human coronavirus HKUI	Rodents	Pneumonia
	Middle east respiratory syndrome-related coronavirus (MERS-CoV or HCoV-EMC). the cause of MERS)	Bats	Severe acute respiratory syndrome, 37% mortality rate
	Severe acute respiratory syndrome coronavirus (SARS-CoV-1), the cause of SARS	Bats	Severe acute respiratory syndrome, 10% mortality rate
	Severe acute respiratory syndrome coronavirus 2 (SARS-CoV-2) or 2019-nCoV, the cause of COVID-19	Bats	Severe acute lower respiratory tract infection, pneumonia, >4% mortality

* Adapted and reproduced from Artika, Dewantari, and Wiyatno (2020) and Chen, Liu, and Guo 2020)

infected person can transmit the disease to 2.56 other persons (Zhao et al. 2020). The droplets that emerged from a human respiratory system such as during talking, coughing, and sneezing are the potential means to transmit the disease (Huang et al. 2020). Another way of the spread of the virus is fomite transmission that is through any surfaces/objects (Otter et al. 2016). Once the droplets from infected humans fall on surfaces, their survivability on those surfaces determines whether further contact transmission is possible or not. The SARS-CoV can persist on surfaces like metal, glass, and plastics for 96 h and other coronaviruses for up to 9 days (Kramer, Schwebke, and Kampf 2006; Kampf et al. 2020; Warnes, Little, and Keevil 2015). The SARS-COV-2 is found more persistently on plastic and stainless steel than on cardboard and copper metal (Van Doremalen et al. 2020). Furthermore, the viable virus was detected on plastic even 72 h after application to these surfaces (see Table 4.3); hence, the transmission can also happen indirectly through touching contaminated surfaces or objects (Carraturo et al. 2020; Fiorillo et al. 2020). Some microbicides are effective against coronaviruses; for example, alcohols, hydrogen peroxide, phenolics, and sodium hypochlorite (Kampf et al. 2020). The virus genome has also been detected in other biological materials, urine, and feces of some patients (Pan et al. 2020). But there is no report about the transmission of SARS-CoV-2 through feces or urine. Also, there is no report about the transmission of this virus from infected pregnant women to their fetuses.

TABLE 4.3
Persistence of SARS-CoV-2 on Different Surfaces*

	Investigated material	Time (hours)	Note on results
SARS-CoV-2	Aerosols	3	Reduction from $10^{3.5}$ to $10^{2.7}$ $TCID_{50}$ per liter of air
	Plastic	72	Reduction from $10^{3.7}$ to $10^{0.6}$ $TCID_{50}$ per liter of a millimeter
	Stainless steel	48	Reduction from $10^{3.7}$ to $10^{0.6}$ $TCID_{50}$ per liter of a millimeter
	Copper	4	No viable SARS-CoV-2
	cardboard	24	No viable SARS-CoV-2

$TCID_{50}$ (50% tissue culture infectious dose)
* Adapted and reproduced from (Fiorillo et al. 2020)

4.7 GENOME STRUCTURE OF CORONAVIRUS AND SARS-COV-2

Coronaviruses are spherical (or pleomorphic) enveloped molecules (80–120 nm diameter) with a positive-sense single-stranded RNA genome of ~30-kilobases (kb). The coronaviruses have the largest genome size among all known RNA viruses. The RNA genome is coated with capsid protein with an outer membranous envelope derived from the host cell having spike protein and glycoprotein embedded (see Figure 4.5). At both ends of the genome, there are two secondary structures, a methylated cap structure at 5′ end and 3′ poly-A tail that contribute to the stability of the genome (Liu, Wimmer, and Paul 2009). These secondary structures are the parts of the untranslated regions (UTR), 5′ UTR, and 3′ UTR. The 5′ cap is formed by the folding of the 5′ UTR and the adjacent sequences encoding the amino-terminus of non-structural protein ORF1a. These UTRs are important for RNA replication and transcription. The coronavirus genome performs multiple functions during infection to the host cell. It acts as a template both for transcription and replication. Initially, the genome of the virus inside the host cell acts as an mRNA, translated directly into two large polyproteins. A polyprotein is later on cleaved by the viral-encoded protease and different structural proteins are produced. These structural proteins that play the role to produce a structurally complete virion include spike (S), envelope (E), membrane (M), and nucleocapsid (N) proteins. Other proteins encoded on the genome of coronaviruses include an RNA-dependent RNA polymerase (RdRP) and an ATPase helicase. Some coronaviruses (HCoV-OC43) also contain a hemagglutinin-esterase (HE) protein (Sturman and Holmes 1983). The genome of SARS-CoV-2 lacks HE. It contains 265 nucleotides at 5′ end and 229 nucleotides at 3′ end constituting two UTRs. The 5′ UTR plays an important role in the synthesis of subgenomic RNAs. 3′ UTR contains all of the *cis*-acting sequences necessary for viral replication. The genome structure of SARS-CoV-2 has the same organization of genes and characteristics as most of the coronavirus's genome shows. It contains an orf1ab gene encoding orf1ab polyprotein, S, E, M, and N structural proteins. The genome also contains ORF3a, ORF6, ORF7a, ORF7b, and ORF8 genes encoding the six accessory proteins as shown in Figure 4.4 and given in Table 4.4.

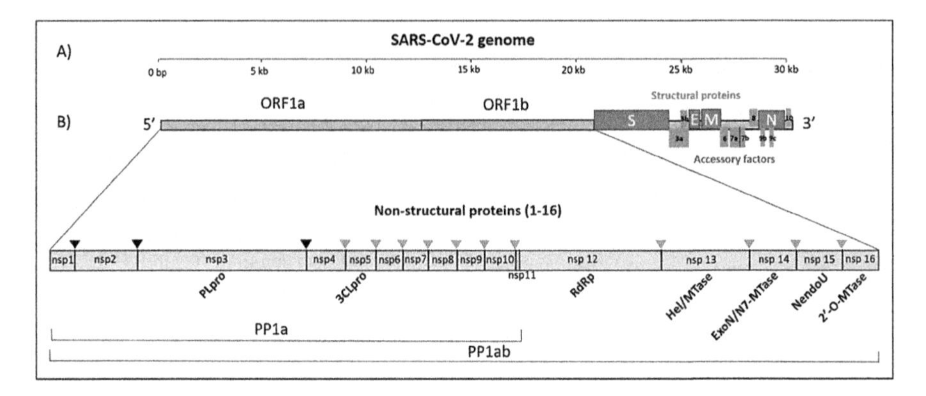

FIGURE 4.4 Genome organization of SARS-CoV-2 (Reprinted under the terms of the Creative Commons Attribution license from Cells, MDPI Publications (Romano et al. 2020)).

TABLE 4.4
List of Genes and the Proteins Expressed by the SARS-CoV-2 Virus

Genes	Proteins
ORF1ab	pp1ab polyprotein
ORF1a	pp1a polyprotein
ORF2 (S)	Spike protein (S protein)
ORF3a	ORF3a protein
ORF4 (E)	Envelope protein (E protein)
ORF5 (M)	Membrane protein (M protein)
ORF6	ORF6 protein
ORF7a	ORF7a protein
ORF7b	ORF7b protein
ORF8	ORF8 protein
ORF9 (N)	Nucleocapsid phosphoprotein (N protein)
ORF10	ORF10 protein

4.8 FUNCTIONS OF SARS-COV-2 PROTEINS

The RNA genome of SARS-CoV-2 and all other coronavirus is translated directly upon entry into the host cell and produces two polyproteins pp1a and pp1ab. These polyproteins are cleaved to produce 11 and 15 non-structural proteins (NSPs), respectively. The viral genome thus acts as a replication and transcription template. RdRP produced by virus then generates a negative-sense RNA using a positive-sense RNA genome. This negative-sense RNA intermediate then serves as a template to produce positive-sense genomic RNA and subgenomic RNAs. These subgenomic RNAs are then translated to produce structural proteins (S, E, M, and N) and at least six accessory proteins (3a, 6, 7a, 7b, 8, and 10).

4.8.1 Non-Structural Proteins (NSPs)

A larger section of the genome, about 2/3 genome located at the 5′ end encodes two long ORFs: ORF1a and ORF1b that together encode the NSPs of the virus. ORF1a and ORF1b are translated first as polyprotein (pp) precursors, pp1a and pp1ab, respectively. The translation for pp1ab is accomplished by a frameshift event at the end of the 1a coding sequence. Virally encoded proteinases cleave these polyproteins into 16 NSPs. pp1a is cleaved to produce 11 NSPs (1–11 NSPs) and pp1ab produces 16 NSPs (1–16 NSPs). These NSPs that are conserved in coronavirus modulate early transcriptional regulation, helicase activity, immunomodulation, gene transactivation, and counter the antiviral response thereby facilitate viral pathogenesis (see Table 4.5). The NSPs of SARS-CoV-2 are found to involve in viral genome replication, protein processing, transcription, proteolysis, RNA-binding, endopeptidase activity, transferase activity, ATP-binding, zinc ion binding, RNA-directed 5′-3′ RNA-polymerase activity, exoribonuclease activity, producing 5′-phosphomonoesters and methyltransferase activity (Naqvi et al. 2020).

4.8.2 Spike Protein (S)

The spike protein is a homotrimeric glycoprotein and facilitates the attachment of the virus to the host cell. This protein binds to the cell receptor ACE2 through its receptor-binding domain (RBD) (Lan et al. 2020a, 2020b) and thus play a significant role in the pathogenesis of virus and the start of the infection cycle. SARS-CoV-2 S protein contains a furin cleavage site (QTQTNSPRRARSVASQSIIA) at the boundary between the two functional subunits, S1 and S2. This furin cleavage site is absent in the S proteins in other SARS-CoV.

4.8.3 Envelope Protein (E)

Envelope protein (E) plays the role in morphogenesis, assembly, and pathogenesis of coronaviruses (Venkatagopalan et al. 2015). The E protein is a small integral membrane protein. This hydrophobic viroporin, although not highly abundant in the envelope, is essential for viral envelope curvature and maturation and can oligomerize to form membrane ion channels.

4.8.4 Membrane Protein (M)

The membrane protein (M) is the most abundant glycoprotein in the infected host cells. It plays a critical role in the packaging of the RNA genome to form a virion. It captures the other structural proteins (S, E, and N) at the budding site and assembles the virus. M protein also interacts with the coronavirus genomic RNA and nucleocapsid (N) (Narayanan et al. 2000).

4.8.5 Nucleocapsid Protein (N)

The nucleocapsid protein (N) is essential for the incorporation of genomic RNA into progeny virus particles. It is a basic protein that binds nonspecifically with ssRNA, ssDNA, and dsDNA. N protein binds with the genomic RNA, protects the genome, and helps it to enter into the host cells, thus facilitate the viral RNA to replicate inside the host (Grunewald et al. 2018). It also overpowers the host defense by acting as a suppressor of the RNA interference mechanism (Mu et al. 2020).

TABLE 4.5
Non-structural Proteins of SARS-CoV-2 and the Functions of Proteins*

Proteins	Functions	References
NSP1	Host mRNA degradation mediates viral RNA replication and processing	(Almeida et al. 2007; Tanaka et al. 2012)
NSP2	Involved in the disruption of intracellular host signaling, interact with two host proteins, prohibitin 1 (PHB1) and prohibitin 2 (PHB2)	(Cornillez-Ty et al. 2009)
NSP3	Binds to viral RNA, nucleocapsid protein, involved in polyprotein processing	(Serrano et al. 2009; Lei, Kusov, and Hilgenfeld 2018)
NSP4	Membrane rearrangements	(Sakai et al. 2017)
NSP5	Mediates processing at 11 distinct cleavage sites of the polyprotein, including its autoproteolysis, and is indispensable for virus replication	(Stobart et al. 2013; Zhu, Fang et al. 2017a; Zhu et al. 2017b)
NSP6	Inhibit autophagosome expansion	(Cottam, Whelband, and Wileman 2014)
NSP7	Cofactor for nsp12 (RdRp)	(Kirchdoerfer and Ward 2019; Gao et al. 2020)
NSP8	Cofactor for nsp12 (RdRp)	(Kirchdoerfer and Ward 2019; Gao et al. 2020)
NSP9	Single-stranded RNA binding protein, play role in replication	(Egloff et al. 2004)
NSP10	regulate nsp14-ExoN activity during virus replication, important for replication fidelity, stimulatory factor to NSP16	(Bouvet et al. 2012; Smith et al. 2015)
NSP11	Unknown	(Fang et al. 2008)
NSP12	RdRP, performs replication of viral RNA	(Kirchdoerfer and Ward 2019; Ahn et al. 2012)
NSP13	Possesses the nucleoside triphosphate hydrolase (NTPase) and RNA helicase activities	(Ivanov et al. 2004; Shu et al. 2020)
NSP14	3′ to 5′ Exoribonuclease involved in proofreading, N7-guanine methyltransferase	(Minskaia et al. 2006; Yu et al. 2009)
NSP15	Endoribonuclease, evade host defense	(Bhardwaj et al. 2006; Deng et al. 2017; Zhang et al. 2018)
NSP16	2′-O-Methyltransferase negatively regulates innate immunity	(Chen et al. 2011; Shi et al. 2019)

* Adapted and reproduced from (Chen, Liu, and Guo 2020)

4.8.6 Accessory Proteins

The genes that encode eight accessory proteins (3a, 3b, 6, 7a, 7b, 8a, 8b, and 9b) of SARS-CoV-2 are dispersed among the structural protein genes. The in vitro studies have shown these accessory proteins unessential for viral replication. But some of these have been reported to modulate virus–host interactions in vivo and perform a role in pathogenesis. These proteins control the interactions involved in cell proliferation, pro-inflammatory cytokine production, programed cell death, and interferon signaling (Narayanan, Huang, and Makino 2008). These accessory proteins perform the following roles for SARS-CoV-2:

a. 3a protein form ion channels that are selective for potassium ions (Lu et al. 2006). It causes necrosis, pyroptosis, apoptosis induction, and cell cycle arrest in the host cells, thus play the role in evading the host immune system, reviewed by (Hartenian et al. 2020).
b. 3b protein localizes to the nucleus and functions as an interferon antagonist. It inhibits interferon production and signaling by inhibiting phosphorylation-dependent activation of IRF3 transcriptional activity (Kopecky-Bromberg et al. 2007).
c. The accessory protein 6 also functions as an interferon antagonist and interferes with cellular DNA synthesis, reviewed by Hartenian et al. (2020).
d. 7a protein is involved in host translation inhibition, apoptosis induction, and cell cycle arrest, while the function of 7b is unknown.
e. The protein 8a is indicated to induce caspase-dependent apoptosis, and 8b has been suggested to have the ability to induce DNA synthesis (Law et al. 2006).
f. 9b protein induces caspase-dependent apoptosis (Sharma et al. 2011).

4.9 CELL ENTRY MECHANISM OF CORONAVIRUSES AND SARS-COV-2

Infection of the coronaviruses and SARS-CoV-2 initiate by the entry of the virus into the host cells that involves the binding of "S" glycoprotein of the virus to the targeted host cell and the subsequent host protease-mediated cleavage of the S protein. The S protein of SARS-CoV-2 is a trimeric protein and each monomer has two subunits, S1 and S2. The entry of the virus depends on the binding of the S1 surface unit to a host cellular receptor, ACE2, which helps the virus in an attachment to the cell surface (see Figure 4.5). The S protein of SARS-CoV-2 has 10–20 times greater affinity to ACE2 than SARS-CoV. A host protease cleaves the S protein at the S1/S2 and the S2′ site and helps the S2 driven virus fusion to the cellular membrane. This cleavage of

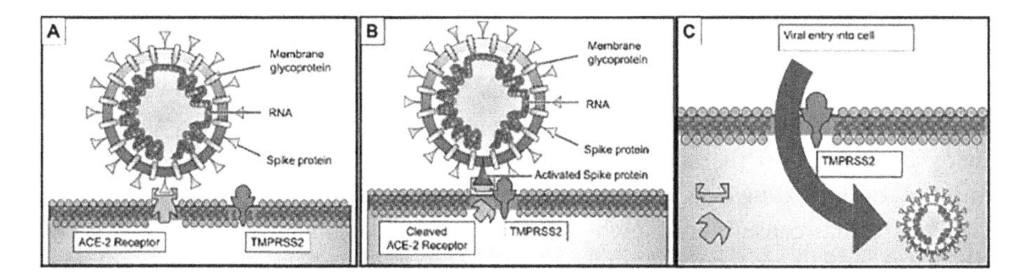

FIGURE 4.5 Clinical presentation of cell entry mechanism of Coronaviruses. A) The coronavirus spike (S) protein binds to the ACE-2 receptor. B) The complex is proteolytically processed by type 2 transmembrane protease (TMPRSS2) leading to cleavage of ACE2 and activation of the spike protein. C; viral entry into the target cell (Reprinted under the terms of the Creative Commons Attribution license from pathogens, MDPI Publications (Rabi et al. 2020)).

S protein into subunits is called priming. Similar to the SARS-CoV, the SARS-CoV-2 interacts with ACE2 for binding and type 2 transmembrane protease (TMPRSS2) for S protein priming. S protein, a class I membrane fusion protein, is cleaved by the furin into S1 and S2 subunits. The subunits remain associated with each other after cleavage (Hoffmann et al. 2020). The S1 subunit carrying the RBD is highly variable within beta coronaviruses whereas the S2 subunit involved in the fusion of virus is conserved within genera. The C-terminal domain in S1 subunit is the ACE2 RBD. The S2 subunit contains a second proteolytic site, S2, the proteolysis of this site triggers the fusion of the virus with the membrane. After fusion, the virus enters the host cell via an endosomal pathway (Coutard et al. 2020). The viral RNA is then released into the cytoplasm of the host cell and serves as mRNA for the translation to produce pp1a and pp1b.

4.10 EMERGENCE AND EPIDEMIOLOGY COVID-19

The present epidemic of coronavirus infection is a third outbreak caused by these viruses. In 2003, WHO announced a global emergency after a severe respiratory disease outbreak in China attributed to the SARS-CoV. The exact origin of SARS was unknown and the death toll was 774 globally. Sequence analysis isolated from Himalayan plain civets revealed a 29 nucleotides deletion in its genome (Muth et al. 2018). While other coronaviruses, from humans and animals, maintain these 29 nucleotides, it is unclear whether these 29 nucleotides' deletion made it more pathogenic compared to other coronaviruses. The data from epidemiological studies revealed that 40% of wild animal traders and 20% of their meat sellers were found positive for SARS. Nevertheless, they did not exhibit any symptoms. The global precautionary measures resulted in the control of the deadly SARS virus during 2003. In September 2012, another coronavirus namely MERS-CoV was identified from Arabian Peninsula (Assiri and Al-Tawfiq et al. 2013). Until January 2017, WHO received the information, which marks identification of MERS-CoV from 17 countries with ~660 deaths (~35% fatality rate). Dromedary camels were considered as a source of animal-to-human transmission. Serological studies indicated that MERS-CoV was present everywhere in Africa and was circulating for three decades (Assiri and Al-Tawfiq et al. 2013; Zumla, Hui, and Perlman 2015). The symptoms of MERS appeared after 2–14 days after the first encounter. However, the severity of symptoms varied from person to person (Li et al. 2016).

The recent outbreak of novel coronavirus SARS-CoV2, with unknown etiology, occurred in Wuhan (China) as early as December 2019. Just like SARS and MERS, there is no antiviral therapy or vaccine available against COVID-19. Due to the absence of any drugs or vaccine, and its highly contagious nature, the novel coronavirus has caused more fatalities than any other coronavirus in human history. The virus, which was limited to China, has become a pandemic. It also resulted in global emergency and trillions of dollar losses to the world economy. The incidence of COVID-19 disease in China rapidly swept across the country due to the holidays of the spring festival. The epidemiological studies revealed that the majority of the infected patients were in their middle age or old age. Most of the patients presented chest abnormalities with a mean incubation period of 5.2 days. The severe cough, breathing difficulties, fever, and fatigue were common symptoms. However, over time, the disease started appearing in young adults, although they were more responsive to treatment as compared to elderly patients (Sun et al. 2020; Chen et al. 2020; Bai et al. 2020). Indeed, the younger population in Italy, Iran, Spain, and the USA caused more spread of disease due to their frequent traveling across the countries. As of September 20, 2020, more than 30.6 million cases have been reported with 953,903 deaths (as of September 20, 2020, the WHO mentioned on its website https://covid19.who.int). Several countries like the USA, Iran, Italy, and Spain were severely plagued by COVID-19 outside its epicenter, China. At present, in the Czech Republic, more than 48,000 cases with 499 deaths have been reported (as of September 20, 2020, the WHO mentioned on its website https://www.worldometers.info/coronavirus/country/czech-republic).

The major phenotype of SARS-CoV-2 in the infected patients is, severe acute distress respiratory syndrome similar to MERS-CoV and SARS-CoV. The infection of coronaviruses is not limited to human beings or birds. The farm animals like pigs in China were also seriously challenged by diarrhea-causing coronaviruses during 2017. Indeed, the sick pigs were found infected by an alphacoronavirus known as novel swine enteric alphacoronavirus (SeACoV) (Gong et al. 2017; Pan et al. 2017). The disease had a 35% fatality rate among young piglets. The sequence analysis suggested that the novel SeACoV was a novel alphacoronavirus with a unique origin. The phylogenetic analysis suggested that SeACoV was closely related to bat alphacoronavirus HKU2 isolated in southern China. The SeACoV shared common ancestors with human coronavirus, namely, 229E/NL63. Nevertheless, further analysis suggested the SeACoV recombination with a beta coronavirus. Surprisingly, the point mutations in the C terminal of spike protein in SeACoV were similar to human coronavirus 229E/NL63, indicating their common receptor binding on the cell surface. Due to extensive bats and other wild animals' diversity coronaviruses can easily jump the species barriers and transmit to the human being. Indeed, wild pigs or other animals can act as mixing vessels for recombination and point mutation and may result in serious viral outbreaks.

4.11 TREATMENT OF SARS-COV-2

Recently on October 22, 2020, US Food and Drug Administration (FDA) has approved Veklury (remdesivir), an antiviral drug to treat adults and children of 12 years age and above with at least 40 kg weight. Remdesivir is a nucleotide analog that competes for the normal nucleotide and once incorporated in viral RNA during replication causes premature termination of replication. Velkury is the first anti-SARS-CoV-2 drug received approval by the FDA to be administrated in a hospital. To date, no vaccine or other antiviral drug or treatment is available to combat the COVID-19. The scientific community around the world is rigorously investigating every aspect of SARS-CoV-2. Several antiviral agents, antibiotics, and herbal treatments have been used to cure and treat the COVID-19. Similar to SARS-CoV and MERS the treatment of SARS-CoV-2 involve the management practices that involve supportive therapy and treatments to prevent respiratory failure. Isolation of the infected patient to avoid exposure to other healthy persons is a prime concern.

Infected persons and the other persons that are suspected to carry the infection must be quarantined. Social distancing and self-isolation are practiced to minimize the spread of disease. People are taking precautionary measures and the guidelines recommended by the WHO to avoid the disease. The drugs designed to treat SARS-CoV-2 are inhibitory agents (against viral polymerases, helicases, and proteinases), monoclonal antibodies, and immunomodulators (interferons and corticosteroids). A rigorous understanding of the structure of coronaviruses greatly helped the scientists to develop 321 candidate vaccine by October, 2020. No candidate vaccine has completed the safety trial to be released to the general public.

4.12 POSSIBLE SOLUTIONS FOR COVID-19 PREVENTION

The development of a vaccine against COVID-19 is a vital requirement of the time and must be started without wasting any time. Of course, for such a global crisis the overall efforts are needed. Indeed, researchers across the different parts of the world need to develop a consortium and start developing the vaccine, targeting different viral proteins. The vaccine trials should be started on a large scale. Every country must develop its vaccine production units, as every single human being on this planet will need the vaccine. Another solution to the COVID-19 is the diagnosis of the virus during the early stages of infection or eclipse period of viral infection. Instead of developing expensive RT-PCR based diagnosis, a strip should be developed for early diagnosis.

Most of the devices used to protect against the transmission of the novel COVID-19 virus are more or less related to textile structures. Hence, it would be advisable for textile professionals (at the academic and industrial research level) to develop "active textile" for manufacturing of efficient, biocompatible, and cost-effective gowns and protective masks and also inform the public in particular about the advantages and disadvantages of individual protective mask materials and the possibilities of sterilizing them. It is highly desirable to address the problem of face masks preventing the penetration of COVID-19 virus into the mouth and nose area of healthy persons and masks preventing the spread of COVID-19 virus from the mouth and nose area of infected persons to surroundings.

4.13 CONCLUSION

The COVID-19, a deadly pandemic, has emerged as a serious threat to humanity, the economy, and the healthcare systems of the whole world. The coronaviruses have regularly emerged as pathogenic and caused life-threatening epidemics in animals and humans. The disease believed to be zoonotic is now being spread to the whole world through human-to-human contact. Despite, the great efforts spent on the molecular biology study of SARS-CoV-2, there is no authentic and straightforward treatment available nor any vaccine has been approved in any part of the world so far. Scientists are working rigorously to find therapies, antiviral drugs, and vaccines to combat this virus. According to the WHO data, there are 26 vaccines under the human trial phaseout of the 169 vaccine candidates. The epidemiological study of the COVID-19 suggests that in the absence of any recommended treatment and vaccine the possible solution is the prevention of epidemic. The prevention of disease has partially been achieved by shifting infected people in quarantine, suggesting that people keep limited social contact with each other and observe lockdowns at national and international levels.

The molecular data analysis revealed that SARS-CoV-2 is more than 97% identical to bat coronavirus, suggesting its zoonotic spillover to humans. The detailed study about the evolution of coronaviruses will help to predict the root cause of the pandemic and the emergence of possible next outbreak. There is a need to isolate and characterize the coronaviruses from bats, pangolins, and other wildlife to precisely predict the possible route of transmission of the disease to humans. Understanding molecular biology, evolution, pathogenesis, and epidemiology will help in controlling the virus thus preventing frequent outbreaks. The frequent emergence of coronavirus related diseases demands the production of broad-spectrum antiviral drugs and vaccine that can provide long-term protection against this group of viruses. It is also speculated that multidisciplinary research is being carried out on the different aspects of COVID-19. and SARS-CoV-2 will soon provide the missing links that are hindering the proper development of a treatment and vaccine.

ACKNOWLEDGEMENT

This work was supported by the Ministry of Education, Youth and Sports of the Czech Republic, and the European Union – European Structural and Investment Funds in the frames of Operational Programme Research, Development, and Education – project Hybrid Materials for Hierarchical Structures (HyHi, Reg. No. CZ.02.1.01/0.0/0.0/16 _019/0000843) and Student Grant Competition Project (SGC, Reg. No. 21407).

REFERENCES

Ahn, Dae-Gyun, Jin-Kyu Choi, Deborah R Taylor, and Jong-Won Oh. 2012. "Biochemical Characterization of a Recombinant SARS Coronavirus Nsp12 RNA-Dependent RNA Polymerase Capable of Copying Viral RNA Templates." *Archives of Virology* 157 (11): 2095–2104.

Almeida, June D, and D. A. J. Tyrrell. 1967. "The Morphology of Three Previously Uncharacterized Human Respiratory Viruses That Grow in Organ Culture." *Journal of General Virology* 1 (2): 175–178.

Almeida, Marcius S, Margaret A Johnson, Torsten Herrmann, Michael Geralt, and Kurt Wüthrich. 2007. "Novel β-Barrel Fold in the Nuclear Magnetic Resonance Structure of the Replicase Nonstructural Protein 1 from the Severe Acute Respiratory Syndrome Coronavirus." *Journal of Virology* 81 (7): 3151–3161.

Artika, I Made, Aghnianditya Kresno Dewantari, and Ageng Wiyatno. 2020. "Molecular Biology of Coronaviruses: Current Knowledge." *Heliyon* 6 (8): e04743. doi:10.1016/j.heliyon.2020.e04743.

Assiri, Abdullah, Jaffar A Al-Tawfiq et al. 2013. "Epidemiological, Demographic, and Clinical Characteristics of 47 Cases of Middle East Respiratory Syndrome Coronavirus Disease from Saudi Arabia: A Descriptive Study." *The Lancet Infectious Diseases* 13 (9): 752–761.

Bai, Yan, Lingsheng Yao, Tao Wei, Fei Tian, Dong-Yan Jin, Lijuan Chen, and Meiyun Wang. 2020. "Presumed Asymptomatic Carrier Transmission of COVID-19." *Jama.*

Baltimore, David. 1971. "Expression of Animal Virus Genomes." *Bacteriological Reviews* 35 (3): 235.

Bhardwaj, Kanchan, Jingchuan Sun, Andreas Holzenburg, Linda A Guarino, and C Cheng Kao. 2006. "RNA Recognition and Cleavage by the SARS Coronavirus Endoribonuclease." *Journal of Molecular Biology* 361 (2): 243–256.

Bouvet, Mickaël, Isabelle Imbert, Lorenzo Subissi, Laure Gluais, Bruno Canard, and Etienne Decroly. 2012. "RNA 3'-End Mismatch Excision by the Severe Acute Respiratory Syndrome Coronavirus Nonstructural Protein Nsp10/Nsp14 Exoribonuclease Complex." *Proceedings of the National Academy of Sciences* 109 (24): 9372–9377.

Carraturo, Federica, Carmela Del Giudice, Michela Morelli, Valeria Cerullo, Giovanni Libralato, Emilia Galdiero, and Marco Guida. 2020. "Persistence of SARS-CoV-2 in the Environment and COVID-19 Transmission Risk from Environmental Matrices and Surfaces." *Environmental Pollution* 265: 115010.

Chan, Jasper Fuk-Woo, Shuofeng Yuan, Kin-Hang Kok, Kelvin Kai-Wang To, Hin Chu, Jin Yang, Fanfan Xing, Jieling Liu, Cyril Chik-Yan Yip, and Rosana Wing-Shan Poon. 2020. "A Familial Cluster of Pneumonia Associated with the 2019 Novel Coronavirus Indicating Person-to-Person Transmission: A Study of a Family Cluster." *The Lancet* 395 (10223): 514–523.

Chen, Yu, Hui Cai, Nian Xiang, Po Tien, Tero Ahola, and Deyin Guo. 2009. "Functional Screen Reveals SARS Coronavirus Nonstructural Protein Nsp14 as a Novel Cap N7 Methyltransferase." *Proceedings of the National Academy of Sciences* 106 (9): 3484–3489.

Chen, Yu, Ceyang Su, Min Ke, Xu Jin, Lirong Xu, Zhou Zhang, Andong Wu, Ying Sun, Zhouning Yang, and Po Tien. 2011. "Biochemical and Structural Insights into the Mechanisms of SARS Coronavirus RNA Ribose 2′-O-Methylation by Nsp16/Nsp10 Protein Complex." *PLoS Pathog* 7 (10): e1002294.

Chen, Yu, Qianyun Liu, and Deyin Guo. 2020. "Emerging Coronaviruses: Genome Structure, Replication, and Pathogenesis." *Journal of Medical Virology* 92 (4): 418–423.

Cornillez-Ty, Cromwell T, Lujian Liao, John R Yates, Peter Kuhn, and Michael J Buchmeier. 2009. "Severe Acute Respiratory Syndrome Coronavirus Nonstructural Protein 2 Interacts with a Host Protein Complex Involved in Mitochondrial Biogenesis and Intracellular Signaling." *Journal of Virology* 83 (19): 10314–10318.

Cottam, Eleanor M, Matthew C Whelband, and Thomas Wileman. 2014. "Coronavirus NSP6 Restricts Autophagosome Expansion." *Autophagy* 10 (8): 1426–1441.

Coutard, Bruno, Coralie Valle, Xavier de Lamballerie, Bruno Canard, N G Seidah, and E Decroly. 2020. "The Spike Glycoprotein of the New Coronavirus 2019-NCoV Contains a Furin-like Cleavage Site Absent in CoV of the Same Clade." *Antiviral Research* 176: 104742.

Deng, Xufang, Matthew Hackbart, Robert C Mettelman, Amornrat O'Brien, Anna M Mielech, Guanghui Yi, C Cheng Kao, and Susan C Baker. 2017. "Coronavirus Nonstructural Protein 15 Mediates Evasion of DsRNA Sensors and Limits Apoptosis in Macrophages." *Proceedings of the National Academy of Sciences* 114 (21): E4251–E4260.

Doremalen, Neeltje Van, Trenton Bushmaker, Dylan H Morris, Myndi G Holbrook, Amandine Gamble, Brandi N Williamson, Azaibi Tamin, Jennifer L Harcourt, Natalie J Thornburg, and Susan I Gerber. 2020. "Aerosol and Surface Stability of SARS-CoV-2 as Compared with SARS-CoV-1."

Egloff, Marie-Pierre, François Ferron, Valérie Campanacci, Sonia Longhi, Corinne Rancurel, Hélène Dutartre, Eric J Snijder, Alexander E Gorbalenya, Christian Cambillau, and Bruno Canard. 2004. "The Severe Acute Respiratory Syndrome-Coronavirus Replicative Protein Nsp9 Is a Single-Stranded RNA-Binding Subunit Unique in the RNA Virus World." *Proceedings of the National Academy of Sciences of the United States of America* 101 (11): 3792 LP–3793796. doi:10.1073/pnas.0307877101.

Fang, Shou Guo, Hongyuan Shen, Jibin Wang, Felicia P L Tay, and Ding Xiang Liu. 2008. "Proteolytic Processing of Polyproteins 1a and 1ab between Non-Structural Proteins 10 and 11/12 of Coronavirus Infectious Bronchitis Virus Is Dispensable for Viral Replication in Cultured Cells." *Virology* 379 (2): 175–180.

Fiorillo, Luca, Gabriele Cervino, Marco Matarese, Cesare D'Amico, Giovanni Surace, Valeria Paduano, Maria Teresa Fiorillo, Antonio Moschella, Alessia La Bruna, and Giovanni Luca Romano. 2020. "COVID-19 Surface Persistence: A Recent Data Summary and Its Importance for Medical and Dental Settings." *International Journal of Environmental Research and Public Health* 17 (9): 3132.

Gao, Yan, Liming Yan, Yucen Huang, Fengjiang Liu, Yao Zhao, Lin Cao, Tao Wang, Qianqian Sun, Zhenhua Ming, and Lianqi Zhang. 2020. "Structure of the RNA-Dependent RNA Polymerase from COVID-19 Virus." *Science* 368 (6492): 779–782.

Gong, Lang, Jie Li, Qingfeng Zhou, Zhichao Xu, Li Chen, Yun Zhang, Chunyi Xue, Zhifen Wen, and Yongchang Cao. 2017. "A New Bat-HKU2–like Coronavirus in Swine, China, 2017." *Emerging Infectious Diseases* 23 (9): 1607.

Grunewald, Matthew E, Anthony R Fehr, Jeremiah Athmer, and Stanley Perlman. 2018. "The Coronavirus Nucleocapsid Protein Is ADP-Ribosylated." *Virology* 517: 62–68.

Guan, Yi, B J Zheng, Y Q He, X L Liu, Z X Zhuang, C L Cheung, S W Luo, P H Li, L J Zhang, and Y J Guan. 2003. "Isolation and Characterization of Viruses Related to the SARS Coronavirus from Animals in Southern China." *Science* 302 (5643): 276–278.

Hamre, Dorothy, and John J Procknow. 1966. "A New Virus Isolated from the Human Respiratory Tract." *Proceedings of the Society for Experimental Biology and Medicine* 121 (1): 190–193.

Hartenian, Ella, Divya Nandakumar, Azra Lari, Michael Ly, Jessica M Tucker, and Britt A Glaunsinger. 2020. "The Molecular Virology of Coronaviruses." *Journal of Biological Chemistry* 295 (37): 12910–12934.

Hoffmann, Markus, Hannah Kleine-Weber, Simon Schroeder, Nadine Krüger, Tanja Herrler, Sandra Erichsen, Tobias S Schiergens, Georg Herrler, Nai-Huei Wu, and Andreas Nitsche. 2020. "SARS-CoV-2 Cell Entry Depends on ACE2 and TMPRSS2 and Is Blocked by a Clinically Proven Protease Inhibitor." *Cell.*

Hu, Tingting, Ying Liu, Mingyi Zhao, Quan Zhuang, Linyong Xu, and Qingnan He. 2020. "A Comparison of COVID-19, SARS and MERS." *PeerJ* 8: e9725.

Huang, Chaolin, Yeming Wang, Xingwang Li, Lili Ren, Jianping Zhao, Yi Hu, Li Zhang, Guohui Fan, Jiuyang Xu, and Xiaoying Gu. 2020. "Clinical Features of Patients Infected with 2019 Novel Coronavirus in Wuhan, China." *The Lancet* 395 (10223): 497–506.

Ivanov, Konstantin A, Volker Thiel, Jessika C Dobbe, Yvonne Van Der Meer, Eric J Snijder, and John Ziebuhr. 2004. "Multiple Enzymatic Activities Associated with Severe Acute Respiratory Syndrome Coronavirus Helicase." *Journal of Virology* 78 (11): 5619–5632.

Kampf, Günter, Daniel Todt, Stephanie Pfaender, and Eike Steinmann. 2020. "Persistence of Coronaviruses on Inanimate Surfaces and Their Inactivation with Biocidal Agents." *Journal of Hospital Infection* 104 (3): 246–251.

Kirchdoerfer, Robert N, and Andrew B Ward. 2019. "Structure of the SARS-CoV Nsp12 Polymerase Bound to Nsp7 and Nsp8 Co-Factors." *Nature Communications* 10 (1): 1–9.

Kopecky-Bromberg, Sarah A, Luis Martínez-Sobrido, Matthew Frieman, Ralph A Baric, and Peter Palese. 2007. "Severe Acute Respiratory Syndrome Coronavirus Open Reading Frame (ORF) 3b, ORF 6, and Nucleocapsid Proteins Function as Interferon Antagonists." *Journal of Virology* 81 (2): 548–557.

Kramer, Axel, Ingeborg Schwebke, and Günter Kampf. 2006. "How Long Do Nosocomial Pathogens Persist on Inanimate Surfaces? A Systematic Review." *BMC Infectious Diseases* 6 (1): 130.

Lan, Jun, Jiwan Ge, Jinfang Yu, Sisi Shan, Huan Zhou, Shilong Fan, Qi Zhang, Xuanling Shi, and Qisheng Wang. 2020a. "Structure of the SARS-CoV-2 Spike Receptor-Binding Domain Bound to the ACE2 Receptor." *Nature* 581 (7807): 215–220.

Lan, Jun, Jiwan Ge, Jinfang Yu, Sisi Shan, Huan Zhou, Shilong Fan, Qi Zhang, Xuanling Shi, Qisheng Wang, and Linqi Zhang. 2020b. "Crystal Structure of the 2019-NCoV Spike Receptor-Binding Domain Bound with the ACE2 Receptor." Biorxiv.

Lau, Susanna K P, Patrick C Y Woo, Kenneth S M Li, Yi Huang, Hoi-Wah Tsoi, Beatrice H L Wong, Samson S Y Wong, Suet-Yi Leung, Kwok-Hung Chan, and Kwok-Yung Yuen. 2005. "Severe Acute Respiratory Syndrome Coronavirus-like Virus in Chinese Horseshoe Bats." *Proceedings of the National Academy of Sciences* 102 (39): 14040–14045.

Law, Pui Ying Peggy, Yuet-Man Liu, Hua Geng, Ka Ho Kwan, Mary Miu-Yee Waye, and Yuan-Yuan Ho. 2006. "Expression and Functional Characterization of the Putative Protein 8b of the Severe Acute Respiratory Syndrome-Associated Coronavirus." *FEBS Letters* 580 (15): 3643–3648.

Lei, Jian, Yuri Kusov, and Rolf Hilgenfeld. 2018. "Nsp3 of Coronaviruses: Structures and Functions of a Large Multi-Domain Protein." *Antiviral Research* 149: 58–74.

Li, Kun, Christine Wohlford-Lenane, Stanley Perlman, Jincun Zhao, Alexander K Jewell, Leah R Reznikov, Katherine N Gibson-Corley, David K Meyerholz, and Paul B McCray Jr. 2016. "Middle East Respiratory Syndrome Coronavirus Causes Multiple Organ Damage and Lethal Disease in Mice Transgenic for Human Dipeptidyl Peptidase 4." *The Journal of Infectious Diseases* 213 (5): 712–722.

Liu, Ying, Eckard Wimmer, and Aniko V Paul. 2009. "Cis-Acting RNA Elements in Human and Animal plus-Strand RNA Viruses." *Biochimica et Biophysica Acta (BBA)-Gene Regulatory Mechanisms* 1789 (9–10): 495–517.

Lu, Wei, Bo-Jian Zheng, Ke Xu, Wolfgang Schwarz, Lanying Du, Charlotte K L Wong, Jiadong Chen, Shuming Duan, Vincent Deubel, and Bing Sun. 2006. "Severe Acute Respiratory Syndrome-Associated Coronavirus 3a Protein Forms an Ion Channel and Modulates Virus Release." *Proceedings of the National Academy of Sciences* 103 (33): 12540–12545.

Lu, Roujian, Xiang Zhao, Juan Li, Peihua Niu, Bo Yang, Honglong Wu, Wenling Wang, Hao Song, Baoying Huang, and Na Zhu. 2020. "Genomic Characterisation and Epidemiology of 2019 Novel Coronavirus: Implications for Virus Origins and Receptor Binding." *The Lancet* 395 (10224): 565–574.

Lwoff, André, and Paul Tournier. 1966. "The Classification of Viruses." *Annual Reviews in Microbiology* 20 (1): 45–74.

Lwoff, André, Robert Horne, and Paul Tournier. 1962. "A System of Viruses." In *Cold Spring Harbor Symposia on Quantitative Biology* 27: 51–55. Cold Spring Harbor Laboratory Press.

Minskaia, Ekaterina, Tobias Hertzig, Alexander E Gorbalenya, Valérie Campanacci, Christian Cambillau, Bruno Canard, and John Ziebuhr. 2006. "Discovery of an RNA Virus 3′→ 5′ Exoribonuclease That Is Critically Involved in Coronavirus RNA Synthesis." *Proceedings of the National Academy of Sciences* 103 (13): 5108–13.

Mu, Jingfang, Jiuyue Xu, Leike Zhang, Ting Shu, Di Wu, Muhan Huang, Yujie Ren, Xufang Li, Qing Geng, and Yi Xu. 2020. "SARS-CoV-2-Encoded Nucleocapsid Protein Acts as a Viral Suppressor of RNA Interference in Cells." *Science China Life Sciences* 1–4.

Muth, Doreen, Victor Max Corman, Hanna Roth, Tabea Binger, Ronald Dijkman, Lina Theresa Gottula, Florian Gloza-Rausch, Andrea Balboni, Mara Battilani, and Danijela Rihtarič. 2018. "Attenuation of Replication by a 29 Nucleotide Deletion in SARS-Coronavirus Acquired during the Early Stages of Human-to-Human Transmission." *Scientific Reports* 8 (1): 1–11.

Naqvi, Ahmad Abu Turab, Kisa Fatima, Taj Mohammad, Urooj Fatima, Indrakant K Singh, Archana Singh, Shaikh Muhammad Atif, Gururao Hariprasad, Gulam Mustafa Hasan, and Md Imtaiyaz Hassan. 2020. "Insights into SARS-CoV-2 Genome, Structure, Evolution, Pathogenesis and Therapies: Structural Genomics Approach." *Biochimica et Biophysica Acta (BBA)-Molecular Basis of Disease*, 1866 (10): 165878.

Narayanan, Krishna, Akihiko Maeda, Junko Maeda, and Shinji Makino. 2000. "Characterization of the Coronavirus M Protein and Nucleocapsid Interaction in Infected Cells." *Journal of Virology* 74 (17): 8127–8134.

Narayanan, Krishna, Cheng Huang, and Shinji Makino. 2008. "SARS Coronavirus Accessory Proteins." *Virus Research* 133 (1): 113–121.

Otter, J A, C Donskey, S Yezli, S Douthwaite, SDea Goldenberg, and D J Weber. 2016. "Transmission of SARS and MERS Coronaviruses and Influenza Virus in Healthcare Settings: The Possible Role of Dry Surface Contamination." *Journal of Hospital Infection* 92 (3): 235–250.

Pan, Yongfei, Xiaoyan Tian, Pan Qin, Bin Wang, Pengwei Zhao, Yong-Le Yang, Lianxiang Wang, Dongdong Wang, Yanhua Song, and Xiangbin Zhang. 2017. "Discovery of a Novel Swine Enteric Alphacoronavirus (SeACoV) in Southern China." *Veterinary Microbiology* 211: 15–21.

Pan, Yang, Daitao Zhang, Peng Yang, Leo L M Poon, and Quanyi Wang. 2020. "Viral Load of SARS-CoV-2 in Clinical Samples." *The Lancet Infectious Diseases* 20 (4): 411–412.

Rabi, Firas A, Mazhar S Al Zoubi, Ghena A Kasasbeh, Dunia M Salameh, and Amjad D Al-Nasser. 2020. "SARS-CoV-2 and Coronavirus Disease 2019: What We Know so Far." *Pathogens* 9 (3): 231.

Romano, Maria, Alessia Ruggiero, Flavia Squeglia, Giovanni Maga, and Rita Berisio. 2020. "A Structural View of SARS-CoV-2 RNA Replication Machinery: RNA Synthesis, Proofreading and Final Capping." *Cells* 9 (5): 1267.

Sakai, Yusuke, Kengo Kawachi, Yutaka Terada, Hiroko Omori, Yoshiharu Matsuura, and Wataru Kamitani. 2017. "Two-Amino Acids Change in the Nsp4 of SARS Coronavirus Abolishes Viral Replication." *Virology* 510: 165–174.

Schalk, A F. 1931. "An Apparently New Respiratory Disease of Baby Chicks." *Journal of the American Veterinary Medical Association* 78: 413–423.

Serrano, Pedro, Margaret A Johnson, Amarnath Chatterjee, Benjamin W Neuman, Jeremiah S Joseph, Michael J Buchmeier, Peter Kuhn, and Kurt Wüthrich. 2009. "Nuclear Magnetic Resonance Structure of the Nucleic Acid-Binding Domain of Severe Acute Respiratory Syndrome Coronavirus Nonstructural Protein 3." *Journal of Virology* 83 (24): 12998–12998.

Sharma, Kulbhushan, Sara Åkerström, Anuj Kumar Sharma, Vincent T K Chow, Shumein Teow, Bernard Abrenica, Stephanie A Booth, Timothy F Booth, Ali Mirazimi, and Sunil K Lal. 2011. "SARS-CoV 9b Protein Diffuses into Nucleus, Undergoes Active Crm1 Mediated Nucleocytoplasmic Export and Triggers Apoptosis When Retained in the Nucleus." *PLoS One* 6 (5): e19436.

Shi, Peidian, Yanxin Su, Ruiqiao Li, Zhixuan Liang, Shuren Dong, and Jinhai Huang. 2019. "PEDV Nsp16 Negatively Regulates Innate Immunity to Promote Viral Proliferation." *Virus Research* 265: 57–66.

Shu, Ting, Muhan Huang, Yujie Ren Di Wu, Xueyi Zhang, Yang Han, Jingfang Mu, Ruibing Wang, Yang Qiu, Ding-Yu Zhang, and Xi Zhou. 2020. "SARS-Coronavirus-2 Nsp13 Possesses NTPase and RNA Helicase Activities That Can Be Inhibited by Bismuth Salts." *Virologica Sinica* 35: 321–329.

Smith, Everett Clinton, James Brett Case, Hervé Blanc, Ofer Isakov, Noam Shomron, Marco Vignuzzi, and Mark R Denison. 2015. "Mutations in Coronavirus Nonstructural Protein 10 Decrease Virus Replication Fidelity." *Journal of Virology* 89 (12): 6418–6426.

Stobart, Christopher C, Nicole R Sexton, Havisha Munjal, Xiaotao Lu, Katrina L Molland, Sakshi Tomar, Andrew D Mesecar, and Mark R Denison. 2013. "Chimeric Exchange of Coronavirus Nsp5 Proteases (3CLpro) Identifies Common and Divergent Regulatory Determinants of Protease Activity." *Journal of Virology* 87 (23): 12611–12618.

Sturman, Lawrence S, and Kathryn V Holmes. 1983. "The Molecular Biology of Coronaviruses." In *Advances in Virus Research*, 28: 35–112. Elsevier.

Sun, Jiumeng, Wan-Ting He, Lifang Wang, Alexander Lai, Xiang Ji, Xiaofeng Zhai, Gairu Li, Marc A Suchard, Jin Tian, and Jiyong Zhou. 2020. "COVID-19: Epidemiology, Evolution, and Cross-Disciplinary Perspectives." *Trends in Molecular Medicine*, 26 (5): 483-495.

Tanaka, Tomohisa, Wataru Kamitani, Marta L DeDiego, Luis Enjuanes, and Yoshiharu Matsuura. 2012. "Severe Acute Respiratory Syndrome Coronavirus Nsp1 Facilitates Efficient Propagation in Cells through a Specific Translational Shutoff of Host MRNA." *Journal of Virology* 86 (20): 11128–11137.

Tang, Daolin, Paul Comish, and Rui Kang. 2020. "The Hallmarks of COVID-19 Disease." *Plos Pathogens* 16 (5): e1008536.

Tyrrell, D. A. J., and M L Bynoe. 1961. "Some Further Virus Isolations from Common Colds." *British Medical Journal* 1 (5223): 393.

Tyrrell, D. A. J., J D Almeida, C H Cunningham, W R Dowdle, M S Hofstad, K McIntosh, M Tajima, L Ya Zakstelskaya, B C Easterday, and A Kapikian. 1975. "Coronaviridae." *Intervirology* 5 (1–2): 76.

Venkatagopalan, Pavithra, Sasha M Daskalova, Lisa A Lopez, Kelly A Dolezal, and Brenda G Hogue. 2015. "Coronavirus Envelope (E) Protein Remains at the Site of Assembly." *Virology* 478: 75–85.

Warnes, Sarah L, Zoë R Little, and C William Keevil. 2015. "Human Coronavirus 229E Remains Infectious on Common Touch Surface Materials." *MBio* 6 (6).

Watson, J D, and F H Crick. 1958. "On Protein Synthesis." *In The Symposia of the Society for Experimental Biology*, 12:138–163.

Wu, Aiping, Yousong Peng, Baoying Huang, Xiao Ding, Xianyue Wang, Peihua Niu, Jing Meng, Zhaozhong Zhu, Zheng Zhang, and Jiangyuan Wang. 2020. "Genome Composition and Divergence of the Novel Coronavirus (2019-NCoV) Originating in China." *Cell Host & Microbe*, 27 (3): 325–328.

Yu, Chen, Hui Cai, et al. 2009. "Functional screen reveals SARS coronavirus nonstructural protein nsp14 as a novel cap N7 methyltransferase". *PNAS*, 106 (9): 3484-3489.

Zhang, Lianqi, Lei Li, Liming Yan, Zhenhua Ming, Zhihui Jia, Zhiyong Lou, and Zihe Rao. 2018. "Structural and Biochemical Characterization of Endoribonuclease Nsp15 Encoded by Middle East Respiratory Syndrome Coronavirus." *Journal of Virology* 92 (22).

Zhang, Tao, Qunfu Wu, and Zhigang Zhang. 2020. "Probable Pangolin Origin of SARS-CoV-2 Associated with the COVID-19 Outbreak." *Current Biology*, 30 (7): 1346–1351.

Zhao, Shi, Salihu S Musa, Qianying Lin, Jinjun Ran, Guangpu Yang, Weiming Wang, Yijun Lou, Lin Yang, Daozhou Gao, and Daihai He. 2020. "Estimating the Unreported Number of Novel Coronavirus (2019-NCoV) Cases in China in the First Half of January 2020: A Data-Driven Modelling Analysis of the Early Outbreak." *Journal of Clinical Medicine* 9 (2): 388.

Zhu, Xinyu, Liurong Fang, Dang Wang, Yuting Yang, Jiyao Chen, Xu Ye, Mohamed Frahat Foda, and Shaobo Xiao. 2017a. "Porcine Deltacoronavirus Nsp5 Inhibits Interferon-β Production through the Cleavage of NEMO." *Virology* 502: 33–38.

Zhu, Xinyu, Dang Wang, Junwei Zhou, Ting Pan, Jiyao Chen, Yuting Yang, Mengting Lv, Xu Ye, Guiqing Peng, and Liurong Fang. 2017b. "Porcine Deltacoronavirus Nsp5 Antagonizes Type I Interferon Signaling by Cleaving STAT2." *Journal of Virology* 91 (10).

Zumla, Alimuddin, David S Hui, and Stanley Perlman. 2015. "Middle East Respiratory Syndrome." *The Lancet* 386 (9997): 995–1007.

5 Disinfection Mechanisms of UV Light and Ozonization

Yuanfeng Wang, Jiri Militký and Aravin Prince Periyasamy
Technical University of Liberec, Czech Republic

CONTENTS

5.1 INTRODUCTION

COVIDs are enveloped RNA viruses that are appropriated extensively among people, different vertebrates, and winged creatures which cause respiratory, enteric, hepatic, and neurologic sicknesses. Two possibly perilous zoonotic Covids have developed in the previous 20 years. The serious intense respiratory disorder Covid (SARS-CoV), starting from China, was answerable for the main episode that stretched out from 2002 to 2003. The subsequent flare-up happened in 2012 in the Middle East and was caused by the Middle East respiratory condition Covid (MERS-CoV) (P. I. Lee and Hsueh 2020). Another type of Covid, assigned as the 2019 novel Covid (2019-nCoV), has outbroken all over the world since the end of 2019 (Zhu et al. 2020). Productive human-to-human transmission is a necessity for a huge scope spread of infection. Therefore, to forestall the spread of novel CoVs, it is crucial to contemplate the endurance of infections in various conditions. Endurance wellness of CoVs in the climate decides the opportunity of spread. CoVs are enveloped, positive-sense single-stranded RNA viruses with surface spike protein projection. A high genomic mutation rate permits CoVs to develop and prompts high variety with potential for transmission. Nonetheless, the presence of viral lipid envelopes renders CoVs touchy to ecological conditions, for example, humidity, temperature, outrageous pH, and UV light. Prolonged exposure to adverse living conditions prompts quick decay and loss of infectivity of CoVs (Wong et al. 2019). Thus, to locate a disinfectant that can be utilized to inactivate the CoVs is of high need.

UV radiation can successfully inactivate different microorganisms, and Henri et al. proposed the first UV light system to sanitize water in 1910 (Hernigou, Auregan, and Dubory 2019). UV radiation has various preferences over traditional purification (e.g., chlorination or ozonation), for example, no compound residue, no unsafe disinfection by-products (DBPs) generation, and no result in the disinfectant-opposition to microbes (Song, Mohseni, and Taghipour 2016). Ozone, as a disinfectant highlighted with high bactericidal impact, has been generally utilized in water and wastewater treatment. The wastewater after ozone treatment turns out to be clear and transparent without scent due to the decoloring and deodorizing nature of ozone sterilization. Likewise, the molecular conformation of ozone is unsteady, which suggests that strong oxidative nuclear oxygen delivered by the deterioration of the ozone atom would rapidly break down microorganisms, for example, microbes and infections, in wastewater (Wang et al. 2020). The two disinfectants have been broadly utilized. This paper will investigate their purification mechanism and further discuss their plausibility for the CoVs' sterilization. The present review is divided into two principal segments. The first segment of this paper will present the purification mechanism of UV light and the effect of UV light on the endurance of the CoVs, and the subsequent segment presents the sterilization system of ozone and the effect of ozone on the CoVs.

5.2 DISINFECTION MECHANISM OF UV AND INFLUENCE OF UV LIGHT ON COVS STATE

5.2.1 UV Light and Photochemistry

Like a wide range of electromagnetic radiation, visible light engenders as waves. Nevertheless, the power presented by the waves is consumed at single locations, exactly how the particles are maintained. The assimilated energy of the EM waves is known as a photon and represents the quanta of light. Photochemistry is the process of absorption of photons by a molecule that raises the molecule to an excited state from which a chemical reaction may happen. On account of photochemical reactions, light gives the activating energy. Briefly, light is one mechanism for giving the activating energy needed for numerous reactions. The assimilation of a photon of light by a reactant molecule may likewise allow a reaction to happen not simply by carrying the molecule to the essential activation energy, but additionally by modifying the evenness of the electronic confirmation of a molecule, achieving a generally unavailable reaction route. Some photochemical reactions are some orders of magnitude quicker than thermal reactions; reactions as quick as 10^{-9} s and related cycles as quick as 10^{-15} ss may be noticed. The photon can be ingested straightforwardly by the reactant or by a photosensitizer, which assimilates the photon and conveys the energy to the reactant. When a photoexcited state is deactivated by a synthetic reagent the contrary cycle is called quenching. Most photochemical conversions happen through a progression of common steps known as primary photochemical processes. One ordinary case of this process is the excited-state proton transfer.

The practicable photochemical reaction by molecules is determined by a few laws of photochemistry. The first law of photochemistry, otherwise called the Grotthuss–Draper law (Albini 2016), asserts that a photochemical reaction occurs only when light is assimilated by a compound substance. This implies molecules that do not ingest light of a specific frequency will not go through a photochemical response when presented to light at that frequency. For instance, eyes do not get hurt when one wears sunglasses because most of the damaging UV light is absorbed by sunglasses and cannot reach the eyes. Very little UV penetrates the sunglass, so no damaging takes place. The second law of photochemistry expresses that for every photon of light assimilated by a synthetic component, just a single molecule is triggered for a photochemical response (Bolton, Mayor-Smith, and Linden 2015). This law implies that each photon of light can lead to the photochemical response of one molecule at the most. A concerning law expresses that the photoreaction is straightforwardly proportional to the absorbed photon flow and the time of exposure. As such, more light delivers more photo products. This is the explanation that UV portion or fluence is so significant in the UV sterilization of microorganisms. The UV dose or fluence is straightforwardly relative to the overall number of UV photons consumed by a specific microbe. Most investigations in photochemistry include the UV ranges (100–400 nm). This range can be divided into three sections: UVA (315–400 nm), UVB (280–315 nm), and UVC (200–280 nm) as shown in Figure 5.1. The UVA leads to changes in the skin that further results in tanned skin, while UVB can cause burning from the sun and can inevitably incite skin cancer. The UVC can be consumed by RNA, DNA, and proteins which causes cell malignancy or cell demise. The UVC range is also called the germicidal reach since it is fruitful for killing microscopic

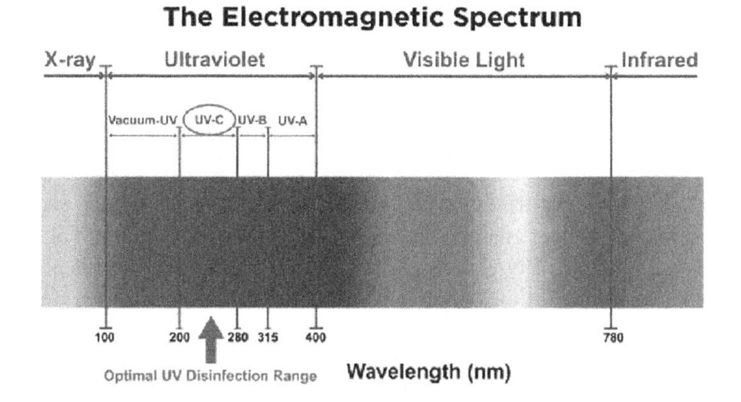

FIGURE 5.1 Schematic diagram of spectral ranges.

organisms and viruses. The vacuum ultraviolet (VUV) range is consumed by practically all materials. Thus, it has no alternative but to be transferred to a vacuum (Bolton and Cotton 2008).

At the point when the target substance was irradiated by UV light, some portion of UV light will be consumed by molecules in the substance (e.g., by DNA in a microorganism). The consumed photon energy makes the molecule ascend in energy from its primary state to an excited state. Now, a few cycles can take place:

- The molecule backtracks to its initial state with the discharge of a photon, which is called fluorescence.
- The excited state can change over to another longer-lived state from which photon discharge can happen – this is called intersystem crossing.
- The molecule can backtrack to its initial state with the overabundance energy being lost as warmth into the medium – this is called inside transformation.
- The molecule can go through a substantial change (modification or splitting) – this is the essence of photochemistry.

The portion of excited states that bring about photochemistry is known as the Quantum yield determined by the proportion of moles of product (Goldstein et al. 2007).

5.2.2 Structure and Photochemistry of DNA and RNA

DNA is a molecule consisting of two polynucleotide chains that curl around one another to shape into a double helix conveying hereditary information for the growth, development, and propagation of every known living being and numerous viruses. DNA and RNA are nucleic acids. Just like proteins, lipids, and complex starches (polysaccharides), nucleic acids are one of the four crucial kinds of macromolecules that are fundamental for all known types of life. The two DNA strands are referred to as polynucleotides as they consist of monomeric units known as nucleotides. Each nucleotide consists of a sugar called deoxyribose, a phosphate group, and one of four nitrogen-containing nucleobases (cytosine [C], guanine [G], adenine [A], or thymine [T]). The nucleotides are joined to each other in an exceeding chain by covalent bonds (i.e., phospho-diester linkage) between the sugar of one nucleotide and the phosphate of the subsequent, forming an alternating sugar–phosphate backbone. According to the base-pairing rule, the nitrogenous bases of two independent polynucleotide chains are bound together with hydrogen bonds to form double-stranded DNA. Complementary nitrogen-containing bases are divided into two groups: pyrimidines and purines. In DNA, the pyrimidines are thymine and cytosine; the purines are adenine and guanine. Similar to DNA, RNA is composed of nucleotide strands, but unlike DNA, RNA has a single strand of nucleotide strands rather than a double helix structure. Cellular organisms use RNA (mRNA) to carry and transmit genetic information (using the nitrogenous bases of guanine, uracil, adenine, and cytosine), which in turn leads to the production of specified proteins. The RNA genome is widely used by viruses to replicate and transmit their genetic information. Within the cell, RNA molecules typically play active roles, specifically including

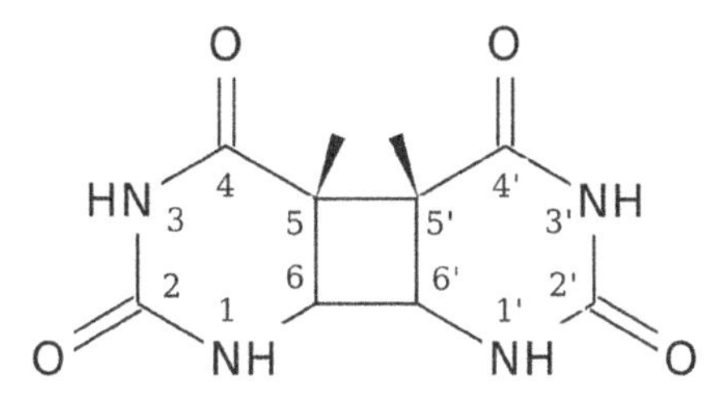

FIGURE 5.2 Thymine cyclobutane dimer. (Reprinted from [Hassanali, Zhong, and Singer 2011], with kind permission of ACS Publications).

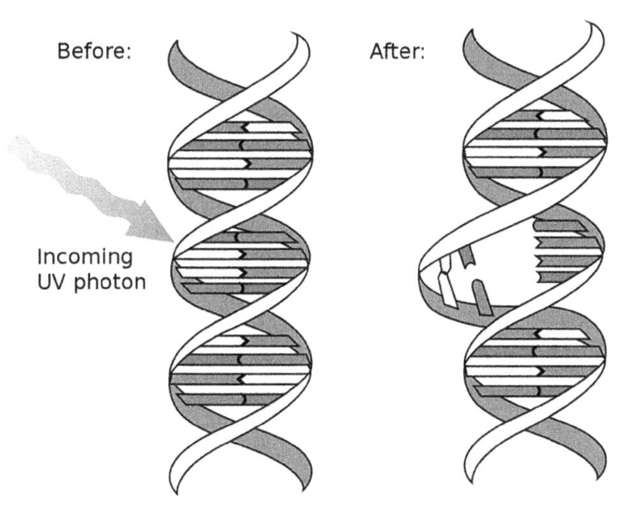

FIGURE 5.3 Formation of thymine dimer lesion in DNA.

catalyzing cellular reactions, delivering genes, and the perception and transmission of cellular signals. One important role is in the control of protein synthesis within the cell, a common function of RNA molecules on the ribosome. In this process, transfer RNA (tRNA) molecules first deliver amino acids to the ribosome, followed by ribosomal RNA (rRNA) that links the amino acids together to ultimately form the coding protein. UV light between 200 and 300 nm is absorbed by all the nucleotides in the DNA or RNA. The so-called thymine dimers, induced by UV light to form covalent junctions between successive bases in the nucleotide chain near their carbon–carbon double bonds, are shown in Figure 5.2. Wrong lesions (Figure 5.3) can disrupt the polymerase, creating errors during transcription or replication that can lead to a stop in replication. This is the basic mechanism of UV disinfection. By inhibiting microbial and viral reproduction, disinfection is ultimately achieved. For viruses containing only RNA, a similar photochemical dimerization reaction occurs between two uracil bases.

The inactivation response of microorganisms to UV light is determined by the ability of UV light to be absorbed. Inactivation of organisms occurs through

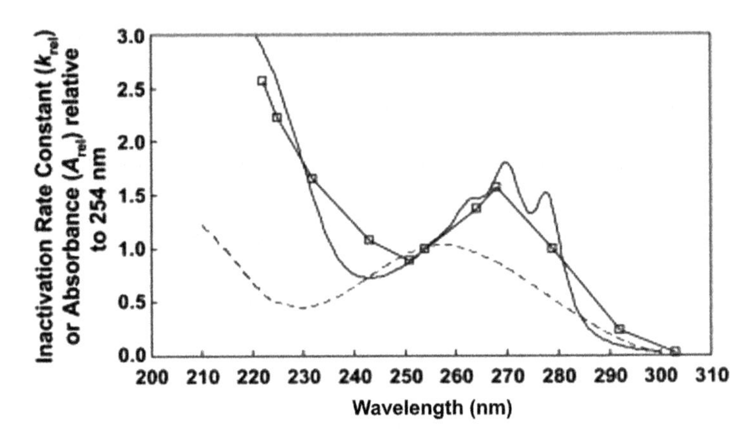

FIGURE 5.4 Comparison of the absorbance spectrum of DNA purified from *Bacillus subtilis* spores, the absorbance spectrum of decoated *B. subtilis* spores, and the spore action spectrum. (Reprinted from [Hassanali, Zhong, and Singer 2011], with kind permission of ACS Publications).

photochemical reactions, such as the dimerization of adjacent thymines in the DNA strand. This relative reaction is referred to as the spectrum of inactivation action. Figure 5.4 shows the spectrum of action of some microorganisms and the relative spectrum of DNA. It is worth noting that the inactivation potency reaches its maximum at about 260 nm. Differences in the composition of nucleotides in DNA and RNA may lead to differences in the action spectra. In many models of the effects of UV light on microorganisms, the relative bactericidal effect of different wavelengths of UV light is indispensable. This is especially true for multicolored light sources such as medium-pressure UV lamps. For this purpose, a germicidal factor (GF) was introduced and defined at the dominant wavelength emitted by low-pressure UV lamps, i.e., GF = 1,000 at 254 nm. If an action spectrum for microorganisms is available, 1,000 at 254 nm can be used as a reference. However, in most cases, the action spectrum is not known; then the relative absorption spectrum of DNA often used as the GF is particularly important for UV equipment using UV lamps that emit multicolored UV light.

5.2.3 Reactivation Mechanisms

Some microorganisms can detect and repair lesions of thymidine dimers on DNA strands during proliferation. Host reactivation enzymes can reactivate some viruses. These activation mechanisms are divided into dark and light mechanisms, with light activation using near-ultraviolet light and short-wavelength visible light. Most bacteria have some reactivation potential in a dark environment. Replacement of UV-damaged nucleotides is one mechanism by which microorganisms repair DNA damage in the dark. In this case, UV light causes the removal of the thymine dimer and adjacent nucleotide sequences from the DNA and the resynthesis of a corresponding sequence. Besides, another mechanism for repairing DNA damage is through the replication and assembly of undamaged regions of the DNA molecule.

An intact, double-stranded DNA that is identical to the original undamaged DNA can be obtained by copying and recombining the undamaged regions of the DNA fragment. This is possible because every single strand of DNA stores complete genetic information. Thus, one strand can function even if the other strand is damaged. UV damage can also be overcome by light induction, which is known as photolytic repair. This mechanism is the most common reactivation mechanism. In this process, the thymidine dimer is split by a special light-activating enzyme, photolyase, which restores the original structure of the DNA. This process is triggered by the absorption of UVA light and is therefore called photoactivation. This repair mechanism can be resisted using a higher UV flux. Photoreactivation must be considered when water treatment systems are exposed to sunlight. Consideration of photoactivation is indispensable for UV treatment of wastewater, which is discharged into natural watersheds. On the other hand, potable water is transported through pipelines and is not exposed to sunlight, so photoactivation is unlikely. Photoreactivation need not be a concern when a chlorinated disinfectant is retained in the water supply system, as the reactivated microorganisms will be inactivated by the disinfectant.

5.2.4 Effects of Different UV Wavelengths on CoVs

Detection of infectious disease viruses such as SARS-CoV-2 during a pandemic can be difficult due to the required biosafety level (BSL) precautions. It is a common practice to study related viruses as a reference for high BSL viruses. Surrogate viruses should respond similarly to UV treatment. Besides, the mechanism of transmission of the SARS-CoV-2 virus is not fully understood, as only a few studies have been performed on it. Therefore, the dependence on surrogate viruses becomes necessary when determining the effectiveness of UV inactivation. In this case, we can use previous CoVs, such as SARS and MERS, as a reference. Duan et al. (2003) first found that the virus was destroyed and reached undetectable levels when SARS-CoV was irradiated for 60 min in a culture medium using UV with an intensity of more than 90 uw/cm^2. UVA (315–400 nm), UVB (280–315 nm), and UVC (200–280 nm) are divided into three classifications: UVA (315–400 nm), UVB (280–315 nm), and UVC (200–280 nm). UVC, when absorbed by RNA and DNA bases, can dimerize two neighboring pyrimidines, which can then lead to base-pairing failure. UVB can also induce pyrimidine dimer photochemical reactions, but with a 20–100-fold lower efficiency than UVC. UVA is weakly taken up by DNA and RNA and does not cause photochemical reactions in terms of pyrimidine dimers. However, UVA may cause base oxidation and strand disruption through the production of reactive oxygen species, resulting in a loss of genetic information (Darnell et al. 2004). Darnell et al. (2004) studied the effects of ultraviolet radiation at different wavelengths on the deactivation of SARS-CoV. The results showed that the virus was completely lost to the test limit after 15 min of exposure to UVC. By contrast, UVA exposure within 15 min had no significant effect on the deactivation of the virus. The deactivation of the virus varied over time, as shown in Figure 5.5.

As shown in Figure 5.6, the efficacy and use of an automatic three-emitter full chamber UV-C disinfection system were reported by Bedell, Buchaklian, and Perlman (2016), resulting in the commercially available MERS-CoV virus

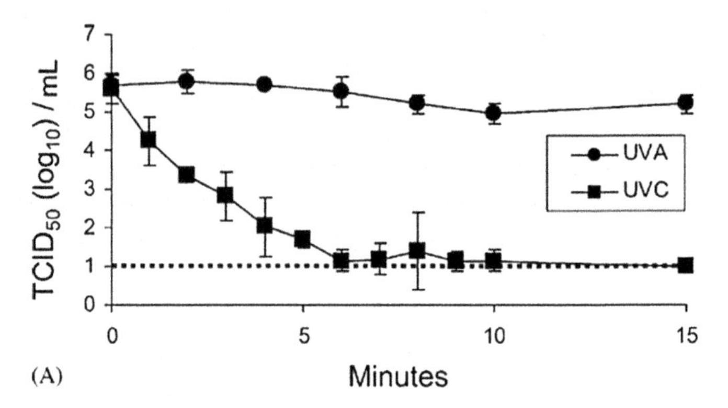

FIGURE 5.5 Effect of UV irradiation on the infectivity of SARS-CoV, (Reprinted from [Darnell et al. 2004], with kind permission of Elsevier Publications).

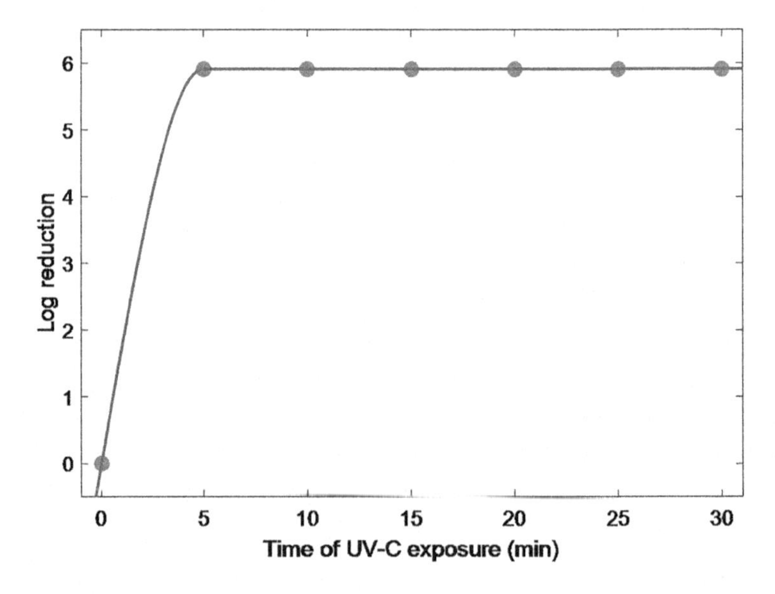

FIGURE 5.6 Log reduction of MERS-CoV during exposure to Surfacide UV-C.

deactivating on the surface and reducing 5log 10. This whole room ultraviolet-C disinfection system can prevent the spread of the virus and protect the staff when the virus explodes. Different wavelengths of UV light also affect the activity of SARS and MERS viruses in the blood. UV-A (Hashem et al. 2019) (Hindawi et al. 2018) (Lin et al. 2005) and UV-B light (Keil, Bowen, and Marschner 2016) can inactivate the nucleic acids of pathogens with amytosin or riboflavin, while a third method uses only UV-C light (Eickmann et al. 2018, 2020). These commercial systems all reduce the activity of SARS-CoV and MERS-CoV in plasma or platelet concentrations to varying degrees. Therefore, the above studies are of great value for the current inactivation of COVID-19 virus. A review of these studies is presented in Table 5.1 (Chang, Yan, and Wang 2020).

TABLE 5.1
Different Methods on Inactivation of Coronavirus in Blood Products

Methods	Commercial Systems	Mechanism of Action	SARS-CoV	MERS-CoV
Amotosalen + UV-A	Intercept Blood System for plasma and platelets	Amotosalen (s-59) intercalates into nucleic acid and induces covalent crosslinking upon UV-A exposure	MEM+10% FBS (reduction of >5.8 log_{10} PFU/mL) (Lin et al. 2005)	Platelet concentrate (reduction of 4.48 ± 0.3 log_{10} PFU/mL) (Hashem et al. 2019) Fresh-frozen plasma (reduction of 4.67 ± 0.25 log_{10} PFU/mL) (Hindawi et al. 2018)
Riboflavin +UV-B	MIRASOL PRT system for plasma and platelets	Riboflavin associates with nucleic acids and mediates an oxygen-independent electron transfer upon UV exposure	N/A	reduction of >4.07 log_{10} PFU/mL for blooded plasma reduction of >4.42 log_{10} PFU/mL for individual donor plasma (Keil, Bowen, and Marschner 2016)
UV-C	THERAFLEX UV-Platelets	UV-C directly interacts with nucleic acids, causing the formation of nucleotide dimers	Platelet concentrates (reduction of ≥3.4 log_{10} $TCID_{50}$/mL) (Eickmann et al. 2020)	Platelet concentrates (reduction of ≥3.7 log_{10} $TCID_{50}$/mL) (Eickmann et al. 2018)

(Reprinted from [Chang, Yan, and Wang 2020], with kind permission of Elsevier Publications)

5.2.5 UV Disinfection System for Masks

Currently, disposable mask products have a major market share. Therefore, during the current COVD-19 outbreak, there is an urgent need for personal protection products with disinfection and reuse capabilities. While there is already guidance on limited reuse and expanded use of masks, the implementation of a masking strategy is a more complex process. For disposable medical devices, the FDA requires validation data on cleanliness, sterilization, and functional performance. Cleaning is typically performed before decontamination to ensure that dirt does not interfere with the decontamination process. Cleaning the mask is a daunting task because the N95 mask is an exposed filter and does not meet standard cleaning techniques. Besides, studies have shown that several models of masks are not effectively cleaned of various contaminants. The Institute of Medicine states that any disposable N95 mask must be decolorized to remove pathogens, be harmless to the user, and not compromise the integrity of the respirator. If the decontamination process removes viable pathogens from a medical device in the presence of other organisms, the question arises as to whether cleaning is still necessary, especially in a public health

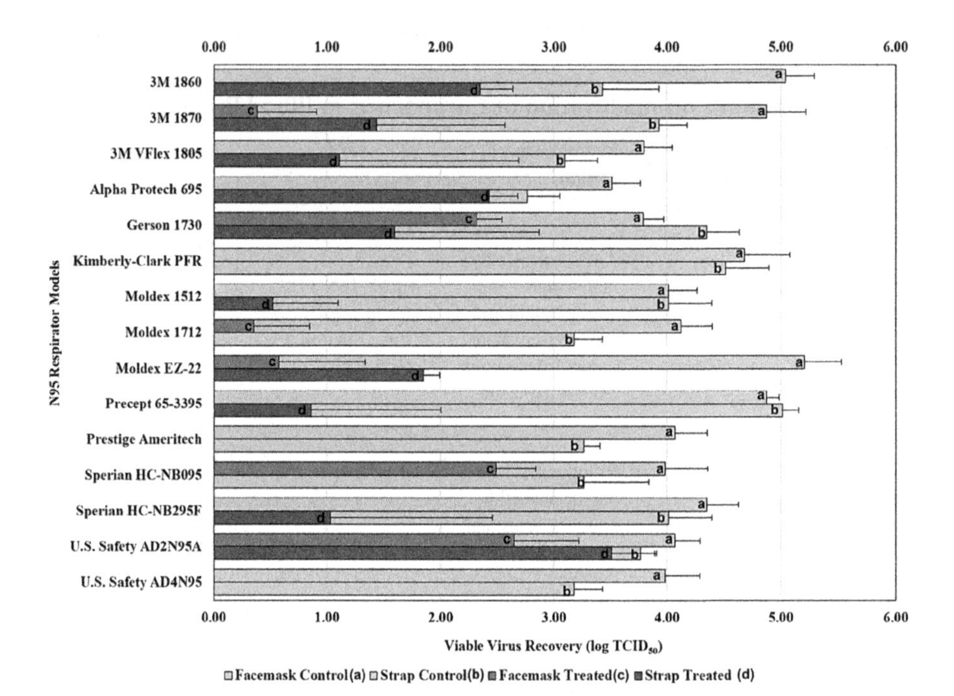

FIGURE 5.7 Viable virus recovered from mucin-soiled N95 respirators. (Reprinted from [Mills et al. 2018], with kind permission of Elsevier Publications).

emergency (Mills et al. 2018). Several studies have been conducted to test the effectiveness of mask decontamination methods, with UV disinfection being one of the most effective methods. Attempts have been made to modify the original UV-emitting device into a device specifically designed for mask disinfection and decontamination. Complete FFRs were evaluated using 15 live FFR models contaminated with the N95 influenza virus. As shown in Figure 5.7, both contamination conditions (artificial saliva and artificial skin oil) significantly reduced the viability of influenza in UV-treated masks (Mills et al. 2018). Decontamination of disposable masks for repeated use has been discussed as a remedial strategy during a disease pandemic, but to date has not been specifically implemented. To evaluate model parameters for UV-C, Fisher and Shaffer (2011) proposed a method to quantify UV-C transmittance for different layers of six N95 FFR models and used it to calculate model-specific a-values – the percentage of available UV-C surface irradiance in the inner filtering medium (IFM). The results show that the log reduction in UV-C is a function of the UV-C dose of a particular IFM. The method described here can be used to assess the potential success of UV-C purification in a given mask. As can be seen above, UV disinfection is a viable method of disinfecting respirators. However, UV radiation simultaneously degrades the polymer, causing accelerated aging of the respiratory protector and weakening the protection of the disposable inhaler for the user. To investigate this issue, William et al. (Lindsley et al. 2015) exposed four N95 mask material coupons and inhaler straps to a UVGI dose of 120–950 J/cm^2 to test the

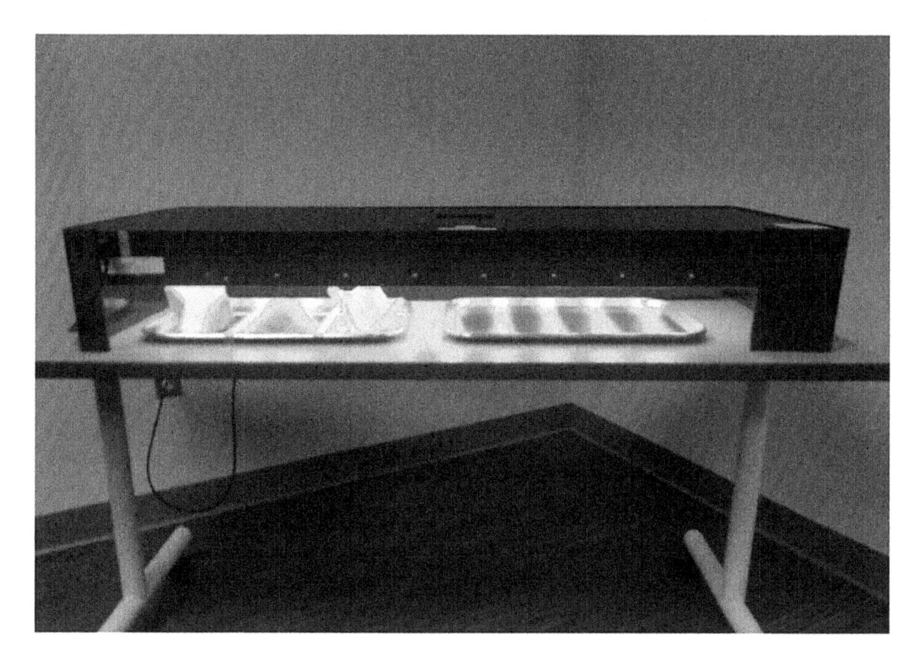

FIGURE 5.8 UVC germicidal disinfection platform for the mask. (Reprinted from [Hamzavi et al. 2020], with kind permission of Elsevier Publications).

permeability, flow resistance, and strength of single-piece mask straps, and the breaking strength of inhaler straps. They found that UVGI radiation had a more significant effect on the strength of the inhaler material. At higher UVGI doses, the strength of each layer of the respirator decreases (some by more than 90%). The strength of the different types of inhaler materials varied considerably. UVGI had a lesser effect on the inhaler backstrap, with a dose of 2,360 J/cm^2, which reduced the fracture strength by 20%–51%. The results indicate that UVGI can be used for effective disinfection and reuse in disposable inhalers, but the maximum cycle frequency will be limited by the type of inhaler and the dose required to inactivate pathogens.

Hamzavi et al. (2020) repurposed UVB units used to treat skin diseases to serve as a platform for UVC germicidal disinfection. In a prototype model that has been developed (Figure 5.8), a UV dose of 1 J/cm^2 can be delivered in 1 min 40 sec at an irradiance of 10 mW/cm^2. The distance from the lamp to the top of the table in Figure 5.8 is approximately 14 cm.

5.3 INFLUENCE OF OZONIZATION ON COVS STATE

5.3.1 Physical–Chemical Properties of Ozone

Oxygen exists in nature in many forms: (1) as a free single atom (O), it is highly reactive and unstable; (2) oxygen (O_2), the most common and stable form, is a colorless gas widely found in the air; (3) ozone (O_3) has a molecular mass of 48, 1.5 times that of oxygen, and is a blue gas; and (4) O_4 is a very unstable, rare, nonmagnetic,

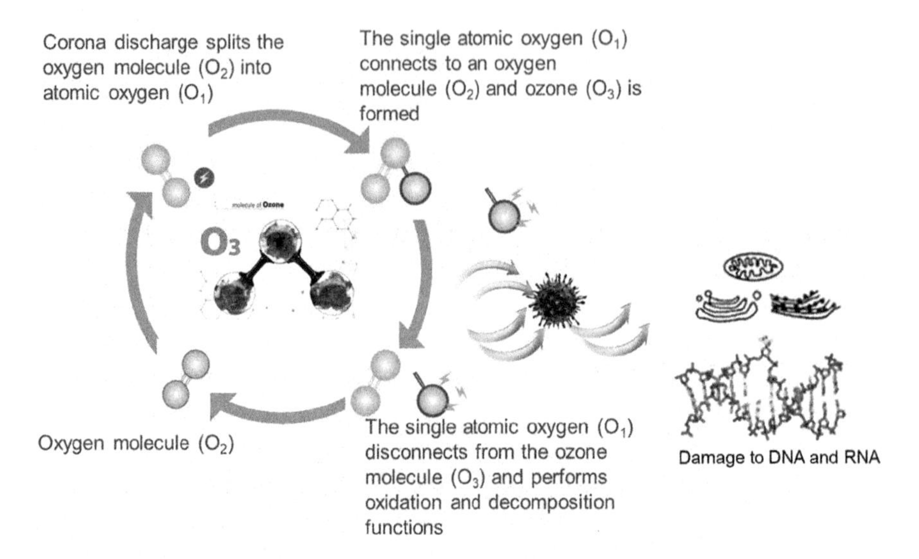

FIGURE 5.9 Schematic diagram of the conversion between the ozone molecule and oxygen molecule.

pale blue gas that breaks down easily into two oxygen molecules. An ozone molecule consists of three oxygen atoms (O_3) and has a molecular weight of 48.00 compared to an oxygen binary molecule (32.00). Ozone has a circular structure with a distance of 1.26 Å between each oxygen atom and is present in several dynamically balanced polymers. The solubility of ozone and oxygen in 100 ml water at 0°C is 49.0 ml and 4.89 ml (10 times lower), respectively. Thus, the large solubility of ozone in water allows it to react immediately with any soluble compounds and biological molecules in the liquid. Ozone is the third most powerful oxidant after fluorine and persulfuric acid, making it highly reactive. Ozone is formed by the process of absorbing oxygen through a high-pressure gradient between electrodes.

This reaction is recyclable (Figure 5.9). It is a spontaneous breakdown of ozone and therefore difficult to store. Besides, the lifetime of the ozone molecule is temperature dependent, with ozone concentration halving in 40 min at 20°C, in 25 min at 30°C, and extending to 3 months at −50°C.

5.3.2 Generation of Ozone and Measurement of Its Concentration

To use ozone as a disinfectant, it is important to understand how ozone is produced and how it is measured. Because of its instability, ozone needs to be produced and used only as soon as it is needed. Ozone can be produced by (1) UV irradiation, (2) electrochemical processes, and (3) corona discharges, but it should be noted that the first two methods produce low yields of ozone and poor production regulation. The ozone therapy instrument must have a safe, nontoxic, reproducible ozone generator. The instrument must use the best ozone-resistant materials, such as Inox 316L stainless steel, pure titanium grade 2, Pyrex glass, Teflon, Viton, and polyurethane, to avoid any substances released by ozone oxidation. It is highly recommended to purchase only one generator to measure the ozone concentration in real-time through a reliable photometer. Unused ozone cannot diffuse into the environment and must be decomposed into oxygen by a catalytic reaction in an indispensable destruction chamber containing heavy metal oxides at a temperature of about 70°C, maintained by an electric thermostat. The medical ozone generator consists of two to four high-voltage tubes connected to an electronic program that determines a voltage difference of 4,000 to 13,000 volts. The discharged energy breaks down the oxygen molecules into oxygen atoms, and when there are too many oxygen molecules, three ozone atoms are formed. The generator supplies medically pure oxygen, a mixture of less than 5% ozone and 95% oxygen at the nozzle, which can be collected with a little pressure. The synthesis of ozone is permitted by the energy released by the discharge, and therefore the decomposition of ozone releases energy. For medical purposes, air cannot be used because, due to the 78%nitrogen content, the final mixture contains, in addition to oxygen and ozone, a variable amount of highly toxic nitrogen oxides. The ozone concentration is determined by three parameters:

- *Voltage:* Eventually, ozone concentration increases with increasing voltage, albeit in a disproportionate manner.
- *The space between the electrodes:* This helps regulate the gradual increase in ozone concentration.
- *The flow of oxygen:* This represents a volume of 1 L/min and can usually be adjusted from 1 to 10 L/min. The final ozone concentration is inversely proportional to the oxygen flow rate, so that at any given time, the greater the flow rate per unit of oxygen, the lower the ozone concentration, and vice versa.

The criteria for calculating the ozone dose are as follows:

- The overall volume of the gas mixture is composed of oxygen and ozone.
- Ozone concentration is expressed as micrograms per ml (mcg/mL).
- Barometric pressure (mm Hg), if different from normal. For safety reasons, we must avoid hyperbaric pressure.

The total ozone dose is equivalent to the gas volume (ml) multiplied by the ozone concentration (mcg/mL). As an example, for a volume of 100 ml blood, we use an equivalent volume of gas (1:1 ratio) with an ozone concentration of 40 mcg/mL, the total ozone dose is 100 × 40 = 4,000 mcg or 4.0 mg. The normal medical generators deliver ozone concentrations from 1 up to 70–100 mcg/mL (Bocci 2011).

5.3.3 Antiviral Action Mechanisms of Ozone

Ozone neutralizes bacteria, viruses, fungi, and parasites in the water. As a result, water purification plants have been established in many major cities around the world. Since then, ozone's unique physical, chemical, and biological properties, as well as its environmental friendliness, have been used in a variety of industrial applications such as pharmaceutical packaging, fumigation of homes and buildings (sick building syndrome), indoor air treatment in operating theatres and nursing homes, and disinfection of air conditioning systems in major hospitals. First, ozone, like other gases, is readily soluble in water, either in blood plasma, in extracellular fluid, or even in a thin layer of water covering the skin, including the mucous membranes of the respiratory and intestinal tracts. At normal temperature and atmospheric pressure, ozone is soluble in water because of its solubility and relative pressure, but unlike oxygen, ozone is unbalanced in the gas phase. This is because ozone, as a strong oxidant, responds immediately to many ions and biological molecules in the liquid, namely antioxidants, proteins, carbohydrates, and polyunsaturated fatty acids, which are preferentially bound to albumin. Phospholipids and cholesterol are present either in cell membranes or in lipids, protected by antioxidants and albumin molecules (Travagli et al. 2010). By this mechanism, ozone destroys viral proteins, lipoproteins, lipids, glycolipids, or glycoproteins. The presence of a large number of double bonds in these molecules makes them susceptible to oxidation by ozone, which provides oxygen atoms and receives electrons in a redox reaction. As a result, the unsaturated bonds are reconfigured, the molecular structure is disrupted, and the envelope is broken. Without the envelope, the virus cannot survive or replicate itself. Ozone itself as well as the peroxide compounds it produces may alter the structure of the viral shell, which is critical for host cell adhesion. Viral glycoprotein protrusions associated with host cell receptors may be the site of ozone action. Even small changes in the integrity of the afferent through peroxidation of the glycoprotein can disrupt adhesion to the host cell membrane and impede viral adhesion and penetration. Fat-coated viruses are the most sensitive to ozone. The viral envelope provides a complex strategy for cell adhesion, penetration, and cell exit. Guided by certain parts of the viral genome, receptors have been carefully tuned to different host cell changes, and new glycoprotein structures are constantly formed to adapt to the host cell's defense. Envelopes are fragile. Ozone and its by-products can destroy them (Gérard & Sunnen, n.d.).

5.3.4 Virus Inactivation by Ozone

Several studies have reported various virus types tested for gaseous ozone disinfection on solid surfaces. Zhang et al. (2004) reported that dissolved ozone solutions inactivate SARS-CoV, commonly known as the SARS virus, which is very similar in structure to the SARS-CoV-2 virus causing the current COVID-19 pandemic. The results showed that the SARS virus was inactivated in 4 min at a high concentration of 27.73 mg/L ozone. The medium (17.82 mg/L) and low (4.86 mg/L) concentrations could also inactivate the SARS virus with different speeds and efficacy (Figure 5.10). Hudson, Sharma, and Vimalanathan (2009) demonstrate the inactivation efficacy of ozone on Murine coronavirus (MCV), which was used at that time as a surrogate for SARS-CoV-1. The literature shows that the envelope proteins in un-enveloped viruses are the most vulnerable. Since SARS-CoV-2 is a virus, the literature supports the hypothesis that ozone will effectively inactivate it. It is important to note that we are not aware of any studies showing that the virus cannot be inactivated by ozone. Of course, the lack of literature does not prove that ozone can inactivate all viruses, but the current evidence suggests that ozone is effective in inactivating SARS-CoV-2. The first task is to determine the concentration of ozone gas and the exposure time,

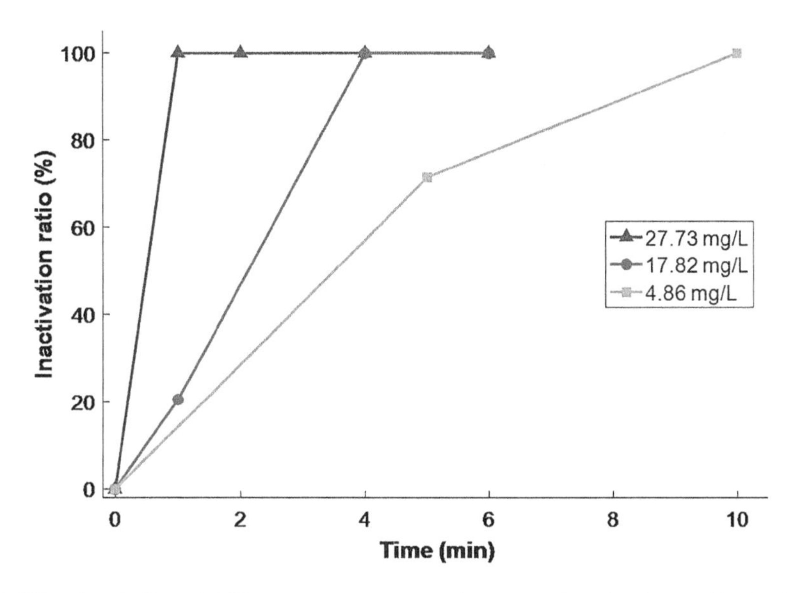

FIGURE 5.10 Effect of different concentrations of ozone on inactivation of SARS virus in water (Reprinted from Zhang et al. 2004).

two variables that are most useful in a basic system such as the one described in this paper. One study has shown that the sterilization effect is primarily influenced by the product of the ozone concentration times the duration, rather than by these two variables alone. This implies that the disinfection effect at higher concentrations over a shorter period of time may not differ significantly from that achieved at lower concentrations over a longer period of time (Tseng and Li 2008). Hudson, Sharma, and Vimalanathan (2009) report viral inactivation data for various surfaces at 45% relative humidity. This data suggests that a dose of 300 min (ppm) would effectively inactivate 99% of all viruses tested on a variety of solid substrate surfaces. In Lee's study (J. Lee et al. 2020), the replacement of SARS-CoV-2 contaminated masks with the human coronary virus (HCoV-229E) was used to show that the virus lost its infectivity against a human cell line (MRC-5) after a short exposure (1 min) to ozone from a dielectric barrier discharge plasma generator (see Figure 5.11). SEM and particle filtration efficiency tests showed that no deterioration in the structure and function of the mask was observed even after five 1-min ozone exposures. These results suggest that rapid disinfection of contaminated masks using plasma generators in well-ventilated areas is possible.

As with UV, ozone is highly destructive and can cause severe damage to latex and masks during the disinfection process. In another study, the durability of N95 mask material (nonwoven polypropylene fiber pads) was tested for 10, 20, and 60 min at ozone concentrations of 10 and 20 ppm. Microscopic examination revealed no significant damage to the fibers and forced-flow filtration efficiency experiments ultimately showed no loss of filtration efficiency after these exposures (Figure 5.12), and there was no evidence of degradation of the filtration material according to the NIOSHN 95 Mask Material Specification Interpretation. Polypropylene mask

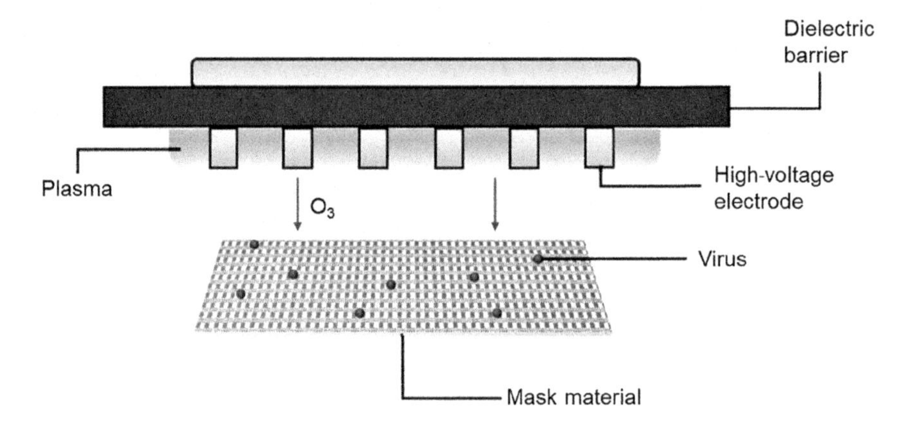

FIGURE 5.11 Schematic diagram of ozone disinfection device for the face mask.

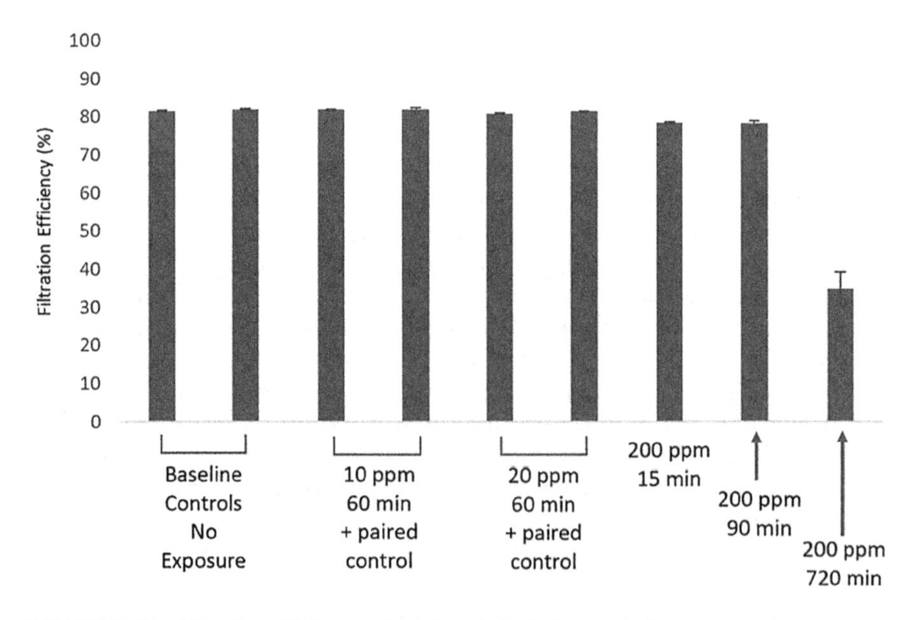

FIGURE 5.12 Filtration efficiency of mask material before and after exposure to ozone gas. (Reprinted under the terms of the Creative Commons Attribution license from *The Journal of Science and Medicine*, JoSaM.org Publications [Dennis, Pourdeyhimi, et al. 2020]).

materials are expected to withstand many disinfection cycles at ozone concentrations well above the concentration and duration of severe (95% to 99%) viral inactivation. Other components of disposable N95 masks may deteriorate upon exposure, but filter effectiveness is not a limiting factor. This means that using a carefully selected combination of ozone concentration and exposure time, virus particles may be inactivated with minimal damage to the disposable filter material (Dennis, Cashion, et al. 2020).

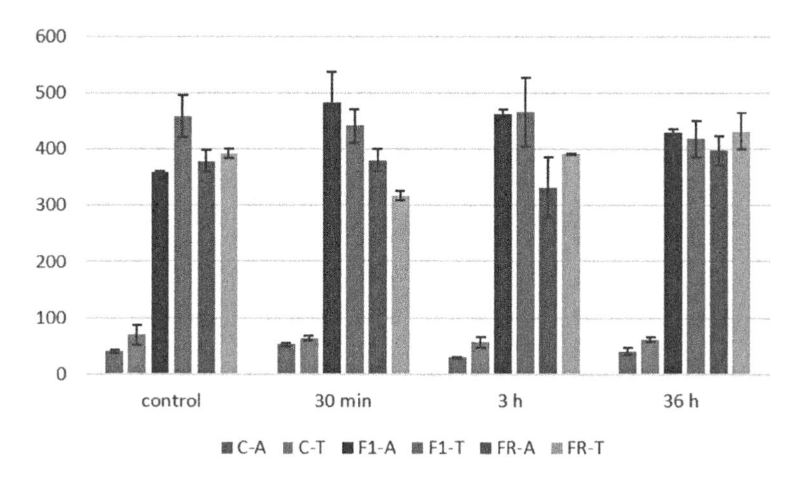

FIGURE 5.13 Strain energy (mJ) from zero to peak load for each of the materials tested, (Reprinted under the terms of the Creative Commons Attribution license from *The Journal of Science and Medicine*, JoSaM.org Publications [Dennis, Pourdeyhimi, et al. 2020]).

To fully verify the mechanical integrity of the filter material throughout the ozone exposure process, basic tensile tests were performed. The test material exhibited strong anisotropy during the preparation, where "along" refers to the tensile properties in the rolling direction and "horizontal" refers to the properties measured orthogonal to the rolling direction. The energy required to reach the peak load is also illustrated (Figure 5.13).

Finally, we conclude that the use of simple methods and convenient and inexpensive doses of antiviral ozone allows the N95 masks to be used repeatedly without compromising filtration performance. Considering the trans viral nature of ozone, it has certain advantages as a gas, while also disinfecting inaccessible spaces. Besides, ozone offers significant advantages in oxygen recovery, whereas liquid disinfectants are likely to damage the surfaces they are used on and leave toxic residues. However, ozone-mediated environmental decontamination requires strict protocols to ensure that the ambient ozone in the target environment being disinfected has time to return to stable parent oxygen without becoming toxic to humans.

5.4 CONCLUSION

Through a review of several studies, this paper first analyzes the mechanisms of disinfection by UV and ozone. The possibility of applying UV and ozone technology to the disinfection of protective equipment is then demonstrated, and finally, the potential damage to protective equipment during the disinfection process is discussed. It was found that UV and ozone technologies can effectively and safely disinfect more than 99% of viral surface protective equipment while causing negligible damage to the protective equipment. In summary, both UV and ozone technologies are experimentally validated to disinfect protective equipment at appropriate doses and concentrations and other parameters.

ACKNOWLEDGEMENT

This work was supported by the Ministry of Education, Youth and Sports of the Czech Republic and the European Union – European Structural and Investment Funds in the Frames of Operational Programme Research, Development and Education – project Hybrid Materials for Hierarchical Structures (HyHi, Reg. No. CZ.02.1.01/0.0/0.0/16 _019/0000843) and the student grant competition No. 21408 at Technical University of Liberec.

REFERENCES

Albini, Angelo. 2016. "Some Remarks on the First Law of Photochemistry." *Photochemical and Photobiological Sciences* 15 (3): 319–324. doi:10.1039/c5pp00445d.

Bedell, Kurt, Adam H. Buchaklian, and Stanley Perlman. 2016. "Efficacy of an Automated Multiple Emitter Whole-Room Ultraviolet-C Disinfection System against Coronaviruses MHV and MERS-CoV." *Infection Control and Hospital Epidemiology* 37 (5): 598–599. doi:10.1017/ice.2015.348.

Bocci, Velio. 2011. *Ozone A New Medical Drug*. 2nd edition. Dordrecht: Springer Netherlands. doi:10.1007/978-90-481-9234-2.

Bolton J R and Cotton, C A. 2008. *The Ultraviolet Disinfection Handbook*. Denver, CO: American Water Works Association.

Bolton, James R., Ian Mayor-Smith, and Karl G. Linden. 2015. "Rethinking the Concepts of Fluence (UV Dose) and Fluence Rate: The Importance of Photon-Based Units – A Systemic Review." *Photochemistry and Photobiology* 91 (6): 1252–1262. doi:10.1111/php.12512.

Chang, Le, Ying Yan, and Lunan Wang. 2020. "Coronavirus Disease 2019: Coronaviruses and Blood Safety." *Transfusion Medicine Reviews*, no. September 2012. Elsevier Inc.: 2–7. doi:10.1016/j.tmrv.2020.02.003.

Darnell, Miriam E.R., Kanta, Subbarao, Stephen M. Feinstone, and Deborah R. Taylor. 2004. "Inactivation of the Coronavirus That Induces Severe Acute Respiratory Syndrome, SARS-CoV." *Journal of Virological Methods* 121 (1). Elsevier: 85–91. doi:10.1016/j.jviromet.2004.06.006.

Dennis, Robert, Avery Cashion, Steven Emanuel, and Devin Hubbard. 2020. "Ozone Gas: Scientific Justification and Practical Guidelines for Improvised Disinfection Using Consumer-Grade Ozone Generators and Plastic Storage Boxes." *The Journal of Science and Medicine* 2 (1). doi:10.37714/josam.v2i1.35.

Dennis, Robert, Behnam Pourdeyhimi, Avery Cashion, Steve Emanuel, and Devin Hubbard. 2020. "Durability of Disposable N95 Mask Material When Exposed to Improvised Ozone Gas Disinfection." *The Journal of Science and Medicine* 2 (1). doi:10.37714/josam.v2i1.37.

Duan, Shu Ming, Xin Sheng Zhao, Rui Fu Wen, Jing Jing Huang, Guo Hua Pi, Su Xiang Zhang, Jun Han, Sheng Li Bi, L. Ruan, and Xiao Ping Dong. 2003. "Stability of SARS Coronavirus in Human Specimens and Environment and Its Sensitivity to Heating and UV Irradiation." *Biomedical and Environmental Sciences* 16 (3): 246–255.

Eickmann, Markus, Ute Gravemann, Wiebke Handke, Frank Tolksdorf, Stefan Reichenberg, Thomas H. Müller, and Axel Seltsam. 2018. "Inactivation of Ebola Virus and Middle East Respiratory Syndrome Coronavirus in Platelet Concentrates and Plasma by Ultraviolet C Light and Methylene Blue plus Visible Light, Respectively." *Transfusion* 58 (9): 2202–2207. doi:10.1111/trf.14652.

Markus Eickmann, Ute Gravemann, Wiebke Handke, Frank Tolksdorf, Stefan Reichenberg, Thomas H. Müller, and Axel Seltsam. 2020. "Inactivation of Three Emerging Viruses – Severe Acute Respiratory Syndrome Coronavirus, Crimean–Congo Haemorrhagic Fever Virus and Nipah Virus – in Platelet Concentrates by Ultraviolet C Light and in Plasma by Methylene Blue plus Visible Light." *Vox Sanguinis* 115 (3): 146–151. doi:10.1111/vox.12888.

Fisher, E. M., and R. E. Shaffer. 2011. "A Method to Determine the Available UV-C Dose for the Decontamination of Filtering Facepiece Respirators." *Journal of Applied Microbiology* 110 (1): 287–295. doi:10.1111/j.1365-2672.2010.04881.x.

Gérard V. Sunnen, M.D. n.d. "SARS and Ozone Therapy: Theoretical Considerations." http://oxygenmedicine.com/2018/05/30/sars-ozone.aspx.

Goldstein, Sara, Dorit Aschengrau, Yishay Diamant, and Joseph Rabani. 2007. "Photolysis of Aqueous H2O2: Quantum Yield and Applications for Polychromatic UV Actinometry in Photoreactors." *Environmental Science and Technology* 41 (21): 7486–7490. doi:10.1021/es071379t.

Hamzavi, Iltefat H., Alexis B. Lyons, Indermeet Kohli, Shanthi Narla, Angela Parks-Miller, Joel M. Gelfand, Henry W. Lim, and David Ozog. 2020. "Ultraviolet Germicidal Irradiation: Possible Method for Respirator Disinfection to Facilitate Reuse during COVID-19 Pandemic." *Journal of the American Academy of Dermatology*, no. PG-(April). Mosby. doi:10.1016/j.jaad.2020.03.085.

Hashem, A. M., A. M. Hassan, A. M. Tolah, M. A. Alsaadi, Q. Abunada, G. A. Damanhouri, S. A. El-Kafrawy, M. Picard-Maureau, E. I. Azhar, and S. I. Hindawi. 2019. "Amotosalen and Ultraviolet A Light Efficiently Inactivate MERS-Coronavirus in Human Platelet Concentrates." *Transfusion Medicine* 29 (6): 434–441. doi:10.1111/tme.12638.

Hassanali, Ali A., Dongping Zhong, and Sherwin J. Singer. 2011. "An AIMD Study of CPD Repair Mechanism in Water: Role of Solvent in Ring Splitting." *Journal of Physical Chemistry B* 115 (14): 3860–3871. doi:10.1021/jp107723w.

Hernigou, Philippe, Jean Charles Auregan, and Arnaud Dubory. 2019. "Vitamin D: Part II; Cod Liver Oil, Ultraviolet Radiation, and Eradication of Rickets." *International Orthopaedics* 43 (3): 735–749. doi:10.1007/s00264-019-04288-z.

Hindawi, Salwa I., Anwar M. Hashem, Ghazi A. Damanhouri, Sherif A. El-Kafrawy, Ahmed M. Tolah, Ahmed M. Hassan, and Esam I. Azhar. 2018. "Inactivation of Middle East Respiratory Syndrome-Coronavirus in Human Plasma Using Amotosalen and Ultraviolet A Light." *Transfusion* 58 (1): 52–59. doi:10.1111/trf.14422.

Hudson, James B., Manju Sharma, and Selvarani Vimalanathan. 2009. "Development of a Practical Method for Using Ozone Gas as a Virus Decontaminating Agent." *Ozone: Science and Engineering* 31 (3): 216–223. doi:10.1080/01919510902747969.

Keil, Shawn D., Richard Bowen, and Susanne Marschner. 2016. "Inactivation of Middle East Respiratory Syndrome Coronavirus (MERS-CoV) in Plasma Products Using a Riboflavin-Based and Ultraviolet Light-Based Photochemical Treatment." *Transfusion* 56 (12): 2948–2952. doi:10.1111/trf.13860.

Lee, Jinyeop, Cheolwoo Bong, Pan K. Bae, Abdurhaman T. Abafog, Seung Ho Baek, Yong-Beom Shin, Moon S. Park, and Sungsu Park. 2020. "Fast and Easy Disinfection of Coronavirus-Contaminated Face Masks Using Ozone Gas Produced by a Dielectric Barrier Discharge Plasma Generator." *MedRxiv*, 2020.04.26.20080317. doi:10.1101/2020.04.26.20080317.

Lee, Ping Ing, and Po Ren Hsueh. 2020. "Emerging Threats from Zoonotic Coronaviruses-from SARS and MERS to 2019-NCoV." *Journal of Microbiology, Immunology and Infection*, no. xxxx. Elsevier Taiwan LLC: 2019–21. doi:10.1016/j.jmii.2020.02.001.

Lin, Lily, Carl V. Hanson, Harvey J. Alter, Valérie Jauvin, Kristen A. Bernard, Krishna K. Murthy, Peyton Metzel, and Laurence Corash. 2005. "Inactivation of Viruses in Platelet Concentrates by Photochemical Treatment with Amotosalen and Long-Wavelength Ultraviolet Light." *Transfusion* 45 (4): 580–590. doi:10.1111/j.0041-1132.2005.04316.x.

Lindsley, William G., Stephen B. Martin, Robert E. Thewlis, Khachatur Sarkisian, Julian O. Nwoko, Kenneth R. Mead, and John D. Noti. 2015. "Effects of Ultraviolet Germicidal Irradiation (UVGI) on N95 Respirator Filtration Performance and Structural Integrity." *Journal of Occupational and Environmental Hygiene* 12 (8): 509–517. doi:10.1080/15459624.2015.1018518.

Mills, Devin, Delbert A. Harnish, Caryn Lawrence, Megan Sandoval-Powers, and Brian K. Heimbuch. 2018. "Ultraviolet Germicidal Irradiation of Influenza-Contaminated N95 Filtering Facepiece Respirators." *American Journal of Infection Control* 46 (7): e49–e55. doi:10.1016/J.AJIC.2018.02.018.

Song, Kai, Madjid Mohseni, and Fariborz Taghipour. 2016. "Application of Ultraviolet Light-Emitting Diodes (UV-LEDs) for Water Disinfection: A Review." *Water Research* 94 (May): 341–349. doi:10.1016/j.watres.2016.03.003.

Travagli, Valter, Iacopo Zanardi, Patrizia Bernini, Stefano Nepi, Leonardo Tenori, and Velio Bocci. 2010. "Effects of Ozone Blood Treatment on the Metabolite Profile of Human Blood." *International Journal of Toxicology* 29 (2): 165–174. doi:10.1177/1091581809360069.

Tseng, Chunchieh, and Chihshan Li. 2008. "Inactivation of Surface Viruse by Gaseous Ozone." *Journal of Environmental Health* 70 (10): 56–63.

Wang, Jiao, Jin Shen, Dan Ye, Xu Yan, Yujing Zhang, Wenjing Yang, Xinwu Li, Junqi Wang, Liubo Zhang, and Lijun Pan. 2020. "Disinfection Technology of Hospital Wastes and Wastewater: Suggestions for Disinfection Strategy during Coronavirus Disease 2019 (COVID-19) Pandemic in China." *Environmental Pollution* 262 (July): 114665. doi:10.1016/j.envpol.2020.114665.

Wong, Antonio C.P., Xin Li, Susanna K.P. Lau, and Patrick C.Y. Woo. 2019. "Global Epidemiology of Bat Coronaviruses." *Viruses* 11 (2): 1–17. doi:10.3390/v11020174.

Zhang, Jia-min, Chong-yi Zheng, Geng-fu Xiao, Yuan-quan Zhou, and Rong Gao. 2004. "Examination of the Efficacy of Ozone Solution Disinfectant in in Activating Sars Virus." *Chinese Journal of Disinfection* 01.

Zhu, Na, Dingyu Zhang, Wenling Wang, Xingwang Li, Bo Yang, Jingdong Song, Xiang Zhao, et al. 2020. "A Novel Coronavirus from Patients with Pneumonia in China, 2019." *New England Journal of Medicine* 382 (8): 727–733. doi:10.1056/NEJMoa2001017.

6 Influence of UV Light and Ozonization on Microbes State

Xiaodong Tan, Qingyan Peng, Kai Yang, Jana Saskova, Aravin Prince Periyasamy, Jiri Militký and Jakub Wiener
Technical University of Liberec, Czech Republic

CONTENTS

6.1 INTRODUCTION

Due to the persistent outbreak of water vector and food vector virus, mechanical understanding of virus disinfection is urgently needed (Wigginton and Kohn 2012). For a long time, scientists have been trying to provide mechanical descriptions of virus inactivation during disinfection of drinking water (Thurman and Gerba 1988). In the 1960s and 1980s, researchers used scintillation spectrum and electron microscopy to detect changes in the virus's genome and protein, and usually reported one of the two findings: i) inactivation due to impaired viral proteins or ii) inactivation due to damaged genomes (Dennis Jr, Olivieri and Krusé 1979). But recent research has focused less on elucidating mechanisms and more on comparing the dynamics of extinction with those of various viral strains, disinfectants, and hydrochemistry (Cromeans, Kahler, and Hill 2010). Overall, the virus inactivation mechanisms of common water disinfectants vary greatly and are often contradictory. For example, chlorine deactivates the poliovirus by RNA degradation and capsid modification (Dennis Jr et al. 1979, O'Brien and Newman 1979). At this point, there is still no certain answer for which modifications lead to or do not lead to inactivation. The reaction between amino acid or nucleotide monomer and common water disinfectants such as ozone and UV radiation has good characteristics (Table 6.1) (Wigginton and Kohn 2012). Therefore, the reaction rate constant of the established amino acid and nucleotide monomer can be summarized according to the known abundance in virus particles to predict the relative reaction rate of the genome and protein target (Davies 2003). When this is done for Poliovirus-1 Mahoney, chemical disinfectants such as chlorine and ozone are much more active in viral protein substances than genomic substances. On the contrary, UVC radiation affects genomic matter more than protein matter. Unfortunately, such predictions are not accurate because of the effect of higher tissue levels of proteins and genomes on response rates. The degradation rate of the genomes of the denatured poliovirus is different from that of the original poliovirus genome. The reactions that occur to genomes and proteins during disinfection

TABLE 6.1

Reported Second-order Rate Constants and Photochemical Constants for the Most Reactive Amino Acid and Nucleoside Monomers with Common Disinfectants in Aqueous Solutions at pH – 7

	Ozone	UV	
Nucleotides and Amino Acids	**k ($M^{-1}S^{-1}$)**	**ε (254 nm) ($M^{-1}cm^{-1}$)**	**Φ_d**
Adenine	$200^{a,b}$	1.2×10^4	4.4×10^{-4c}
Cytosine	1.4×10^{3a}	3.5×10^3	5.3×10^{-4c}
Guanine	5.0×10^{4a}	1×10^4	2.1×10^{-4c}
Uracil	650	7.8×10^3	1.4×10^{-3d}
Thymine	1.6×10^{4a}	6.3×10^3	9.6×10^{-4c}
Cysteine	1.0×10^{9b}	–	–
Histidine	4.0×10^{5b}	–	–
Methionine	6×10^{6b}	–	–
Phenylalanine	–	140	0.019
Tryptophan	1×10^{7b}	2.8×10^3	9.0×10^{-3}
Tyrosine	4×10^{6d}	340	0.022

Source: Reprinted from (Wigginton and Kohn 2012), with Kind Permission of Elsevier Publications.

[a] Values for dAMP, dCMP, dGMP, and Dtmp.

[b] Values may include radical pathways.

[c] Quantum yield for nucleotide destruction in air/O_2 saturated solution.

[d] Data for chromophore loss in air/O_2 saturated solution.

may form a by-product of further reactions with amino acids and nucleotides; this makes the response prediction more complex.

Unlike chemical oxidizers, UVC radiation directly causes photolytic virus components, regardless of whether their solvents are available or not. Similar to genomic-based methods, the prevalence of pyrimidines in virus genomes is considered as a framework for predicting UVC susceptibility. There is a correlation between the number of potential dimerization sequences in the virus genome and the effective UV dose. However, the presence of abnormal virus strains suggests that alternative pathways play a role in some mechanisms of UV inactivation. Overall, these differences suggest that information on viral components (i.e., genomes and protein sequences alone cannot accurately predict susceptibility), so previous understanding of the crust and genomic structure should help explain and predict the response of the virus particles and help identify the most sensitive areas of the particle.

6.2 UV LIGHT

6.2.1 State-of-the-Art of UV Light

Usually, UV light is produced by mercury lamps, including low pressure (LP) mercury lamps, which emit near-monochromatic UV light at 254 nm, and medium-pressure lamps, which emit a multicolor spectrum of different wavelengths (Bolton and Cotton 2011). However, UV mercury lamps are wrapped in glass, which is vulnerable. UV mercury lamps can easily cause mercury leaks in the external environment and release harmful gases into the air within weeks or years. In recent years, a new, friendly approach has been developed to produce UV light, known as UV-LEDs. Vilhunen uses UV radiation at wavelengths of 269 and 276 nm as the best way to sterilize bacteria. Other studies support the idea that UV radiation can be effectively sterilized (Vilhunen, Särkkä, and Sillanpää 2009). However, the development of UV-LED chips continues to make progress, although it is difficult to develop short-wavelength epitaxial slices and the lattice failure in the chips is difficult to manage. Chips with wavelengths below 280 nm have been developed, of which 200–280 nm are the most useful disinfection belts (Li et al. 2019). Besides, the output efficiency of the chip encapsulation module is improved under the chip encapsulation mode, including a single chip. At present, the multichip array module is widely used, which greatly improves the output power of the whole device. UV-LED packaging disinfection application products are constantly emerging in the market. UV radiation lamps have several advantages over UV mercury lamps; UV radiation lamps with adjustable wavelength from 210 nm of radiation to visible light, is a very promising UV radiation source. UV-LED does not cause disposal problems (mercury-free), leaving only a small footprint (flexible structure), mechanical strength, immediate shutdown (high-frequency response), low voltage, low power requirements, and long life (reduced frequency of replacement). Unmanned aircraft and e-commerce have many advantages, but are still underdeveloped in all respects due to the prevailing technical difficulties. Most UV-LED applications are still under development, and there are no standard measurements and descriptions to determine how UV-LEDs can be sterilized. At the same time, the traditional measuring technology of mercury lamp has been developed. As early as 2003, Bolton and Linden developed a standardized protocol and obtained accurate and consistent results from microbiological and photochemical experiments by measuring accurate UV flux (Bolton and Linden 2003). Through the study of different parameters, a measuring method for UV mercury lamp's output was developed. More practical programs have since been introduced, and there has been strong interest among different industry participants, including the International UV society. However, these measurement protocols are designed for UV mercury lamps and are not suitable for UV LEDs with different structures and output specifications. Therefore, it is necessary to start a measuring and analyzing system for UV-LEDs (Chevremont, Farnet, Coulomb, Boudenne, 2012). Therefore, a standard protocol should be established for accurate control and monitoring of doses of UV radiation in different experiments and equipments and for assessing the consistency of doses of UV radiation. Agreements must also be developed to measure and control UV-LED outputs, radiation profiles, radiation power, and UV doses for water disinfection applications. Other researchers studied the synergy of two wavelengths to compare log inactivation (Chevremont, Farnet, Sergent, Coulomb, Boudenne 2012). Scholars have also expressed interest in adding assistive devices, such as ultrasonic pretreatment systems, to achieve higher antimicrobial rates (Bowker et al. 2011). The study of disinfection of air surfaces, such as automatic disinfection stethoscope made with UV-LED lights, solves the problem of hospital pathogen propagation through a stethoscope. UV-LED systems will improve step-by-step, but forming a comprehensive, standardized, and standardized disinfection mechanism of UV-LED is still urgent (Chin and Bérubé 2005).

6.2.2 Influence Rule of Disinfection Methods

Here, the UV-LED sterilization system was divided into two parts: UV-LED light source system, which is a key part of the system efficiency and the most important factor; the water treatment system, which receives the UV-LED light source and acts as a reaction site to kill microbes(Chen et al. 2019).

6.2.3 Influence of UV-LED Light Source System Parameters on Inactivation Mechanism

The UV emission light system is the main factor affecting disinfection. Microorganisms in UV water treatment systems can be sterilized to prevent the replication and transcription of microbial genetic material by absorbing UV photons and producing polymers based on the mechanism of deactivation. In UV water disinfection systems, the most common light source transmitter systems include traditional mercury lamps and emerging UV-LEDs. However, large and bulky UV mercury lamps may be replaced by compact UV-LEDs, which is very marketable and promising because of its compact appearance. The UV-LED light source system includes a basic circuit, power supply, and four UV-LED chips in 1, 2, 3, and 4 locations, respectively. The wavelength of the UV-LED chip can be selected as required. According to the analysis, researchers can choose the same or different wavelength chip to set the UV-LED chip number or wavelength (Ding et al. 2019).

6.2.4 Wavelength of UV-LEDs

Wavelength is the key factor of microbiological disinfection in the UV system. From the perspective of UV source disinfection, 253.7 nm is considered the most effective when used with conventional mercury lamps (Bolton and Cotton 2011). However, mercury lamps come in two ways: LP and MP. MP is mixed with multiple emission peaks. Determining the most effective wavelength for MP is difficult. The emission peaks of LP are around 254 nm, whereas the optimal bactericidal wavelength is 253.7 nm. The development of UV-LED will make microbiological deactivation more effective because it can emit light with adjustable wavelength and provide insight into its effect on disinfection (Bowker et al. 2011). As shown in Figure 6.1 and Figure 6.2, the absorption of 260 nm wavelength is the largest for the DNA absorption curve. However, in the actual experiments of UV-LEDs on microbe inactivation, different microorganisms showed different absorption curves and different disinfection ability. Absorption values of 260 and 280 nm show strong inactivation characteristics (Chen, Craik, and Bolton 2009). When the inactivation ability of different microorganisms was compared with a UV dose-response, the logarithmic inactivation dose-response of each species at 260 nm was always lower than that at 280 nm (Bolton and Cotton 2011).

At current technology levels, 280 nm is the best option to achieve high inactivation efficiency with minimal energy consumption. Oguma observed that 280 nm of UV-LED could achieve a high survival rate constant, with minimum energy consumption, and that three logarithmic inactivation values could be achieved at 265,

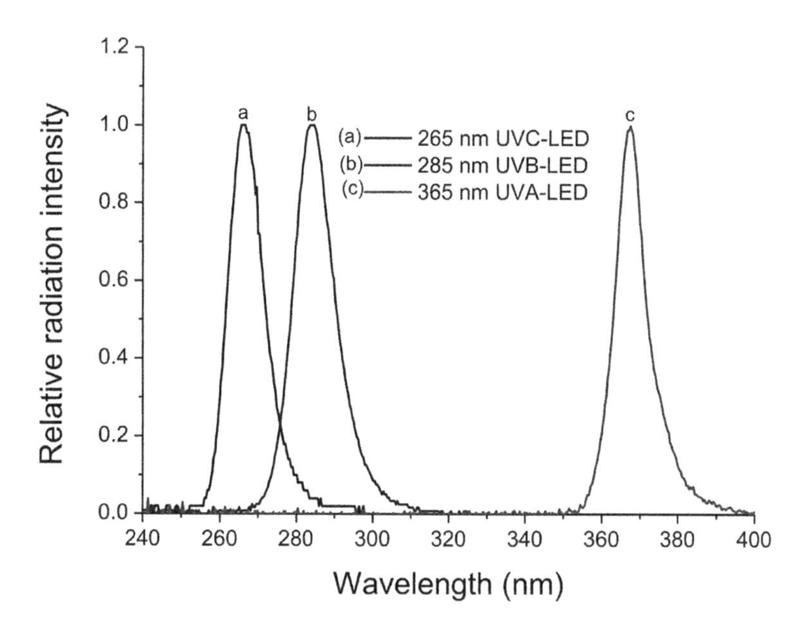

FIGURE 6.1 Emission spectra of 265 nm UVC-LED, 285 nm UVB-LED, and 365 nm UVA-LED. [Reprinted from (Song, Taghipour, and Mohseni 2019), with kind permission of Elsevier Publications]

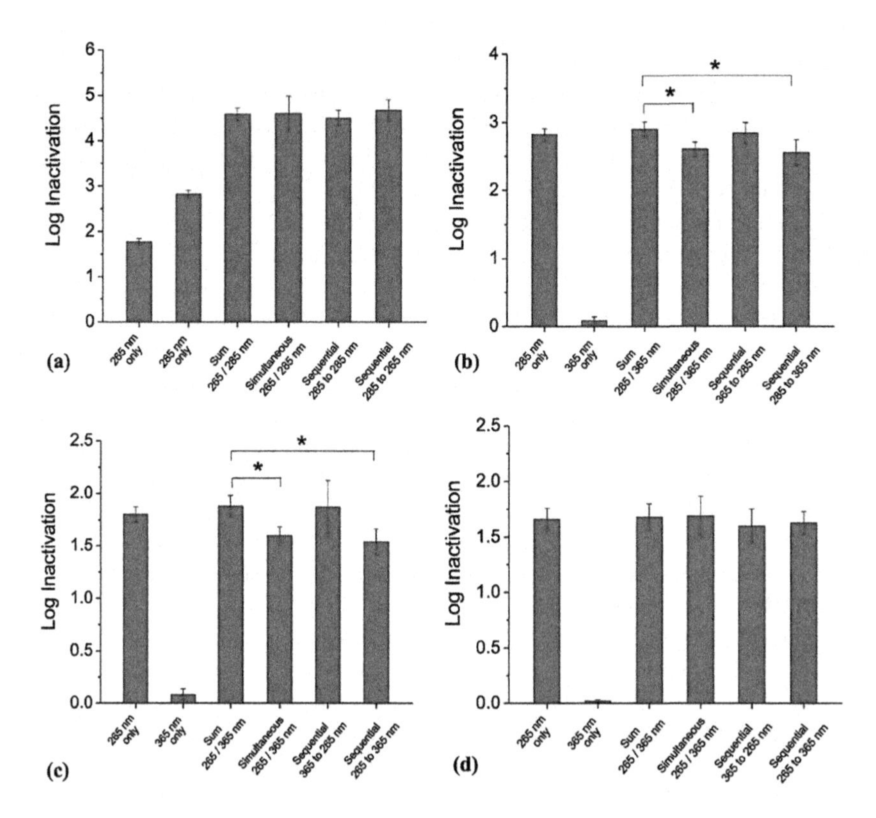

FIGURE 6.2 Comparisons of microbe's inactivation by various UV-LEDs' wavelength combinations: *E. coli* inactivation by UVC 265 nm/UVB 285 nm combinations (a), by UVB 285 nm/UVA 365 nm combinations (b), by UVC 265 nm/UVA 365 nm combinations (c); and MS2 inactivation by UVC 265 nm/UVA 365 nm combinations (d). [Reprinted from (Song, Taghipour, and Mohseni 2019), with kind permission of Elsevier Publications].

280, and 300 nm for tested microbial species (Oguma et al. 2013). By comparing disinfection efficiency, another group also noted that the wavelength of 280 nm was better suited to the water purification system than the wavelength of 255 nm, as the external quantum efficiency of 280 nm generated more than 10 times the power generated by 255 nm. Würtele reported light output and inactivation at 282 nm higher than at 269 nm (Würtele et al. 2011). While 269 nm UV-LED showed better bactericidal effects, 282 nm UV-LED showed significant improvement in spore inactivation due to the high photon output of 282 nm UV-LED at the same period and input power. In the band of UV disinfection, higher wavelength light power can improve the disinfection ability of bacteria. Lower wavelength peaks are close to the DNA absorption curve, but lower wavelength UV-LED is more efficient when considering light power (Song, Taghipour, and Mohseni 2019). The selection of UV-LED wavelength shows the potential for future application in water disinfection systems equipped with UV-LED. The coupling wavelength has been widely used in recent years, but it shows different results. The total power of coupling wavelength is lower than that of single UV-LEDs. Although further research is needed, other wavelength combination applications produce better disinfection effects and optimize the advantages of different wavelength types (Zhou et al. 2017).

6.2.5 UV Dose

UV dose is another key factor in the water disinfection system of UV-LED. UV doses are not only produced directly by UV post-radiation doses but may also be influenced by parameters and associated instrumentations. UV radiation dose is the product of UV radiation flux, exposure time, attenuation factor, and radiation surface. UV radiation flux, also known as radiation power, means that when light radiates outward in this formula, it passes through a certain portion of radiation energy per unit time, expressed in watts (W); UV dose is defined as the radiant energy received per unit area of receiving ionizing radiation (mJ/cm^2) and represents the absorbed dose of the water sample from the UV transmitter to the water treatment reactor (Bolton and Linden 2003). Complex attenuation factors are involved in the system, including water, reflection, divergence, and sensor factors. Thus, the UV dose is not a true value, but an approximation. If the UV radiation flux was in control, the experiment results will vary greatly due to the different radiation time, intensity, and radiation distance (Sinha and Häder 2002). The radiation flux of LP is high while the dose is constant, but between LP and UV-LED, there are many differences in radiation time, radiation distance, and radiation intensity (Würtele et al. 2011). LP sends out a parallel light, while the UV-LED lamp sends out a point light source, emitting a diffuse light from an observation angle. Unlike LP, radiation intensity refers to the radiation flux that leaves the point source in a given direction cube in the UV-LED light source system. Given the greatest influence factors, UV doses can be controlled in specific cases, considering the same attenuation factor. Exposure times are closely related to flux and dose, i.e., larger radiation flux will result in shorter exposure times. Besides, some researchers have found that high radiation flux and short exposure times can lead to higher logarithmic inactivation, while low radiation flux and long exposure times can lead to microbial aggregation, thereby reducing inactivation (Vilhunen, Särkkä, and Sillanpää 2009, Zhou et al. 2017). Because the output power of 275 nm UV-LEDs is higher, it is better to inactivate *Escherichia coli* with high flux and short exposure time than with low flux and long exposure time (Bowker et al. 2011, Sommer et al. 1998). Therefore, when the same doses of UV radiation were controlled, a larger radiation flux may produce better disinfection effects. Overall, UV doses that optimize disinfection effects must be identified and therefore require extensive investigation.

6.2.6 Inactivation Rate Constant (K)

The inactivation rate constant (K) (cm^2/mJ) is equal to the sensitivity of microorganisms to UV light (Pirnie, Linden, and Malley 2006). Table 6.2 shows the required UV dose to achieve 4log inactivation for the listed microorganisms in three wavelengths.

Overall, 260 nm is generally more sensitive than 280 nm. Under current conditions, however, the cost of producing UV-LED chips at a range of 260 nm is considerable

TABLE 6.2
UV Dose Required to Achieve 4log Inactivation

Microorganism	Wavelength (nm)	UV dose (mJ/cm²)
B. subtilis	265	28
E. coli K12 IFO3301	265	16.4
	265	10.8
	265	6
L. pneumophila	265	4.5
Qβ	265	3.9
B. subtilis	280	48
E. coli K12 IFO3301	280	25.5
	280	13.8
	280	9
HAdV2	280	89
L. pneumophila	280	9
P. aeruginosa	280	8.5
Qβ	280	75
B. subtilis	LP	58
E. coli K12 IFO3301	LP	7.5
HAdV2	LP	112
L. pneumophila	LP	6.5
P. aeruginosa	LP	9
T1	LP	19.3
PRD1	LP	36
PhiX174	LP	8.9
MS2	LP	64.4

Source: Reprinted from (Li et al. 2019), with Kind Permission of Elsevier Publications.

due to considerable research and development difficulties and a lack of optical strength and external quantum output (Oguma, Katayama, and Ohgaki 2002). On the other hand, UV-LED at a wavelength of 280 nm is more effective in actual inactivation experiments, because the wavelength of 280 nm produces more external output and is efficient in lighting. That is, the external quantum efficiency of 280 nm of UV-LED is almost five to ten times, although the disinfection efficiency of 255 nm is twice that of 280 nm, the total disinfection efficiency of the UV-LED operating at 280 nm is almost five times (Mamane-Gravetz and Linden 2005). The damage induced by 280 nano-UV rays may inhibit the activity of proteins, and the detection of the sensitive site of enzyme cutting may cause less damage. A chip with a higher output power of 260 nm will result in higher inactivation efficiency. As a result, the wavelength of 280 nm is still more efficient and relatively inexpensive, but with advances in UV-LED chip development, ~260 nm in the future may have a more competitive wavelength range of light (Qiu et al. 2018).

6.3 OZONIZATION

The mechanism of ozone disinfection has been shown in Figure 6.6. There is ample evidence in the scientific literature that a mixture of ozone and oxygen deactivates

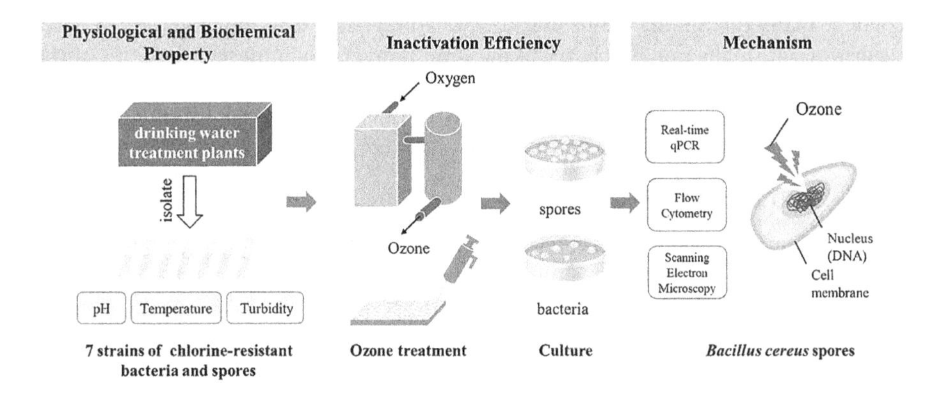

FIGURE 6.3 Mechanism of ozonization disinfection. [Reprinted from (Ding et al. 2019), with kind permission of Elsevier Publications].

microbes such as bacteria, fungi, and viruses (Hoff 1986). In the ongoing examination, exact ozone fixation (in a fluid medium) was utilized to inactivate the infection's infectivity. Some information proposes that an assortment of infection types can be inactivated in a realized ozone uncovered climate (Murray et al. 2008). Ozone is known for its function in air biological parity (Figure 6.3). The special organic properties of ozone are being concentrated to decide its likely application in different clinical orders. Since the late nineteenth century, ozone has been studied in response to a range of unsaturated carbon compounds called carbon bonds (Razumovskii and Zaikov 1984). Ozone (O_3) and atomic oxygen (O_2) are both receptive oxygen (RO), just as extra responsive oxygen and responsive nitrogen (NO) and cushion arrangement delivered in water. These incorporate countless essential ROs and NOs: superoxide free radicals (O_2-), ozone free radicals (O_3-), hydroxyl radical (OH), hydrogen peroxide (H_2O_2), nitrite (NO_2-), and nitrite peroxide (NOOO-). Auxiliary RO and NO may result from the holding oxidation of lipids, proteins, and amino acids, creating different responsive revolutionaries. Through double-bond oxidation, ozone has the unique ability to eliminate toxic or toxic industrial impurities and inactivate biological (viral and bacterial) pollutants (Bruice 2016). The cleansing and sanitization properties of ozone have been perceived since the First World War when ozone was utilized medically to disinfect wounds. After clarifying the biochemical components associated with ozone-interceded detoxification, the likely utilization of ozone has extended extraordinarily, including water cleansing.

6.4 OZONE INACTIVATION

6.4.1 Lipid Peroxidation

The data suggests that there are two main ways to inactivate the virus: lipid peroxidation and protein peroxidation (Carbonneau et al. 1991, Friedman and Stromberg 1993). Lipid peroxidation is a component of tissue harm brought about by different ROs, which delivers free revolutionaries and causes numerous obsessive sequelae (Esterbauer, Schaur, and Zollner 1991). The unsaturated fat creation of phospholipid contains many point immersions in its hydrocarbon chain, and the oxidation of these securities will make genuine harm to the lipid bilayer structure and capacity of the film. A generally acknowledged rule for deciding lipid peroxidation is malondialdehyde (MDA), a steady peroxide side effect (Esterbauer, Schaur, and Zollner 1991).

6.4.2 Protein Peroxidation

Besides, protein peroxidation is characterized as the covalent adjustment of a protein by direct association with RO or circuitous communication with optional collaboration by oxidative pressure items (Shacter 2000). Uncommonly, protein peroxidation assumes a significant part in the inactivation of nonwrapped infections, for example, adenovirus, poliovirus, and different enteroviruses (Shacter 2000).

6.4.3 Ozone for Disinfection

As an efficient and promising disinfectant, O_3 shows better performance than chlorine disinfectants (such as free chlorine, dioxide, and chloramines) or hydroxyl radicals (Gottschalk, Libra, and Saupe 2009). Chlorine disinfectants have been successfully used to control pathogen contamination (not including parasitic organisms), and there are serious defects in their application. Its potential to react to natural organic matter (NOM) in the water environment can easily form a sterile by-product (DBP, especially trihalomethane [THM], and haloacetic acid [HAA]), which may have negative influences on human health (Chin and Bérubé 2005). Ozone, even at low concentrations, is more effective at controlling many microbes, including parasites, bacteria and viruses, and some NOMs, compared to chlorine (Xu et al. 2002). Besides, because of its properties, O_3 is not persistent in water and therefore may pose minimal health risks. An important issue related to ozone disinfection (or ozone oxidation) is the formation of bromate, a so-called carcinogen (Von Gunten and Hoigne 1994). Since bromine (Br-) is usually contained in various water environments, bromate can be produced between Br- and O_3 (or OH) in a multistep process (shown in Figure 6.4) (Wang and Chen 2020). Strategies for reducing bromate

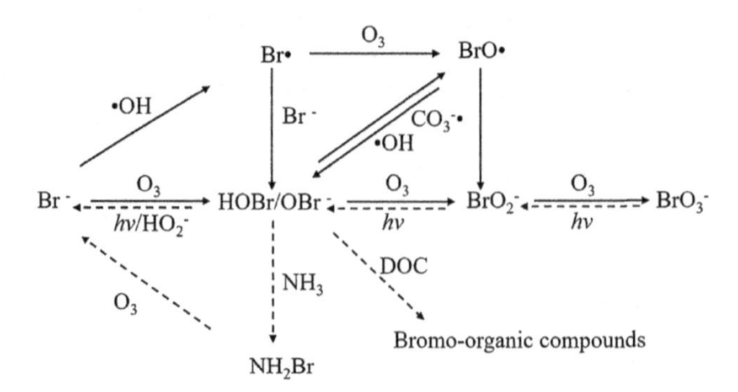

FIGURE 6.4 Formation of bromate in ozonation/ozone-based processes and mitigation strategies, [Reprinted from (Wang and Chen 2020), with kind permission of Elsevier Publications].

formation were studied: if O_3 concentration or pH (<6.0) is reduced, if ammonia or H_2O_2 is added, or if organic carbon is combined with UV radiation or is dissolved (DOC) (Fischbacher et al. 2015, Von Gunten and Oliveras 1997).

The virus inactivation caused by ozone transfer strongly indicates the damage of lipid virus membrane and protein capsid. To confirm that the integrity of the membrane and the protein membrane of the virus have been compromised, samples were prepared for electron microscope observation and negative staining with phosphotungsten acid. VACV morphology was severely damaged and HSV-1 fusion was destroyed (Chin and Bérubé 2005).

6.4.4 Air Disinfection by Ozone

O_3 is a gas that serves as a strong oxidizer for biocides. It has the broad-spectrum antibacterial ability and is active against bacteria, fungi, viruses, protozoa, and fungal spores. For this reason, ozone has been used in water treatment for decades. In recent years, ozone has been widely accepted as an eco-friendly "green" technology. Due to the growing interest in ozone applications, the opinion of the Italian Ministry of Health favored the use of gaseous ozone for the disinfection of empty cheese ripening and storage facilities. Portable ozone generators are now available. They have discharge devices and fans that produce ozone at different concentrations, and catalytic converters that then break it down into oxygen. The benefits associated with ozone use are ease of access to hidden sites in the gaseous state. It also has the advantage of not having a by-product, as it quickly breaks down into oxygen without leaving bad residues on food or food surfaces. This technology can both save water and improve the quality of wastewater, for example, by avoiding the presence of harmful chlorinated compounds, as compared to other biocides. Besides, ozone is generated locally as needed and does not need to be stored. On the other hand, some of the disadvantages are high capital costs (i.e., corona discharge generators). Nevertheless, ozone treatment remains more cost-effective than alternative treatment technologies (Masotti et al. 2019).

Similar to the Fenton process, adding catalysts to the ozone oxidation process promotes the decomposition of oxidizers (O_3) into active free radicals, such as OH. Catalytic ozone can reduce operating costs compared with other ozone-based treatments, as it does not require other energy costs (such as UV or pH adjustment costs), as it is highly effective in various pH values. Besides, catalytic ozone treatment systems exhibit good performance in water treatment, which has many advantages over single ozone treatment (Ding et al. 2019). Generally, based on the catalyst used in the system, the catalytic ozone process can be divided into two types: (1) homogeneous catalytic oxidation, in which metal ions are commonly used as catalysts and (2) heterogeneous catalytic oxidation based on ozone activation of the carrier by a heterogeneous catalyst (such as a metal oxide or a metal) (see Figure 6.5).

6.4.5 Ozone-Based Water Treatment Processes (Ozonization)

Ozone treatment technology is a developed technology and has been widely used in the world. Due to its strong oxidizing properties, O_3 can be used in many fields,

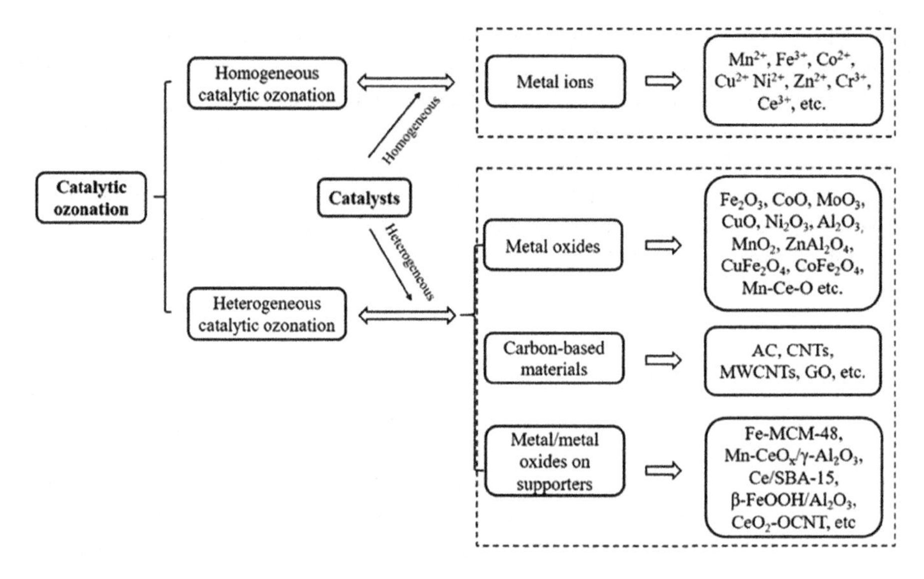

FIGURE 6.5 A variety of homogenous and heterogeneous catalysts applied in catalytic ozonation process, [Reprinted from (Wang and Chen 2020), with kind permission of Elsevier Publications].

such as oxidizing and sterilizing inorganic–organic matter (or pathogen control). As more and more microbes and chemical pollutants are found in the water environment (water source, wastewater, etc.), attention to the efficient and economical removal of pollutants (and disinfection of by-products) is increasing. They also lead to renewed attention to ozone treatment and ozone water treatment processes (Gehr et al. 2003).

6.4.6 Perspectives of Ozone-Based Water Treatment

The catalytic oxidation processes and possible reaction mechanisms of homogeneous and heterogeneous catalysts are analyzed. There are many studies on catalytic oxidation, but some results or conclusions on the catalytic mechanism still seem to be inconsistent and need further study. Besides, the following aspects should be considered to enhance the understanding and application of the catalytic ozone depletion mechanisms (Gottschalk, Libra, and Saupe 2009).

- The effect of pH
  The pH value of the solution has a significant effect on the catalytic oxidation process. In most cases, however, the pH of the reaction solution is not controlled or changes in pH are not monitored during ozone oxidation, for example, when catalysts are added to the solution. In addition to laboratory studies, consideration should also be given to the effect of pH on the catalytic ozone process in the wastewater matrix. Besides, the optimization of other operational parameters is important for the application of the process (Murray et al. 2008).

- The role of active species
  Many studies focus only on the hydroxyl radical that is produced in the system. In some cases, due consideration should also be given to the role of other active species formed within the system. Besides, quenching experiments combined with other detection techniques, such as electron paramagnetic resonance (EPR), will be more conducive for the study of catalytic mechanisms (Murray et al. 2008).
- The active center of the catalyst
  Conclusions about catalyst activity centers are sometimes inconsistent. Some researchers believe that Lewis is an active site for the decomposition of O_3 and the formation of free radicals, while others suggest that Lewis plays an important role in catalyzing the ozone process (especially when carbon is used as a catalyst). Genuine centers need further clarification (Murray et al. 2008).
- The performance of the catalyst
  To date, ozone catalysis processes have been studied for various catalysts. However, the synthesis technology of some catalysts is very complex and the conditions are very strict, which are not favorable for large-scale production and industrial application. Besides, despite the good performance of certain materials in catalyzing ozone systems, research into the reuse and stability of catalysts is insufficient. Therefore, low-cost, stable, and reusable catalysts need to be developed. On the other hand, a systematic method for evaluating catalyst performance should be established (Razumovskii and Zaikov 1984).
- Mitigation of DBPs formation
  DBP may form during ozone treatment, which is considered to be a serious threat to human health and ecosystem, and possible mitigation strategies are required for both ozone and ozone water treatment applications.
- Combination with other treatment processes (Xu et al. 2002).
  The combination of ozone and other treatments has been extensively studied. However, how the characteristics of an integrated system improve or limit individual and global efficiency, and the costs of the entire treatment system in practice require further study (Ding et al. 2019).

6.5 CONCLUSION

UV-LED water disinfection system involves a wide range of topics. These systems include not only microelectronic encapsulation technology or modular design of UV-LED chips but also the deactivation of pathogens in water environment engineering analysis. In this paper, the problems of UV-LED in water treatment are discussed through the comprehensive analysis and study of various wavelength disinfection and energy consumption of UV-LED. Current data suggest that it is possible to accurately deliver, monitor, and record ozone concentrations and that the toxic effects of ozone can be minimized while maximizing the sterilization and sterilization effects of ozone. Real-time monitoring of the flow of ozone gases, ozone absorption, and ozone–oxygen exposure times provide an unprecedented level of control for the ozone–oxygen transport system. Significant improvements can be made in water disinfection programs, in the disinfection of serological products, and the treatment of atherosclerosis and viremia. These results provide a basis for further research into ozone-mediated virus inactivation and an opportunity to study other infectious substances in biological materials. Overall, UV and odorous oxidation have proved to be a good way to inactivate and sterilize microbes, including bacteria and viruses. The development of ozone and UV radiation provides feasible and cost-effective solutions for decontamination of selected areas of the facility. Air purification can reduce the deposition of microorganisms on surfaces of frequent contact and thus prevent the risk of their spreading.

ACKNOWLEDGEMENT

This work was supported by the Ministry of Education, Youth and Sports of the Czech Republic and the European Union – European Structural and Investment Funds in the Frames of Operational Programme Research, Development and Education – project Hybrid Materials for Hierarchical Structures (HyHi, Reg. No. CZ.02.1.01/0.0/0.0/16_019/0000843) and the research project of Student Grant Competition of Technical University of Liberec no. 21406/2020 granted by Ministry of Education Youth and Sports of Czech Republic.

REFERENCES

Bolton J R, Cotton C A. 2011. *The UV disinfection handbook.* American Water Works Association, Denver, CO, USA.

Bolton, James R, and Karl G Linden. 2003. "Standardization of methods for fluence (UV dose) determination in bench-scale UV experiments." *Journal of Environmental Engineering* 129 (3):209–215.

Bowker, Colleen, Amanda Sain, Max Shatalov, and Joel Ducoste. 2011. "Microbial UV fluence-response assessment using a novel UV-LED collimated beam system." *Water Research* 45 (5).

Bruice, Paula Yurkanis. 2016. *Essential Organic Chemistry*, University of California, Santa Barbara, USA.

Carbonneau, MA, E Peuchant, D Sess, P Canioni, and M Clerc. 1991. "Free and bound malondialdehyde measured as thiobarbituric acid adduct by HPLC in serum and plasma." *Clinical Chemistry* 37 (8):1423–1429.

Chen, Lizhu, Felipe Caro, Charles J. Corbett, and Xuemei Ding. 2019. "Estimating the environmental and economic impacts of widespread adoption of potential technology solutions to reduce water use and pollution: Application to China's textile industry." *Environmental Impact Assessment Review* 79:106293. doi: 10.1016/j.eiar.2019.106293.

Chen, Ren Zhuo, Stephen A Craik, and James R Bolton. 2009. "Comparison of the action spectra and relative DNA absorbance spectra of microorganisms: Information important for the determination of germicidal fluence (UV dose) in an UV disinfection of water." *Water Research* 43 (20):5087–5096.

Chevremont, A-C, A-M Farnet, B Coulomb, and J-L Boudenne. 2012. "Effect of coupled UV-A and UV-C LEDs on both microbiological and chemical pollution of urban wastewaters." *Science of the Total Environment* 426:304–310.

Chevremont, A-C, A-M Farnet, M Sergent, B Coulomb, and J-L Boudenne. 2012. "Multivariate optimization of fecal bioindicator inactivation by coupling UV-A and UV-C LEDs." *Desalination* 285:219–225.

Chin, August, and PR Bérubé. 2005. "Removal of disinfection by-product precursors with ozone-UV advanced oxidation process." *Water Research* 39 (10):2136–2144.

Davies, Michael J. 2003. "Singlet oxygen-mediated damage to proteins and its consequences." *Biochemical and Biophysical Research Communications* 305 (3):761–770. doi: 10.1016/s0006-291x(03)00817-9.

Ding, Wanqing, Wenbiao Jin, Song Cao, Xu Zhou, Changping Wang, Qijun Jiang, Hui Huang, Renjie Tu, Song-Fang Han, and Qilin Wang. 2019. "Ozone disinfection of chlorine-resistant bacteria in drinking water." *Water Research* 160:339–349.

Esterbauer, Hermann, Rudolf Jörg Schaur, and Helmward Zollner. 1991. "Chemistry and biochemistry of 4-hydroxynonenal, malonaldehyde and related aldehydes." *Free Radical Biology and Medicine* 11 (1):81–128.

Fischbacher, Alexandra, Katja Löppenberg, Clemens von Sonntag, and Torsten C Schmidt. 2015. "A new reaction pathway for bromite to bromate in the ozonation of bromide." *Environmental Science & Technology* 49 (19):11714–11720.

Friedman, LI, and RR Stromberg. 1993. "Viral inactivation and reduction in cellular blood products." *Revue française de transfusion et d'hémobiologie* 36 (1):83–91.

Gehr, Ronald, Monika Wagner, Priya Veerasubramanian, and Pierre Payment. 2003. "Disinfection efficiency of peracetic acid, UV and ozone after enhanced primary treatment of municipal wastewater." *Water Research* 37 (19):4573–4586. doi: 10.1016/s0043-1354(03)00394-4.

Gottschalk, Christiane, Judy Ann Libra, and Adrian Saupe. 2009. *Ozonation of water and waste water: A practical guide to understanding ozone and its applications*. John Wiley & Sons, Hoboken, New Jersey, United States.

Hoff, JC. 1986. "Inactivation of microbial agents by chemical disinfectants EPA 600 S2-86 067." Office of Water, US Environmental Protection Agency, Washington, DC: 1-7.

Li, Xiaoling, Miao Cai, Lei Wang, Fanfan Niu, Daoguo Yang, and Guoqi Zhang. 2019. "Evaluation survey of microbial disinfection methods in UV-LED water treatment systems." *Science of the Total Environment* 659:1415–1427.

Mamane-Gravetz, H, and KG Linden. 2005. "Relationship between physiochemical properties, aggregation and uv inactivation of isolated indigenous spores in water." *Journal of Applied Microbiology* 98 (2):351–363.

Masotti, Fabio, Stefano Cattaneo, Milda Stuknytė, and Ivano De Noni. 2019. "Airborne contamination in the food industry: An update on monitoring and disinfection techniques of air." *Trends in Food Science & Technology* 90:147–156. doi: 10.1016/j.tifs.2019.06.006.

Murray, Byron K, Seiga Ohmine, David P Tomer, Kendal J Jensen, F Brent Johnson, Jorma J Kirsi, Richard A Robison, and Kim L O'Neill. 2008. "Virion disruption by ozone-mediated reactive oxygen species." *Journal of Virological Methods* 153 (1):74–77.

O'Brien R T, Newman J. 1979. "Structural and compositional changes associated with chlorine inactivation of polioviruses." *Applied and Environmental Microbiology* 38 (6):1034–1039.

Oguma, Kumiko, Hiroyuki Katayama, and Shinichiro Ohgaki. 2002. "Photoreactivation of Escherichia coli after low-or medium-pressure UV disinfection determined by an endonuclease sensitive site assay." *Applied Environmental Microbiology* 68 (12):6029–6035.

Oguma, Kumiko, Ryo Kita, Hiroshi Sakai, Michio Murakami, and Satoshi Takizawa. 2013. "Application of UV light emitting diodes to batch and flow-through water disinfection systems." *Desalination* 328:24–30.

Pirnie Malcom, Karl G Linden, and JPJ Malley. 2006. "UV disinfection guidance manual for the final long term 2 enhanced surface water treatment rule." *US Environmental Protection Agency* 2 (11):1–436.

Qiu, Y., Q. Li, B. E. Lee, N. J. Ruecker, N. F. Neumann, N. J. Ashbolt, and X. Pang. 2018. "UV inactivation of human infectious viruses at two full-scale wastewater treatment plants in Canada." *Water Research* 147:73–81. doi: 10.1016/j.watres.2018.09.057.

R.B. Thurman, C.P. Gerba. 1988. "Molecular mechanisms of viral inactivation by water disinfectants." *Advances in Applied Microbiology* 33:75–105.

Razumovskii, SD, and Gennadiĭ Efremovich Zaikov. 1984. "Ozone and its reactions with organic compounds." *Studies in Organic Chemistry* 15.

Shacter, Emily. 2000. "Quantification and significance of protein oxidation in biological samples." *Drug Metabolism Reviews* 32 (3–4):307–326.

Sinha, Rajeshwar P, and Donat-P Häder. 2002. "UV-induced DNA damage and repair: a review." *Photochemical & Photobiological Sciences* 1 (4):225–236.

Sommer, R, Th Haider, A Cabaj, W Pribil, and M Lhotsky. 1998. "Time dose reciprocity in UV disinfection of water." *Water Science and Technology* 38 (12):145–150.

Song, Kai, Fariborz Taghipour, and Madjid Mohseni. 2019. "Microorganisms inactivation by wavelength combinations of UV light-emitting diodes (UV-LEDs)." *Science of The Total Environment* 665:1103–1110. doi: doi:10.1016/j.scitotenv.2019.02.041.

Cromeans Theresa L., Amy M. Kahler, Vincent R. Hill. 2010. "Inactivation of adenoviruses, enteroviruses, and murine norovirus in water by free chlorine and monochloramine." *Public Health Microbiology* 76 (4):1028–1033.

Vilhunen, Sari, Heikki Särkkä, and Mika Sillanpää. 2009. "UV light-emitting diodes in water disinfection." *Environmental Science and Pollution Research* 16 (4):439–442.

Von Gunten, Urs, and Juerg Hoigne. 1994. "Bromate formation during ozonization of bromide-containing waters: interaction of ozone and hydroxyl radical reactions." *Environmental Science & Technology* 28 (7):1234–1242.

Von Gunten, Urs, and Yvonne Oliveras. 1997. "Kinetics of the reaction between hydrogen peroxide and hypobromous acid: implication on water treatment and natural systems." *Water Research* 31 (4):900–906.

W.H. Dennis Jr, V.P. Olivieri, C.W. Krusé. 1979. "Mechanism of disinfection: Incorporation of Cl-36 into f2 virus." *Water Research* 13 (4):363–369.

Wang, J., and H. Chen. 2020. "Catalytic ozonation for water and wastewater treatment: Recent advances and perspective." *Science Total Environment* 704:135249. doi: 10.1016/j.scitotenv.2019.135249.

Wigginton, K. R., and T. Kohn. 2012. "Virus disinfection mechanisms: the role of virus composition, structure, and function." *Current Opinion Virology* 2 (1):84–89. doi: 10.1016/j.coviro.2011.11.003.

Würtele, MA, T Kolbe, M Lipsz, A Külberg, M Weyers, M Kneissl, and M Jekel. 2011. "Application of GaN-based UV-C light emitting diodes–UV LEDs–for water disinfection." *Water Research* 45 (3):1481–1489.

Xu, Pei, Marie-Laure Janex, Philippe Savoye, Arnaud Cockx, and Valentina Lazarova. 2002. "Wastewater disinfection by ozone: main parameters for process design." *Water Research* 36 (4):1043–1055.

Zhou, Xiaoqin, Zifu Li, Juanru Lan, Yichang Yan, and Nan Zhu. 2017. "Kinetics of inactivation and photoreactivation of Escherichia coli using ultrasound-enhanced UV-C light-emitting diodes disinfection." *Ultrasonics Sonochemistry* 35:471–477.

7 Photocatalysis and Virus Spreading

Photocatalysis

Aamir Mahmood, Jiri Militký and Miroslava Pechociakova

Technical University of Liberec, Czech Republic

CONTENTS

7.1 INTRODUCTION

Numerous diseases including fever, heart issues, hepatitis, loss of motion, and respiratory infections are brought about by infections. Infections have a less irresistible amount of $<10–10^3$ particles in examination with bacterial microbes and a critical high ailment danger of 10–10,000 times under a comparative degree of presentation (Gibson 2003). This prompts extraordinary interest from the scientific community to discover an effective and minimal effort disposal/inactivation technique for water-borne infections. Granular enacted carbon adsorption as a typical barrier in water treatment was accounted for the viable expulsion of protozoan (oo) growths and microbes, however not for viruses (Hijnen et al. 2010). Customary cleansing strategies depend on the use of heat, radiation, or chemical compounds Chlorine, ozone, and UV radiation are among the most utilized agents at present used to sanitize water, air, or fomites (Magalh and Andrade 2017). Chlorination as a cleansing procedure is predominantly established by the use of gaseous chlorine or potentially hypochlorite. Chlorine gas (Cl_2) is the elemental form of chlorine at standard temperature and pressure. Chlorine gas is around 2.5 times heavier than air and is exceptionally harmful. Hypochlorite (ClO^-) is generally acquired from sodium hypochlorite and calcium hypochlorite (Rutala and Weber 1997). Chlorination on the other hand increases the formation of possibly mutagenic and cancer-causing disinfection by-products (DBPs) which can prompt the issues of repeated pollution and salting of freshwater sources (Nieuwenhuijsen 2000). Ozone is delivered when oxygen particles are separated by an energy source into oxygen atoms and hence collide with the nonseparated oxygen molecules. Ozone is one of the most impressive oxidizing agents ($E^0 = 2.07$ V) and it is generally used to annihilate organic (Agustina, Ang, and Vareek 2006). Despite its profoundly effective inactivation of all microorganisms, ozonation can also deliver DBPs, such as aldehydes, carboxylic acids, and ketones, within the sight of dissolved organics (Huang, Fang, and Wang 2005).

7.2 PHOTOCATALYSIS

Photocatalysis is defined as a process where light is used to activate a substrate (photocatalyst) that in turn triggers the kinetics of a chemical reaction remaining unencumbered by itself (usually semiconducting metal oxides) (Keane, McGuigan, Ibáñez, Inmaculada Polo-López, Byrne, Dunlop, O'Shea, Dionysiou, and Pillai, 2014). According to the general description of thermal catalysis, photocatalysis is a process in which the speeding up of a photoreaction takes place by the presence of a catalyst, which shows that both light and a catalyst are needed to cause or to accelerate a chemical transformation. Since the photoreaction occurs in more than one homogeneous medium, it is commonly referred to as "heterogeneous photocatalysis." Photocatalysis is a key mechanism for developing possible disinfection strategies to combat harmful microorganisms. The water source contains many bacteria, viruses, fungi, protozoa, etc. These microorganisms present in water have an extremely complicated nature. Therefore, understanding microbial morphology is of keen importance in achieving advanced photocatalysis disinfection procedures. In recent years, germicidal UV has gained more attraction due to the fact that it does not create almost any DBPs. UV rays, depending on the wavelength of radiation, are divided into UV-A (315–400 nm), UV-B (280–315 nm), UV-C (200–280 nm), and vacuum UV radiation (VUV) (100–200 nm). Among them, UV-C is very efficient against microbial disinfection. UV-C destroys illuminated DNA by directly inducing pyrimidine and purine dimers and pyrimidine adducts. UVC intensity of 7 mJ/cm^2 inactivates bacterial cells by 99% for water disinfection. Protozoa is vulnerable, like bacteria to UV-C damage and consequently, 99% of cryptosporidium are subject to inactivation at 5 mJ/cm^2.

7.3 COUPLING OF TITANIA WITH SEMICONDUCTORS

For achieving effective visible-light photocatalytic activity, the technique of integrating titania with narrow hole semiconductors is obvious. To achieve this, the valence and conduction band potentials of coupled semiconductors must be more negative and less positive than those of titania. This technique ensures separating photoinduced electron–hole pairs through carrier transfer pathways. Reducing the reduction capacity of conduction band electrons of coupled semiconductor and underutilization of high oxidation capacity of valence band holes of titania are two disadvantages of this technique (Gratzel 2001). The critical issues for semiconductor coupling are to locate a narrow band hole semiconductor with appropriate band electronic structure and subsequently keep it in close contact with titania. PbS and CdS are the most normally utilized coupled semiconductor. However, they have light corrosion and excessive charge carrier recombination issues (Wu, Jimmy, and Xianzhi, 2006). Including an electron contributor, for example, sulfide or sulfite to the reaction framework can adequately take care of the erosion issues of the coupled semiconductor.

7.4 MODIFICATIONS OF TIO_2 WITH METAL DEPOSITION

To effectively function under visible light, modified titanium dioxide has been carried out by metal doping. Nature of dopant ion, its level, the method used, type of titanium dioxide used, the reaction for which catalyst is used, and reaction conditions play an important role in photoactivity of metal-doped titanium dioxide photocatalysts. Deposited nanoparticles on the surface of titanium dioxide play an important role in improving the photocatalytic behavior of the material. Figure 7.1 describes the interfacial cycle of electron move from the photoexcited semiconductor to the deposited metal caused by the contact between the surface of a semiconductor and the metal. An increase in the photocatalytic performance of the system is caused by the Schottky barrier which limits the likelihood of recombination of photogenerated charge carriers. Localized surface plasmon resonance (LSPR) properties of the metal present are responsible for shifting the absorption into the visible region (Rodríguez-gonzález et al. 2020).

Researchers believe that metal doping of titanium dioxide causes an overlap of Ti 3d orbitals with the d levels of metals resulting in a shift in the absorption spectrum to longer wavelengths, thus, supporting the utilization of visible light to photoactivate titanium dioxide (Moma and Baloyi 2019). It has been reported that doping of

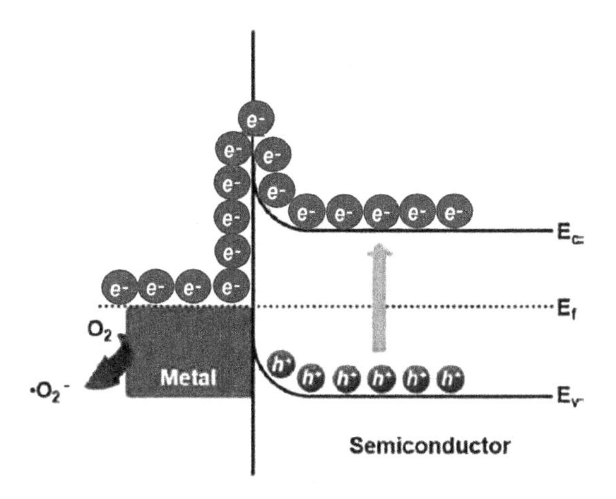

FIGURE 7.1 Illustrative diagram depicting electron shifting in metal–semiconductor interface junction (Rodríguez-gonzález et al. 2020), reprinted with permission from Elsevier.

titanium dioxide with Fe, Mg, Co, Na, and Li enlarges the titanium dioxide visible light response range when ball-milled with high energy and with metal nitrates. It has been learned that in Na doping, Ti was found as both Ti^{4+} and Ti^{3+}. Furthermore, the prevention of recombination of electrons (e^-) and gaps (h^+) was caused by the conversion between Ti^{4+} and Ti^{3+}. The crystal phase changes that caused electrons and holes were improved by metal particle doping (Q. Zhao et al. 2018). To achieve enhanced catalytic performance in photocatalytic water splitting for Pt-doped, mesoporous titanium dioxide doped with various levels of Pt in the range of 1–5 wt% nominal loading and made by the sol–gel method led to a high surface area in titanium dioxide and consequently resulting in an increase in catalytic performance. The optimum catalytic performance was observed at 2.5 wt% for Pt–TiO_2 with a minimum electron–hole recombination rate (Guayaquil-sosa et al. 2017). It has been reported that due to the Schottky barrier at the metal–TiO_2 interface, noble metal nanoparticles also demonstrate effective deterring electron–hole recombination and thus have been used to modify titanium dioxide for photocatalysis purposes. The function of noble metal nanoparticles is to keep and transfer photogenerated electrons from the surface of titanium dioxide to an accepter thus increasing photocatalytic activity and decreasing charge carrier recombination rate. Recently Low, Cheng, and Yu (2017) found that depositing Au on the surface of titanium dioxide resulted in electron shifting from photoexcited Au particles (>420 nm) to the conduction band of titanium dioxide with a decrease in their absorption band (~550 nm). Furthermore, the band recovery was observed by the addition of electron donors like Fe^{2+} and alcohols. In a study conducted by Zhang et al. (2016), it was revealed that the electric field improvement close to the metal nanoparticles could ascribe the coupled Au/TiO_2 visible light activity. Besides, coupling Au and Ag nanoparticles on the surface of titanium dioxide to utilize their properties of localized surface plasmonic resonance (LSPR) in photocatalysis purposes were carried out by various scientists (Sousa-Castillo et al. 2016). Wang et al. (2017) revealed that Pt nanoparticles increased the electron-shifting rate to oxidant thus improving the photocatalytic performance. Various parameters like solution pH, platinum loading (wt%) on titanium dioxide, and light intensities including UV, visible, and sunlight affect the photocatalytic sacrificial hydrogen generation (Chowdhury, Gomaa, and Ray 2015). Besides, utilizing Pt/TiO_2 as a catalyst gives complete staining and dye mineralization and the results are subject to the fact that higher Pt content of photocatalyst prepared with the highest deposition time is guaranteed. The best results for Pt/TiO_2 catalysts were achieved when the catalyst was manufactured by photochemical deposition technique with 120 min deposition time.

Huang et al. (2018) developed Pt/TiO_2 nanoparticles from titanium dioxide (produced at different hydrolysis pH values) and discovered that the titanium dioxide phase acquired relied generally upon the hydrolysis pH. It was revealed that in the Pt/TiO_2 sample, the anatase/rutile intersection showed a lower recombination rate in comparison to the anatase phase of Pt/TiO_2 because of the more extended recombination pathways. Whereas the anatase phase of Pt/TiO_2 was found with more effective degradation efficacy. It was further found that the anatase composition in Pt/TiO_2 framework was important in increasing the photocatalytic degradation of Acid Red 1 dye.

Several noble metals advanced TiO_2 (P25) by wet impregnation were developed by Repousi et al. (2017). They discovered that the distribution of small metal crystallites had no influence on the optical band hole of titanium dioxide. Under sunlight Pt-promoted catalyst most effectively degraded the bisphenol-A. Furthermore, the presence of humic acid also contributed noticeably to improving the reaction rate of Rh/TiO_2. On the other hand, humic acid had a bad effect on the P25 TiO_2 catalyst. It is believed that humic acid is suitable for photosensitizing Rh/TiO_2 catalysts. For decreasing photocatalytic CO_2, Indium doped with TiO_2 results in increased surface area due to the suppression of TiO_2 particle development during the titanium dioxide synthesis. Introducing the impurity level below the conduction band level of titanium dioxide improves the light absorption ability of In-TiO_2. As a result of high surface area and broadened light absorption range, photocatalytic carbon dioxide reduction activity of In-TiO_2 was around eight times that of pure titanium dioxide.

Various research works reveal that transition metals doping with titanium dioxide enhances the photocatalytic activity. As it causes a change in electronic structure resulting in the absorption region that shifts from UV to visible light. The charge transfer transition between d electrons of the transition metals and the valence or conduct band of titanium dioxide nanoparticles cause this shift. Nagi et al. (2020) carried out a comparison of titanium dioxide nanoparticles doping with Fe, Mn, Co, Cr, V, Cu, Ce, Mo, Y, Zr, and Ni and it was revealed that V, Cr, and Fe indicated improved conversions in the visible area whereas inclusion of remaining transition metals, i.e., Zr, Cu, Mn, Ni, Mo, Y, Co, and Ce showed a restraint impact on the photocatalytic activity. For Cr-doped TiO_2 the rate constant was about 8–19 times greater than the rest of the metal-doped catalysts thus showing superior catalytic performance. Furthermore, it was found that reduction peaks in Chromium doped Titanium dioxide moved to much lower temperatures because of the increase in reduction capability of chromium and titania. So, the strong interaction, i.e., the formation of Cr–O–Ti bonds causes the higher effectiveness of Cr/TiO_2 in visible light. For visible light degradation of para-nitrophenol, Fe-doped titanium dioxide nanoparticles were reported to be used. It was discovered that the concentration of Fe-dopant was critically significant in deciding the catalyst activity. It was found that with a 0.05 mol% $Fe^{(3+)}$ molar concentration and without any addition of oxidizing reagents, the maximum para-nitrophenol degradation rate could be as maximum as 92% in 5 h. Maximum separation of photogenerated charge carriers and an increase in threshold wavelength response led to excellent photocatalytic activity (Sood et al. 2015). On the contrary, for humic acid degradation, Fe-doped TiO_2 demonstrated a slowdown effect for the doped catalysts in comparison with bare TiO_2 specimens when assessed for sunlight photocatalytic activity. This could be ascribed to surface complexation reactions instead of the reactions occurring in an aqueous medium. The quicker removal rates achieved by the utilization of bare titanium dioxide could be explained as substrate-specific instead of being related to inefficient visible light-activated catalytic performance (Birben et al. 2017). Ola and Maroto-valer (2015) found that the properties of V-doped titanium dioxide were compatible with visible light due to the replacement of Ti^{4+} by V^{4+} or V^{5+} particles as the V^{4+} is focused at 770 nm whilst the absorption band of V^{5+} is lower than 570 nm. Moradi et al. (2018) studied the impacts of Fe^{3+} doping content on the band hole and size of the nanoparticles and found the high photocatalytic activity of Fe doped titanium dioxide. They also discovered that with an increase in doping content there is a decrease in band hole energy and particle size from 3.3 eV and 13 nm for bare TiO_2 to 2.9 eV and 5 nm for Fe_{10}-TiO_2 accordingly.

The good electron catching properties of rare earth metals doped titanium dioxide can bring stronger absorption edge transfer toward longer wavelength thus acquiring a narrow bandgap. Bhethanabotla, Russell, and Kuhn (2017) discovered that at low loadings under simulated solar irradiation with improvements differing by catalyst composition, the rare earth dopants caused an improvement in the aqueous phase photodegradation of phenol. The improved exhibition of rare-earth-doped samples in contrast to pure titania can be explained by the dissimilarities in defect chemistry on key kinetic steps. Reszczy (Reszczy 2016) arranged a series of Pr^{3+}, Eu^{3+}, Y^{3+} and Er^{3+}, and Eu^{3+} adjusted titanium dioxide nanoparticles photocatalysts and it was found that fusion of RE^{3+} ions into titanium dioxide nanoparticles and caused a blue shift of absorption edges of titanium dioxide nanoparticles. This can be explained by the movement of the conduction band edge above the first excited state of RE^{3+}. Furthermore, fused RE^{3+} ions interact with the electrons of the conduction band of titanium dioxide which results in a higher move from the TiO_2 to RE^{3+} ions at the first excited state. On the other hand, a decrease in crystallite size of RE^{3+}-TiO_2 in contrast with titanium dioxide can be credited by the observed blue shift. Under visible light irradiation the Pr^{3+}, Eu^{3+}, Y^{3+}, and Er^{3+} modified titanium dioxide nanoparticles showed higher activity and could be excited in the range from 420 to 450 nm in contrast with pure P25 TiO_2. In a work similar to this on rare earth's titania nanotubes, (Mazierski et al. 2017) It was discovered that the RE^{3+} species are located at the crystal boundaries instead of inside the titanium dioxide unit cell. Furthermore, an excitation into the titanium dioxide absorption band was observed with subsequent RE^{3+} discharge confirming energy movement between the TiO_2 matrix and RE^{3+}. It was found that the presence of rare earth elements served in reducing the recombination of electrons and holes effectively by getting them and as well by advancing their fast development along the surface of the titanium dioxide nanoparticles. Lanthanide improved the light absorption of the catalyst instead of affecting the energy gap of titanium dioxide nanoparticles. The La^{3+} particle doping tends to increase the surface range of titanium dioxide nanoparticles and lessening the crystallite size thus increasing the adsorption limit of the doped titanium dioxide nanoparticle. It was postulated in view of some theoretical computations that during the electrochemical process new surface vacancies and Ho-f states took place that might decrease the photon excitation energy from valence band to conduction under visible light irradiation. Under visible light irradiation, the photocatalytic movement was ascribed not to OH but to different types of reactive oxygen species, for example, O_2, HO_2, H_2O_2.

7.5 MODIFIED TIO_2 WITH NONMETAL DEPOSITION

Doping titanium dioxide with nonmetals like carbon, iodine, nitrogen, sulfur, and so forth have indicated excessive improvement in the new visible light active photocatalyst. Nitrogen with little ionization energy and high stability is generally used for doping. Furthermore, the atomic size of nitrogen is analogous to oxygen and thus can be introduced in titanium dioxide structure. Asahi et al. (2001) used N-TiO_2 for the photocatalytic degradation of methylene blue and gaseous acetaldehyde in visible light. Subsequently, several research methodologies have been carried out aiming at introducing nitrogen into titanium dioxide structure. Among the techniques used for N-TiO_2 synthesis are the physical methods such as pulsed laser deposition, sputtering, atomic layer deposition, and ion implementation have been used by the researchers for nitrogen-doped titanium dioxide. Yet sol-gel technique is the common procedure for nitrogen-doped titanium dioxide synthesis. The main benefits of this technique include a simple process and low cost. This method facilitates to easily combine controlled morphology and porosity.

Highly ordered nitrogen-doped titanium dioxide nanotube arrays using electrochemical anode oxidation of Ti foil were developed by Liu et al. (2011). Samples were further treated with nitrogen-plasma and subsequently annealed under Ar atmosphere. XPS analysis showed that all infused nitrogen was interstitial nitrogen-doped titanium dioxide. UV-vis DRS indicated the bandgap energy of both pure titanium dioxide nanotubes and nitrogen-doped titanium dioxide as ~3.24 and ~3.03 eV, respectively. Furthermore, for the deterioration of methylene blue under visible light, the photocatalytic activity of titanium dioxide nanotubes and nitrogen-doped titanium dioxide nanotubes was observed 68 and 98% with 90 min activity. Dye concentration 10 mg/L and 350 W Xe lamp were chosen for the experiments. The sol-gel technique can be used for interstitial doped titanium dioxide with three kinds of nitrogen precursors namely diethanolamine, urea, and triethylamine (Ananpattarachai, Kajitvichyanukul, and Seraphin 2009). Diethanolamine was found with the highest visible light absorption capacity of interstitial nitrogen-doped titanium dioxide with the least bandgap energy of 2.85 eV and the smallest anatase crystal size of 4.86 nm. The highest degradation of 2-chlorophenol was observed for nitrogen-doped titanium dioxide against the urea doped and triethylamine doped. Yet there are various perspectives about the nitrogen-doped titanium dioxide photocatalytic activity under visible light. Many researchers think that interstitial nitrogen atoms or NO_X contaminations lead to higher photocatalytic activities (Ananpattarachai, Kajitvichyanukul, and Seraphin 2009) whereas some think that improved photocatalytic activity is caused by oxygen vacancies (Liu et al. 2011). Additionally, donor states situated below the conduction band that permits photoreactions under visible light can also be caused by oxygen vacancies in nitrogen-doped titanium dioxide (Liu et al. 2011). X-ray Photo Spectroscopy (XPS) is a good tool for distinguishing the N doping as a substitutional element on the O lattice or at interstitial lattice sites. The binding energy of N 1 s at around 397 eV represents the substitutional N whereas for exposing interstitial N peaks at >400 eV is given 400 eV and 405 eV to NO and NO_2, respectively (Ananpattarachai, Kajitvichyanukul, and Seraphin 2009). It was found that for the synthesis of nitrogen-doped titanium dioxide, the choice of doping method is of crucial importance and can impact the state of nitrogen in titanium dioxide and its efficiency as well in visible light absorption. Usually, N doping into the titanium dioxide lattice forms bulks oxygen vacancies that do the job of recombination hubs for carriers and result in diminishing photocatalytic efficiency (Jinlong Zhang et al. 2010). For visible light activity in titanium dioxide, phosphorous, carbon, and sulfur as dopants are found to be very effective. Narrowing the bandgap of titanium dioxide, nonmetal dopants are also found to be effective. Bandgap narrowing is caused by factors like change of lattice parameters and the presence of trap states inside conduction and valence bands from electronic disturbances (Hamal and Klabunde 2007). This not only increases the lifetime of photogenerated charge carriers by the presence of trap sites but also permits visible light absorption.

The bigger ionic radius of sulfur makes effective inclusion of it into the titanium dioxide lattice when compared with nitrogen. Compared to sulfur ionic structure (S^{2-}) lattice, the inclusion of cationic sulfur (S^{6+}) represents chemically advantageous behavior. Utilizing a simple sol-gel method, cationic sulfur and ionic nitrogen co-doped with titanium dioxide has been produced from a single source as well (Periyat et al. 2009). Development of S-doped titanium dioxide with modification of titanium isopropoxide with sulfuric acid has been reported by Periyat et al. According to them, at higher temperatures equal or greater than 800°C titanyl oxysulfate forms which in turn results in retention of anatase. They also found that increasing visible light photocatalytic activity of the synthesized materials was caused by the presence of sulfur (Periyat et al. 2008). Based on a unique sol-gel procedure that employed a self-assembly method with a nonionic surfactant to control nanostructure and H_2SO_4 as an inorganic sulfur source, visible light-activated sulfur-doped titanium dioxide was effectively produced in recent times (Han et al. 2011).

7.6 TIO_2 FRAMEWORKS WITH GRAPHENE AND DIFFERENT CARBONEOUS MATERIALS

Research interests for photocatalytic self-cleaning applications based on graphene have been noticed in recent years. Graphene has several advantages including high surface area, quick electron transfer capabilities, brilliant conductivity, and high transparency as well as the good capacity of accepting electrons from the conduction bands of metal oxides. The graphene/metal oxide hybrids offer numerous favorable circumstances, for example, the two-dimensional nature of grapheme that facilitates an expanded surface area, an improved charge separation capacity and decreased electron–hole pair recombination, and enhanced adsorption of contaminants due to the strong π–π interaction. It is a well-known fact that various investigations based on titanium dioxide as a photocatalyst have been reported in the literature. But, the quick recombination of photoexcited electron–hole pair and the absence of visible light absorption restrict the titanium dioxide efficacy. To overcome this problem, loading titanium dioxide with noble metals has been carried out by different researchers. Loaded titanium dioxide with noble metals serves the purpose of diminishing oxygen to H_2O, peroxides, and other species thus expanding recombination time (Lu et al. 2013). As the noble metal-doped titanium dioxide is an expensive process, the advancement of TiO_2/graphene hybrids has got more attention from the scientific society. Doping titanium dioxide with graphene ensures high electron acceptor abilities of graphene thus expanding recombination time.

Anandan et al. have researched the self-cleaning properties of TiO_2/graphene hybrids (Anandan et al. 2013). According to them, the resistivity of titanium dioxide was observed high as compared to TiO_2/graphene hybrid. It was also observed that the resistivity of TiO_2/graphene hybrid was analogous to that of a graphene layer. Improvement in photocatalytic activity of graphene-doped titanium dioxide was observed to be double compared to the corresponding titanium dioxide film. It was also observed that increasing graphene concentration caused an improvement in photocatalytic efficacy of the material and could be explained by the electron transfer abilities of the graphene layer. Besides, higher hydrophilic conversion efficacy of hybrid was also observed compared to titanium dioxide film (Anandan et al. 2013).

It was reported that the presence of highly conductive graphene expanded the capacity of photogenerated holes in titanium dioxide which in turn increased the oxidative capacity to deteriorate the surface contaminants. Titanium dioxide nanorods decorated with graphene sheets have been produced followed by photocatalytic studies under visible light irradiation for the elimination of contamination. Akhavan and Ghaderi (2009) reported on the preparation of graphene–titanium dioxide films for killing over 99.9% of *Escherichia coli* bacteria under sunlight irradiation in an aqueous solution. These films were produced through depositing GO platelets on TiO_2 thin films. It was also observed that the graphene platelets remained chemically stable after photoinactivation of bacteria and that the antibacterial activity of graphene–TiO_2 films increased by about 6 and 7.5 times related to the activity of annealed graphene–titanium dioxide and the simple titanium dioxide thin film correspondingly. The necessary properties of individual semiconductors can be additionally upgraded by utilizing them in combination with different materials to make new composites or hybrids. Such materials are those built from carbon nanotubes (CNTs) and TiO_2 or ZnO. CNTs enable enhancement of electron flow from the semiconductor oxide to the CNT, thus becoming appealing for the manufacturing of semiconductor hybrids. Negatively charged species are created on the surface of CNT positively charged species in semiconductor material due to the movement of electrons from the metal oxide. A depletion layer is formed by the surplus negative and positive charges formed at the interface that conserves charge separation in the hybrid materials (Woan, Pyrgiotakis, and Sigmund 2009). The huge surface area of CNTs and charge transfer ability increase the electron–hole pair recombination time by catching electrons produced during the photocatalysis process. It was observed that each of the 32 carbons could store one electron leading to the fact of the nanohybrids' good capacity for storing numerous such electrons resulting in increasing recombination time. Utilizing laser flash photolysis, it was estimated that the recombination time of electron–hole pairs in CNT-metal oxides had approximately 10^{-9} s whereas the time required for the degradation of contaminants was recorded in the range of 10^{-8}–10^{-3} s (Kongkanand and Kamat 2007). The composition of TiO_2 and CNT in the preparation of a hybrid plays an important role in photocatalytic activity. Effect of the multiwalled nanotube (MWNT)/TiO_2 hybrids under ultraviolet and visible light utilizing model dyes have been observed in recent years. Additional amounts of MWNTs disperse the photons and protect the titanium dioxide from UV absorption. On the other hand, an extra amount of titanium dioxide decreases the electron catching capacity of CNTs by yielding an enormous gap between titanium dioxide particles and CNTs. Hence it is necessary to make some ideal percentage arrangement to achieve the highest photocatalytic activity.

7.7 COMMERCIAL APPLICATIONS OF TIO_2

For improved antifogging and self-cleaning commercial applications, photoinduced hydrophilic change of titanium dioxide can be exploited. Examples include paints, cement, fabrics, and glass. Recently in the construction industry, the use of tiles and glass windows with photocatalytic self-cleaning properties can be observed. Utilization of titanium dioxide photocatalysis thin films are reported in such materials (Radeka, Markov, Loncar, Rudic, Vucetic, and Ranogajec, 2014). A spray of a liquid suspension of titanium dioxide onto a surface is used to produce photocatalytic goods. For this purpose, surfaces are compacted without melting at 600–800°C to firmly fix the titanium dioxide layer on the surfaces (Shimohigoshi and Saeki 2007). Self-cleaning coating items for business uses require high treating temperatures. Therefore, dynamic anatase titanium dioxide having high-temperature permanency is inevitable. Thermal stability of the anatase phase can be enhanced by nonmetallic doping (Periyat et al. 2008). On the other hand, it encourages electron–hole recombination and decreases photocatalytic activity. In recent times, stable anatase phase up to 900°C can be made possible by developing visible light active oxygen-rich titanium dioxide (Etacheri et al. 2011). In self-cleaning construction materials, the high thermal stability of anatase–titanium dioxide is very beneficial. On the surface of China National Opera Hall, self-cleaning glass prepared by coating titanium dioxide nanoparticles has been applied. Misericordia Church in Italy was applied with white cement having titanium dioxide nanoparticles. For indoor and outdoor applications in the building industry, titanium dioxide represents enormous potential due to its photoinduced antimicrobial activity. The development of ceramics having a thin film of semiconductor photocatalysts inside has been produced by some organizations with photoinduced antimicrobial and cleaning properties. Due to the self-cleaning properties of titanium dioxide, the focus of researchers has gained attention toward textile-related applications (Radeti 2013). Cellulose fibers, polyesters, polyamides, etc. covered with titanium dioxide nanoparticles have been developed for photocatalytic and self-cleaning textile fibers. For accomplishing functionalization of textile fibers with titanium dioxide, the carboxylic acid is used that can coordinate to titanium atom and can tie to titanium dioxide as well via H-bonding with lattice oxygen or surface hydroxyl group (Radeti 2013). For UV protection of fabric, adding nanocrystalline titanium dioxide plays an important factor. It was also observed that it could bear several home-washing cycles (Daoud, Xin, and Zhang 2005). Antibacterial properties and high stain removal were reported for fibers coated with titanium dioxide (Daoud, Xin, and Zhang 2005). For the design of self-cleaning cotton fibers, succinic acid is used as a crosslinking agent when cotton fibers with MWNTs and titanium dioxide are treated with cotton (Karimi, Zohoori, and Amini 2014). Under sunlight and UV, the photocatalytic activity of fiber cotton can be considerably enhanced by a simultaneous coating of fiber with titanium dioxide and MWNTs Abrasion resistance and UV blocking capacity of cotton fibers can also be improved by titanium dioxide-multiwalled CNTs. Development of self-cleaning textiles using hydrophilic titanium dioxide-coated surfaces has also been produced.

Advanced antimicrobial and self-cleaning activity can be achieved by the consolidated treatment of polyester fabric with titanium dioxide nanoparticles and colloidal silver nanoparticles (Vodnik et al. 2011). Polyester fabric showed excellent effectiveness for photochemical annihilation of methylene blue and antimicrobial action when treated with $AgCl/TiO_2/AgI$ nanocomposites (Rehan et al. 2013). Sensible designing of photocatalysts using indoor irradiation and sunlight gives rise to self-cleaning materials.

7.8 CONCLUSION

This chapter represents an overview of the fundamentals of photocatalysis and reviews briefly the most relevant strategies to improve photocatalytic activity. TiO_2 is currently the most used photocatalyst for disinfection purposes. TiO_2 is attractive in disinfecting microorganisms due to its high stability, good location of the band edges, low charge transport resistance, high photocatalytic activity, high thermal and chemical stability, low toxicity, and low price. On the other hand, efforts are being made to increase its photocatalytic ability to absorb visible light. TiO_2 doping with other materials was studied as well as the use of TiO_2/grapheme composite photocatalysts. Doping of TiO_2 with metals also decrease charge recombination and reduce the redox overpotentials. Besides the limitations of photocatalysts, the absence of knowledge on the long-time effect of photoinactivation on microorganisms is a matter of concern.

ACKNOWLEDGEMENT

This work was supported by the Ministry of Education, Youth and Sports of the Czech Republic and the European Union – European Structural and Investment Funds in the frames of Operational Programme Research, Development and Education under project Hybrid Materials for Hierarchical Structures [HyHi, Reg. No. CZ.02.1.01/0.0/0.0/16_019/0000843].

REFERENCES

Agustina, T E, H M Ang, and V K Vareek. 2006. "A Review of Synergistic Effect of Photocatalysis and Ozonation on Wastewater Treatment." *Journal of Photochemistry and Photobiology C: Photochemistry Reviews* 6: 264–273. doi: 10.1016/j.jphotochemrev.2005.12.003.

Akhavan, O, and E Ghaderi. 2009. "Photocatalytic Reduction of Graphene Oxide Nanosheets on TiO_2 Thin Film for Photoinactivation of Bacteria in Solar Light Irradiation." *The Journal of Physical Chemistry C* 113 (47): 20214–20220. doi: 10.1021/jp906325q.

Anandan, Srinivasan, Tata Narasinga Rao, Marappan Sathish, Dinesh Rangappa, Itaru Honma, and Masahiro Miyauchi. 2013. "Superhydrophilic Graphene-Loaded TiO_2 Thin Film for Self-Cleaning Applications." *ACS Applied Materials & Interfaces* 5 (1): 207–212. doi: 10.1021/am302557z.

Ananpattarachai, Jirapat, Puangrat Kajitvichyanukul, and Supapan Seraphin. 2009. "Visible Light Absorption Ability and Photocatalytic Oxidation Activity of Various Interstitial N-Doped TiO_2 Prepared from Different Nitrogen Dopants." *Journal of Hazardous Materials* 168: 253–261. doi: 10.1016/j.jhazmat.2009.02.036.

Asahi, R, T Morikawa, T Ohwaki, K Aoki, and Y Taga. 2001. "Visible-Light Photocatalysis in Nitrogen-Doped Titanium Oxides." *Science* 293 (July): 269–272. www.sciencemag.org.

Bhethanabotla, Vignesh C, Daniel R Russell, and John N Kuhn. 2017. "Assessment of Mechanisms for Enhanced Performance of Yb/Er/Titania Photocatalysts for Organic Degradation: Role of Rare Earth Elements in the Titania Phase." *Applied Catalysis B, Environmental* 202: 156–164. doi: 10.1016/j.apcatb.2016.09.008.

Birben, Nazmiye Cemre, Ceyda Senem Uyguner-demirel, Sibel Sen Kavurmaci, Yelda Yalc, Nazli Turkten, Zekiye Cinar, and Miray Bekbolet. 2017. "Application of Fe-Doped TiO_2 Specimens for the Solar Photocatalytic Degradation of Humic Acid." *Catalysis Today* 281: 78–84. doi: 10.1016/j.cattod.2016.06.020.

Chowdhury, Pankaj, Hassan Gomaa, and Ajay K Ray. 2015. "Sacrificial Hydrogen Generation from Aqueous Triethanolamine with Eosin Y-Sensitized Pt/TiO_2 Photocatalyst in UV, Visible and Solar Light Irradiation." *Chemosphere* 121 (February): 54–61. doi: 10.1016/j.chemosphere.2014.10.076.

Daoud, Walid A, John H Xin, and Yi-he Zhang. 2005. "Surface Functionalization of Cellulose Fibers with Titanium Dioxide Nanoparticles and Their Combined Bactericidal Activities." *Surface Science* 599: 69–75. doi: 10.1016/j.susc.2005.09.038.

Etacheri, Vinodkumar, Michael K Seery, Steven J Hinder, and Suresh C Pillai. 2011. "Oxygen Rich Titania: A Dopant Free, High Temperature Stable, and Visible-Light Active Anatase Photocatalyst." *Advanced Functional Materials*. 21: 3744–3752. doi: 10.1002/adfm.201100301.

Gibson, Kristen E. 2003. "Viral Pathogens in Water : Occurrence, Public Health Impact, and Available Control Strategies." *Current Opinion in Virology* 4: 50–57. doi: 10.1016/j.coviro.2013.12.005.

Gratzel, Michael. 2001. "Photoelectrochemical Cells." *Nature* 414: 338–344. doi: 10.1038/35104607.

Guayaquil-sosa, J F, Benito Serrano-rosales, P J Valadés-pelayo, and H De Lasa. 2017. "Applied Catalysis B: Environmental Photocatalytic Hydrogen Production Using Mesoporous TiO_2 Doped with Pt." *Applied Catalysis B, Environmental* 211: 337–348. doi: 10.1016/j.apcatb.2017.04.029.

Hamal, Dambar B, and Kenneth J Klabunde. 2007. "Synthesis , Characterization , and Visible Light Activity of New Nanoparticle Photocatalysts Based on Silver, Carbon, and Sulfur-Doped TiO_2." *Journal of Colloid and Interface Science* 311: 514–522. doi: 10.1016/j.jcis.2007.03.001.

Han, Changseok, Miguel Pelaez, Vlassis Likodimos, Athanassios G Kontos, Polycarpos Falaras, Kevin O Shea, and Dionysios D Dionysiou. 2011. "Applied Catalysis B : Environmental Innovative Visible Light-Activated Sulfur Doped TiO_2 Films for Water Treatment." *Applied Catalysis B, Environmental* 107 (1–2): 77–87. doi: 10.1016/j.apcatb.2011.06.039.

Hijnen, W A M, G M H Suylen, J A Bahlman, A Brouwer-hanzens, and G J Medema. 2010. "GAC Adsorption Filters as Barriers for Viruses , Bacteria and Protozoan (Oo) Cysts in Water Treatment." *Water Research* 44 (4): 1224–1234. doi: 10.1016/j.watres.2009.10.011.

Huang, Winn-jung, Guor-cheng Fang, and Chun-chen Wang. 2005. "The Determination and Fate of Disinfection By-Products from Ozonation of Polluted Raw Water." *Science of the Total Environment* 345 345: 261–272. doi: 10.1016/j.scitotenv.2004.10.019.

Huang, Bing-Shun, En-Chin Su, Yun-Ya Huang, and Hui-Hsin Tseng. 2018. "Tailored Pt/TiO_2 Photocatalyst with Controllable Phase Prepared via a Modified Sol-Gel Process for Dye Degradation." *Journal of Nanoscience and Nanotechnology* 18 (3): 2235—2240. doi: 10.1166/jnn.2018.14237.

Karimi, Loghman, Salar Zohoori, and Atefeh Amini. 2014. "Multi-Wall Carbon Nanotubes and Nano Titanium Dioxide Coated on Cotton Fabric for Superior Self-Cleaning and UV Blocking." *New Carbon Materials* 29 (5): 380–385. doi: 10.1016/S1872-5805(14)60144-X.

Keane, Donal A., Kevin G. McGuigan, Pilar Fernández Ibáñez, M. Inmaculada Polo-López, Anthony J. Byrne, Patrick S.M. Dunlop, Kevin O'Shea, Dionysios D. Dionysiou and Suresh C. Pillai. 2014. "Solar Photocatalysis for Water Disinfection: Materials and Reactor Design." *Catalysis Science and Technology*, 5: 1211–1226.

Kongkanand, Anusorn, and Prashant V Kamat. 2007. "Electron Storage in Single Wall Carbon Nanotubes. Fermi Level Equilibration in Semiconductor–SWCNT Suspensions." *ACS Nano* 1 (1): 13–21. doi: 10.1021/nn700036f.

Liu, Xu, Zhongqing Liu, Jian Zheng, Xin Yan, Dandan Li, Si Chen, and Wei Chu. 2011. "Characteristics of N-Doped TiO_2 Nanotube Arrays by N_2-Plasma for Visible Light-Driven Photocatalysis." *Journal of Alloys and Compound* 509: 9970–9976.

Low, Jingxiang, Bei Cheng, and Jiaguo Yu. 2017. "Applied Surface Science Surface Modification and Enhanced Photocatalytic CO_2 Reduction Performance of TiO_2 : A Review." *Applied Surface Science* 392: 658–686. doi: 10.1016/j.apsusc.2016.09.093.

Lu, Qipeng, Zhenda Lu, Yunzhang Lu, Longfeng Lv, Yu Ning, Hongxia Yu, Yanbing Hou, and Yadong Yin. 2013. "Photocatalytic Synthesis and Photovoltaic Application of Ag-TiO_2 Nanorod Composites." *Nano Letters* 13: 5698–5702.

Magalh, Pedro, Luísa Andrade, Olga C Nunes, and Adélio Mendes. 2017. "Titanium Dioxide Photocatalysis : Fundamentals and Application on Photoinactivation." *Reviews on Advanced Materials Science*. 51: 91–129.

Mazierski, Paweł, Wojciech Lisowski, Tomasz Grzyb, Michał Jerzy, Tomasz Klimczuk, Alicja Mikołajczyk, Jakub Flisikowski, et al. 2017. "Enhanced Photocatalytic Properties of Lanthanide-TiO_2 Nanotubes : An Experimental and Theoretical Study." *Applied Catalysis B, Environmental* 205: 376–385. doi: 10.1016/j.apcatb.2016.12.044.

John Moma and Jeffrey Baloyi. 2019. "Modified Titanium Dioxide for Photocatalytic Applications." In *Photocatalysts – Applications and Attributes*, edited by Sher Bahadar Khan. Intech Open. https://www.intechopen.com/books/photocatalysts-applications-and-attributes/modified-titanium-dioxide-for-photocatalytic-applications.

Moradi, Vahid, Martin B G Jun, Arthur Blackburn, and Rodney A Herring. 2018. "Significant Improvement in Visible Light Photocatalytic Activity of Fe Doped TiO_2 Using an Acid Treatment Process." *Applied Surface Science* 427: 791–799. doi: 10.1016/j.apsusc.2017.09.017.

Nagi, Siva, Reddy Inturi, Thirupathi Boningari, and Makram Suidan. 2020. "Environmental Visible-Light-Induced Photodegradation of Gas Phase Acetonitrile Using Aerosol-Made Transition Metal (V, Cr, Fe, Co, Mn, Mo, Ni, Cu, Y, Ce, and Zr) Doped TiO_2." *Applied Catalysis B, Environmental* 144 (2014): 333–342. doi: 10.1016/j.apcatb.2013.07.032.

Nieuwenhuijsen, Mark J, Mireille B Toledano, Naomi E Eaton, John Fawell, and Paul Elliott. 2000. "Chlorination Disinfection Byproducts in Water and Their Association with Adverse Reproductive Outcomes: A Review." *Occupational and Environmental Medicine* 57: 73–85.

Ola, Oluwafunmilola, and M Mercedes Maroto-valer. 2015. "General Transition Metal Oxide Based TiO_2 Nanoparticles for Visible Light Induced CO_2 Photoreduction." *Applied Catalysis A, General* 502: 114–121. doi: 10.1016/j.apcata.2015.06.007.

Periyat, Pradeepan, Suresh C Pillai, Declan E McCormack, John Colreavy, and Steven J Hinder. 2008. "Improved High-Temperature Stability and Sun-Light-Driven Photocatalytic Activity of Sulfur-Doped Anatase TiO_2." *The Journal of Physical Chemistry C* 112 (20): 7644–7652. doi: 10.1021/jp0774847.

Periyat, Pradeepan, Declan E Mccormack, Steven J Hinder, and Suresh C Pillai. 2009. "One-Pot Synthesis of Anionic (Nitrogen) and Cationic (Sulfur) Codoped High-Temperature Stable, Visible Light Active, Anatase Photocatalysts." *The Journal of Physical Chemistry C* 113: 3246–3253. doi: 10.1021/jp808444y.

Radeka, Miroslava, S. Markov, E. Loncar, O. Rudic, S. Vucetic, J. Ranogajec. 2014. "Photocatalytic Effects of TiO_2 Mesoporous Coating Immobilized on Clay Roofing Tiles." *Journal of the European Ceramic Society* 34: 127–136. doi: 10.1016/j.jeurceramsoc.2013.07.010.

Radeti, Maja. 2013. "Functionalization of Textile Materials with TiO_2 Nanoparticles." *Journal of Photochemistry and Photobiology C: Photochemistry Reviews* 16: 62–76. doi: 10.1016/j.jphotochemrev.2013.04.002.

Rehan, Mohamed, Andreas Hartwig, Matthias Ott, Linda Gätjen, and Ralph Wilken. 2013. "Surface & Coatings Technology Enhancement of Photocatalytic Self-Cleaning Activity and Antimicrobial Properties of Poly (Ethylene Terephthalate) Fabrics." *Surface & Coatings Technology* 219: 50–58. doi: 10.1016/j.surfcoat.2013.01.003.

Repousi, Vasiliki, Athanasia Petala, Zacharias Frontistis, Maria Antonopoulou, Ioannis Konstantinou, Dimitris I Kondarides, and Dionissios Mantzavinos. 2017. "Photocatalytic Degradation of Bisphenol A over Rh /TiO_2 Suspensions in Different Water Matrices." *Catalysis Today* 284: 59–66. doi: 10.1016/j.cattod.2016.10.021.

Reszczy, Joanna. 2016. "Photocatalytic Activity and Luminescence Properties of RE^{3+} – TiO_2 Nanocrystals Prepared by Sol–Gel and Hydrothermal Methods." *Applied Catalysis B: Environmental J* 181: 825–837. doi: 10.1016/j.apcatb.2015.09.001.

Rodríguez-gonzález, Vicente, Sergio Obregón, Olga A Patrón-soberano, and Chiaki Terashima. 2020. "An Approach to the Photocatalytic Mechanism in the TiO_2-Nanomaterials Microorganism Interface for the Control of Infectious Processes." *Applied Catalysis B: Environmental* 270 (January): 118853. doi: 10.1016/j.apcatb.2020.118853.

Rutala, W A, and D J Weber. 1997. "Uses of Inorganic Hypochlorite (Bleach) in Health-Care Facilities." *Clinical Microbiology Reviews* 10 (4): 597–610. doi: 10.1128/CMR.10.4.597.

Shimohigoshi, Mitsuhide, Yoshimitsu Saeki. 2007. "Research and Applications of Photocatalyst Tiles." In *International RILEM Symposium on Photocatalysis, Environment and Construction Materials – TDP 2007*, edited by P. Baglioni and L. Casssar, 291–297. RILEM Publications SARL, Paris, France.

Sood, Swati, Ahmad Umar, Surinder Kumar Mehta, and Sushil Kumar Kansal. 2015. "Highly Effective Fe-Doped TiO_2 Nanoparticles Photocatalysts for Visible-Light Driven Photocatalytic Degradation of Toxic Organic Compounds." *Journal of Colloid and Interface Science* 450 (July): 213—223. doi: 10.1016/j.jcis.2015.03.018.

Sousa-Castillo, Ana, Miguel Comesaña-Hermo, Benito Rodríguez-González, Moisés Pérez-Lorenzo, Zhiming Wang, Xiang-Tian Kong, Alexander O Govorov, and Miguel A Correa-Duarte. 2016. "Boosting Hot Electron-Driven Photocatalysis through Anisotropic Plasmonic Nanoparticles with Hot Spots in Au–TiO_2 Nanoarchitectures." *The Journal of Physical Chemistry C* 120 (21): 11690–11699. doi: 10.1021/acs.jpcc.6b02370.

Vodnik, Vesna, Branislav Potkonjak, Darka Mihailovic, Jovan M Nedeljkovic, Maja Radetic, and Petar Jovanc. 2011. "Multifunctional PES Fabrics Modified with Colloidal Ag and TiO_2 Nanoparticles." *Polym. Adv. Technol.* 22: 2244–2249. doi: 10.1002/pat.1752.

Wang, Fenglong, Roong Jien Wong, Jie Hui Ho, Yijiao Jiang, and Rose Amal. 2017. "Sensitization of Pt/TiO_2 Using Plasmonic Au Nanoparticles for Hydrogen Evolution under Visible-Light Irradiation." *ACS Applied Materials & Interfaces* 9 (36): 30575–30582. doi: 10.1021/acsami.7b06265.

Woan Karran, Georgios Pyrgiotakis, and Wolfgang Sigmund. 2009. "Photocatalytic Carbon-Nanotube–TiO_2 Composites." *Advanced Materials* . 21: 2233–2239. doi: 10.1002/adma.200802738.

Wu, Ling, Jimmy C Yu, and Xianzhi Fu. 2006. "Characterization and Photocatalytic Mechanism of Nanosized CdS Coupled TiO_2 Nanocrystals under Visible Light Irradiation." *Journal of Molecular Catalysis A: Chemical* 244: 25–32. doi: 10.1016/j.molcata.2005.08.047.

Zhang, Jinlong, Yongmei Wu, Mingyang Xing, Sajjad Ahmed Khan Leghari, and Shamaila Sajjad. 2010. "Development of Modified N-Doped TiO_2 Photocatalyst with Metals, Nonmetals, and Metal Oxides." *Energy Environ. Sci.* 3 (6): 715–726. doi: 10.1039/B927575D.

Zhang, Jianming, Xin Jin, Pablo I Morales-Guzman, Xin Yu, Hong Liu, Hua Zhang, Luca Razzari, and Jerome P Claverie. 2016. "Engineering the Absorption and Field Enhancement Properties of Au–TiO_2 Nanohybrids via Whispering Gallery Mode Resonances for Photocatalytic Water Splitting." *ACS Nano* 10 (4): 4496–4503. doi: 10.1021/acsnano.6b00263.

Zhao, Qianfei, Mei Wang, He Yang, Dai Shi, and Yuzheng Wang. 2018. "Preparation, Characterization and the Antimicrobial Properties of Metal Ion-Doped TiO_2 Nano-Powders." *Ceramics International* 44 (5): 5145–5154. doi: 10.1016/j.ceramint.2017.12.117.

8 Kinetic Model for Disinfection by Using of Photooxidation

Qingyan Peng,
Mohanapriya Venkataraman and Jiri Militký
Technical University of Liberec, Czech Republic

CONTENTS

8.1 INTRODUCTION

Phylogenetic analysis indicated that novel coronavirus NKS belonged to the subgenus Sarbecovirus of the genus Betacoronavirus. Due to the similarity between NKS and SARS-CoV, putting on masks and respirators and self-quarantining at home are used as guidelines. As for public practice medical masks, also known as surgical masks, protect wearers from the spread of the virus, especially when exposed face-to-face and in large droplets and sprays. However, no matter which protective equipment is used, it is almost impossible to sterilize the public indoor area of a large mobile population every minute, using traditional disinfection techniques. Common methods of indoor pollutant control include controlling the source of the virus, diluting the virus by ventilation, and using air purification. While source control is considered to be the most effective of these measures, in many cases source control is either impractical or costly. Ventilation, combined with proper air treatment, is an effective way to create an indoor environment that is acceptable for temperature, humidity, and airflow, and to eliminate indoor viruses and pollutants. However, traditionally, building ventilation systems are designed to maintain thermal comfort while controlling carbon dioxide concentrations and odors. Effective removal of viruses and contaminants is generally not a design criterion in nonindustrial buildings. One possible solution to indoor air quality problems is to increase ventilation rates. However, dilution of the mean pollution concentration does not solve the problem of local high pollutant emissions or poor air circulation. Besides, increased ventilation may not be economically feasible, and energy can be wasted. Other general techniques used for air virus cleaners include mechanical filters, electronic (electrostatic) precipitators, mixing filters, ozone generators, and so on. But these technologies move the virus to another stage rather than eliminate it; additional disposal or disposal steps are required. Electronic filters are usually low in energy costs, and deposition batteries are reusable, avoiding the cost of replacing the filter periodically. However, they are often less efficient to use, and sediment cells need to be cleaned frequently. They are more expensive than mechanical filters and can produce ozone as a by-product. Ozone in indoor air can be used to control virus growth and activity. However, ozone is of concern when concentrations in humans reach a certain level (Boeniger 1995). Thus, another remediation technique with many advantages over conventional techniques is the use of photocatalytic oxidation (PCO). Operating at room temperature, the PCO can degrade a large number of pollutants into harmless end products, such as CO_2 and H_2O_2, without requiring significant energy input. In this chapter, we summarize the current knowledge and dynamics of PCO virus disinfection based on the latest literature to find a possible way to limit NKS activity.

8.2 FUNDAMENTAL OF PCO

Photogenic decomposition of titanium dioxide (TiO_2) has been first observed by Fujishima and Honda in the early 1970s (Fujishima and Honda 1972). In the last three decades, PCO technology has attracted great attention for removal of gaseous pollutants at low concentrations (i.e., parts per billion [ppb] level), due to its superior properties, such as room temperature, activity against various pollutants, and benign final products (CO_2 and H_2O) (Goswami, Trivedi, and Block 1997; Zhao and Yang 2003). PCO technology is based on the use of semiconductor catalysts such as TiO_2 and UV to convert toxic viruses or compounds into less hazardous products. In PCO, a key step is to form electron and hole pairs (e^- and h^+), which require semiconductor lighting, photons that absorb enough energy, and electron valence bands to rise to the conduction band (Chen, Nanayakkara, and Grassian 2012). Light is used to trigger photocatalysis by firing photons at the surface of the photocatalyst (Lammel et al. 2013). For photocatalytic surface radiations with compatible wavelength (larger than the bandgap of photocatalytic material) excited electrons (e^-) of the photocatalyst conduction band, see Figure 8.1 (Teta et al. 2007). When the e^- leaves the valence band and is absorbed by the conduction band (e^-_{CB}), a positive hole is formed on the valence band (h^+_{VB}) (Equation (8.1) and Figure 8.1).

$$Photocatalyst + hv \Rightarrow b^+_{VB} + e^-_{CB} \tag{8.1}$$

$$e^-_{CB} + O_2 \Rightarrow O_2^- \tag{8.2}$$

$$h^+_{VB} + H_2O \Rightarrow OH + H^+ \tag{8.3}$$

The e^-reacts with oxygen (O_2) causing the formation of superoxide radicals (O_2^-) or hydroperoxide radicals (HO_2) in the conduction band (e^-_{CB}), (Equation (8.2) and Figure 8.1) (Garrido and Kroemer 2004). The contaminants and viruses are photodegraded by the reactive oxygen species (ROS) into water (H_2O) and carbon dioxide (CO_2). Ulterior degradation can arise by using superoxide radicals (Zhao et al. 2011). The oxidation reaction produces hydroxyl radicals (OH) and hydrogen ions (H^+) (Equation (8.3) and Figure 8.1).

Some semiconductors and materials have enough bandgap energy to stimulate the action of the PCO (Zhao and Yang 2003), and the most widely used photocatalyst is TiO_2 formulated based on the anatase and rutile crystal phases. TiO_2 is a semiconductor with an interband gap of 3.2 eV or more that is produced on the TiO_2 surface of an electron–hole pair once the illumination is less than 385 nm. In the valence band, the hole reacts with H_2O or hydroxide ions adsorbed on the surface to produce hydroxyl

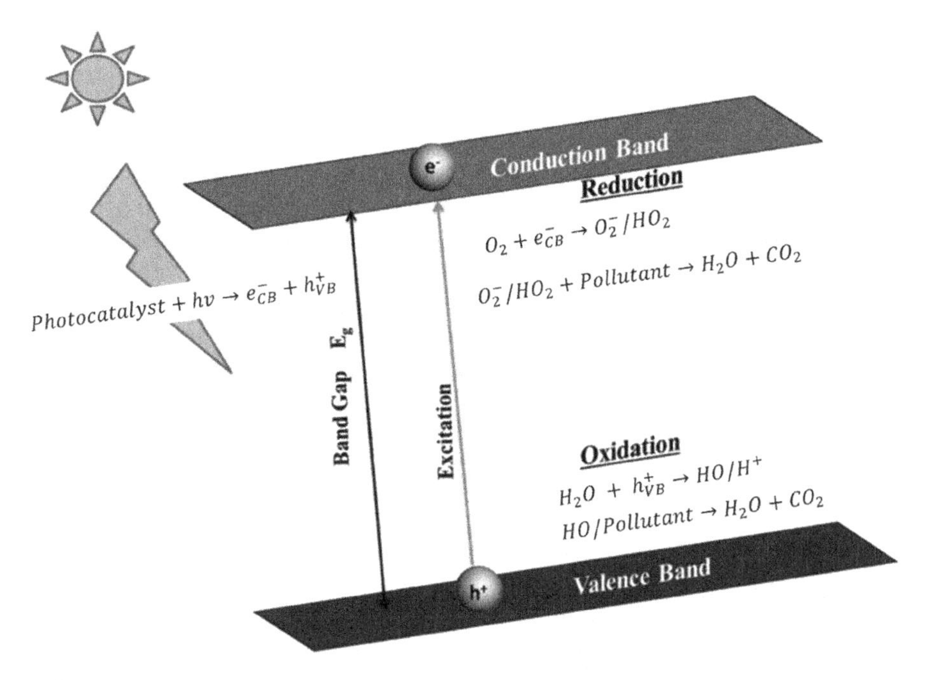

FIGURE 8.1 Mechanism of photocatalytic (reprinted from (Teta et al. 2007), with kind permission of Elsevier Publications).

radicals. In the conduction band, the electron reduces O_2 to produce superoxide ions. The contact reaction between the hydroxyl radical and organic compound is strong. There are also reports of the formation of monofilament and hydrogen peroxide on irradiated TiO_2 (Minero et al. 2000). Except for TiO_2, Fe_2O_3, ZnS, ZnO, SnO_2, CeO_2, WO_3, and ZrO_2 are used for PCO technology. ROS can cause a variety of damage to the cell or viral functions or structures because they contain a large number of organic compounds.

8.2.1 Biological Effects of PCO

The basic condition of PCO is that photons, water molecules, and pollutants are present on the surface of the photocatalyst. With the proper light source design and arrangement, we can ensure that photons reach the photocatalyst surface. The higher the relative humidity of water molecules in the air, the better the PCO effect that can be achieved. However, very high RH causes photocatalysts to be covered by too many water molecules so that pollutants and viruses in the airflow are less likely to contact the catalyst surface. Therefore, the RH value of air should be optimal to achieve maximum PCO effect, especially for air disinfection. Since microorganisms are biological entities with self-defense and self-healing mechanisms, partial or incomplete damage to them can be repaired. Very low RH decreases the possibility of killing the hydroxyl group of the microbe, and very high RH decreases the possibility of contact between the microbe and a hydroxyl group. In 1985, Matsunaga and his colleagues reported for the first time that microbial cells could be killed by a photocatalytic reaction with TiO_2 (Matsunaga et al. 1985). The photocatalytic bactericidal sensitivity of P25 and a metal halide lamp for four microorganisms was studied. *Escherichia Coli* ($Gram^-$), *Lactobacillus acidophilus* ($Gram^+$), *Chlorella vulgaris* (algae), and *Saccharomyces cerevisiae* (yeast) were deactivated to the extent of 20%, 100%, 45%, and 54%, respectively. The first three's inactivation time was 60 min and the last required 120 min. This is roughly related to the thickness of the cell wall. The basic condition of PCO is that photons, water molecules, and pollutants should be present on the surface of the photocatalyst. The photons can reach the photocatalyst surface by reasonable light source design and arrangement. The higher the relative humidity of water molecules in the air, the more effective the PCO effect. However, when RH reaches a high level, resulting in the photocatalyst being covered by too many water molecules, the chance of air pollutants reaching the catalyst's surface decreases. Therefore, the RH value of air should be optimal to achieve maximum PCO effect, especially for air disinfection. Since microorganisms are biological entities with self-defense and self-healing mechanisms, partial or incomplete damage to them can be repaired. Very low RH decreases the possibility of killing the hydroxyl group of the microbe and decreases the possibility of contact between the microbe and a hydroxyl group. Very high RH also makes the surface moist, and microbes regenerate and survive.

The choice of the light source has been highly variable. Usually, experiments have shown that TiO_2 suspended in water has no significant effect on the survival of bacteria in the darkness, and the effect of PCO is greater than the effect of light alone. However, with some light sources alone there is still a great reduction in colony-forming units, this situation usually happens when the wavelength of the light source is below 365 nm (UV) and light can be transmitted through the reactor. The possible mechanism for disinfection of ultraviolet light, which kills only microbes, is that photons cause photochemical reactions in the bacteria DNA (the RNA of the virus) that do not replicate themselves, proving that ultraviolet radiation leads to the formation of thymidine dimers, which interconnects two thymic dimers in the DNA. Besides, the large number of photons absorbed damages the cell, including proteins, and causes cell rupture. The mechanism by which UV/PCO killed microbes includes ultraviolet destruction (same as above), along with photocatalytic oxidation caused by hydroxyl radicals and ROS, which leads to the oxidation of DNA bases and other damage to cell structure.

The most commonly used light source is fluorescent black lights with an emission maximum at about 365 nm, which possesses a pretty match with the bandgap of anatase. A previous study used platinized P25 TiO_2 as the photocatalyst, comparing the effect on *S. cerevisiae* of metal halide, xenon, and white fluorescent lamps with the same photon flux and discovered that the survival ratios were 27%, 46%, and 58%, respectively (Matsunaga et al. 1985). As a photobiological reactor, light irradiation can be classified as internal or external. Reflector or optical fiber has been used in the internal type to enhance light availability (Safapour and Metcalf 1999). Direct delivery of light to the photocatalyst minimizes losses due to absorption and scattering by the reactor wall and solution. During utilization of optical fibers, TiO_2 has been immobilized onto low-cost polymeric optical fibers after scratching cladding polymers to enhance the lateral diffusion of light. However, pollution from absorption by cells and by-products reduces light diffusion. Organisms have defensive and self-healing mechanisms to deal with PCO damage. Phototherapy breaks down at the start of the procedure in favor of the bacterial self-defense mechanism, thus increasing the time it takes to complete disinfection. It is important to determine the lighting time (intermittent or continuous) necessary for each microbe to produce irreversible bacterial deactivation. This is relevant because of intermittent lighting effects, to varying degrees, post-irradiation events of different microorganisms (Rincón and Pulgarin 2003).

8.2.2 Mechanism of Cell and DNA Damage

Some studies have shown that cell wall damage followed after the destruction of the cytoplasmic membrane promotes the cell permeability and the outflow of intracellular contents that lead to cell death finally (Huang et al. 2000). In other words, cell membrane may be an important target of photocatalytic activity. Besides, it has been proved that ROS including hydroxyl radical can cause DNA damage. The synthesis of supercoiled plasmid DNA has been used to demonstrate the activity of free radicals on titanium dioxide. The concentration of 0.05–0.15 mL of titanium dioxide particles at 37°C results in rupture of plasmid DNA after 8 h of culture (Donaldson, Beswick, and Gilmour 1996). Investigators (Dunford et al. 1997) displayed that sunlight-illuminated TiO_2 catalyzes DNA destruction both in vitro and in vivo. As for vitro studies, they showed that supercoiled plasmid DNA with TiO_2 was converted first to the relaxed form and later to the linear form under UV illumination, demonstrating a chain breakage; they also studied human skin fibroblast cells in vivo and found a significant level of hydroxylation of cellular RNA, but no hydroxylation of cellular DNA was discovered. The detection of oxidative damage to cellular DNA is a tough analytical problem that requires highly sensitive and targeted methods. Over the past 20 years, levels of oxidative DNA damage have been widely used as an indicator of the occurrence of oxidative activation, namely 8-oxo-7, 8-dihydro-2-deoxyguanosine(8-oxodGuo), which is a basic and ubiquitous oxidizing product of DNA (Helbock, Beckman, and Ames 1999). Ashikaga et al. (2000) under UVA illumination, tested the direct effects of the PCO activities of different types of TiO_2 on plasmid DNA, they also detected the effects of several radical cleaners on the photodynamic DNA chain-breaking activity of TiO_2, indicating that photodynamic DNA chain-breaking activity is owing to active oxygen species, particularly hydroxyl radical emitted by UVA-irradiated TiO_2. Ying et al. 2005, in a relatively short period of time, studied the effect of PCO on plasmid DNA under ultraviolet irradiation, which will facilitate the application of PCO technology in a PCO reactor with a shorter microbial residence. They detected and compared the DNA destruction by two types of photocatalysts, P25, and the latest developed nanocatalyst. Due to its reduced activation energy and maximum active surface area, the new catalyst is expected to trigger a larger photocatalytic DNA split reaction than the P25. In the presence of photocatalyst, 5 millivolt UV light (germicidal lamp, G25T8, UVC 254 nm, 25W, Sankyo Denki) was used to illuminate the ultra-disc pBR322 DNA solution and the product was analyzed by gel electrophoresis.

In one study, a laboratory-scale UV-TiO_2-PCO reactor was tested for monitoring antibacterial and DNA chain rupture activity (PCO), as shown in Figure 8.2a and Figure 8.2b, respectively (Kim et al. 2013). The reactor used for the ultraviolet antibacterial experiment consisted of a stainless-steel chamber with a volume of 3.5 L and an ultraviolet lamp either surrounded by a coated TiO_2 Silica tube (36 mm in diameter and 570 mm in length) or a silica tube without a titanium dioxide coating. A magnetic mixer is placed at the bottom of the reactor for sufficient mixing. The reactor used to study DNA fracture activity consisted of a lid, an ultraviolet lamp, and a silica tube with an inner diameter of 25 mm and an altitude of 50 mm, used as a source for preparing titanium dioxide films. A TiO_2 solution was made by dissolving $TiCl_4$, HCl, $2(NH_4)HCO_3$, and H_2O_2 in distilled water. This solution was deposited into a quartz tube and completely dried at 250°C for 24 h. The control experiments using TiO_2 alone without UV illumination were also performed. Kim et al. chose supercoiled plasmid DNA as a model of DNA destruction caused by ROS on TiO_2 surfaces. Plasmid DNA pUC19, extracted using an Exprep® plasmid SV mini kit (Gene ALL, Seoul, Korea), subsequently transferred to TiO_2-coated and clear quartz tubes (Sankyo Denki Co., Japan) and illuminated using different UV lamps. Each sample was transferred to an electronic test tube for a 2-min interval of a total 10-min PCO exposure time. Bam H1, a restriction enzyme isolated from bacteria that identify a specific sequence of polyclonal sites and cuts DNA, uses a Bam H1 cutting plasmid to obtain a linear standard (Ogawa et al. 2012). The PUC19 DNA was isolated by BAMH1 from BSA, 10× buffer, and dH_2O_2. The separated DNA hatches 24 h at 37°C. Cleavage results in the relaxation of the supercoiled DNA molecules. Using the linear form of plasmid DNA as control and size marker. So, run a small alkali on

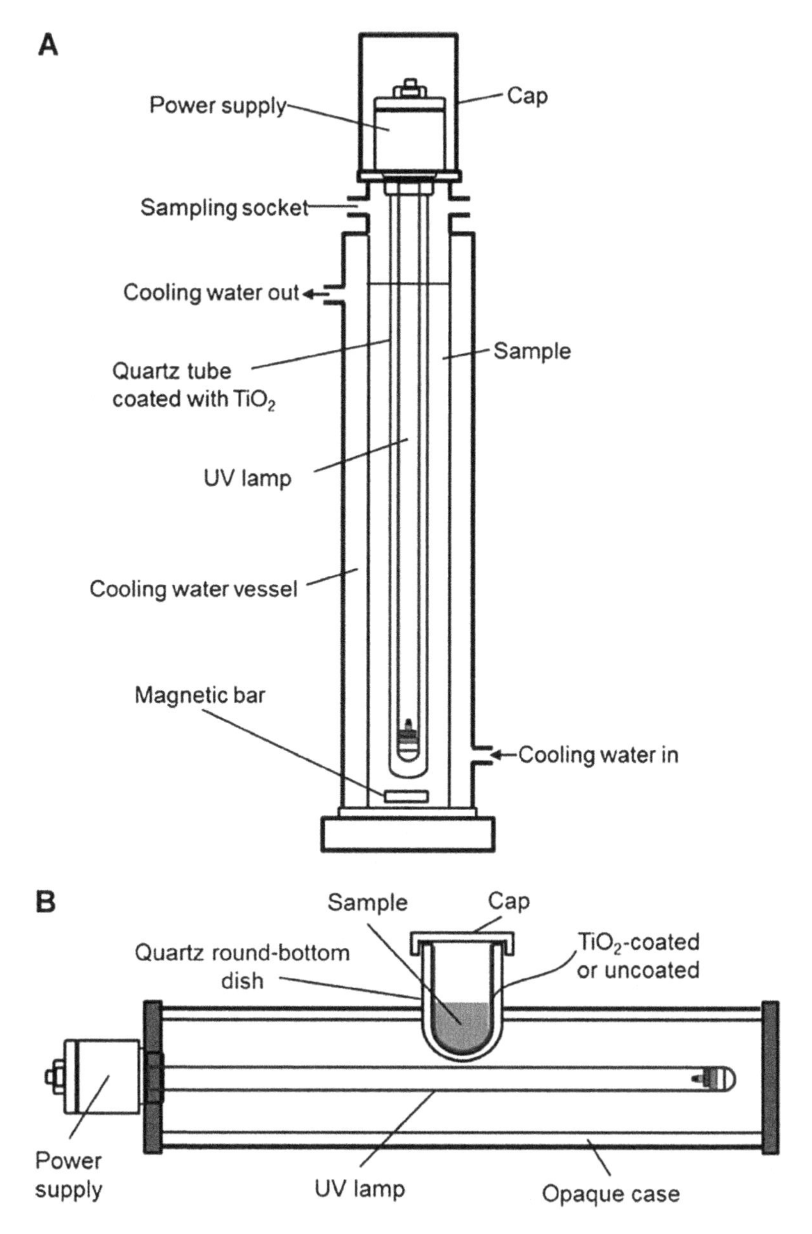

FIGURE 8.2 Schematic diagram of the lab-scale UV-assisted TiO_2-PCO reactor for antibacterial (a) and DNA strand breaking (b) activities (Reprinted from (Kim et al. 2013), with kind permission of Elsevier Publications).

the gel to check the digestion. The genome DNA of *E. coli* was prepared by CTAB genome DNA and treated with TiO_2-UVC at 30 min intervals (Doyle and Doyle 1987). Bacterial cells were captured after a 5-min rotation in a tabletop centrifuge at 1,650 × *g* for 5 min and then washed in dH_2O. CTAB extraction buffer was then instilled and the mixture was incubated at 65°C for 30 min. And 7 mL of the aqueous phase was transferred to a tube containing 7 mL of isopropanol after adding 10 mL of chloroform, and the sediment was acquired by centrifuging at 1,650 × *g* for 5 min. The genomic DNA sediment was washed in 70% ethanol and dissolved in 500 μL of TE buffer. Separation and visualization of DNA fragments by agarose gel electrophoresis based on their size topology. The gel was prepared by dissolving 1% agarose in 1× TAE buffer containing 5 μL/mL of ethidium bromide to stain the DNA, then heated until dissolved, followed by casting into a flat gel dish and setting (Serpone et al. 2006). Samples were subjected to electrophoresis in agarose gel. Agarose constitutes a porous lattice in the buffer solution and the DNA must pass through the holes in the lattice to move toward the positive pole. The DNA migration pattern was recorded using WiseDoc® WGD-20 Gel Documentation. Plasmid DNA and genomic DNA chain-breaking activities were detected as an incline in the ratio of linear to supercoiled plasmid DNA, and the concentration of the segments.

8.3 KINETICS

8.3.1 Kinetics of PCO

Chemical kinetics is necessary for comprehending the rate and the factors affecting the chemical process to acquire equilibrium at any given time (Ollis 2005). An increase in interest over heterogeneous photocatalysis has turned up in several reports on different kinetic models (Satuf et al. 2011).

8.3.1.1 Langmuir–Hinshelwood (L-H) Model

Photocatalysis is a kind of heterogeneous catalysis, which is always explained by the classical Langmuir–Hinshelwood (L-H) model (Yoshihara et al. 2004). It is based on the assumption that the adsorption of the reactants occurs on the surface of the catalyst (Hoffmann et al. 1995). The reaction is carried out between the adsorbed species and the product is removed from the surface. Herrmann (2010), in an important contribution, has tried to state several myths related to photocatalysis, and he also explained this model with a more prevalent approach. An improved L-H model is formed by considering several modifications to the model. In a bimolecular reaction;

$$A + B \rightarrow C + D \tag{8.4}$$

The velocity of the chemical process is proportionate to the surface coverage (θ) of the reactant

$$r = k\theta_A\theta_B \tag{8.5}$$

where the coverage θi differs as

$$\theta i = \frac{KiXi}{1 + KiXi} \tag{8.6}$$

where Ki: adsorption constant (in dark), Xi: concentration in liquid phase, or partial pressure Pi in gas phase; hence Equation (8.5) changes to

$$r = \frac{kK_AK_BX_AX_B}{\left(1+K_AX_A\right)\left(1+K_BX_B\right)} \tag{8.7}$$

K: true rate constant.

In general, one of the two chemical substances has a high concentration and the other B has a low concentration. Therefore, the concentration change in A is constant, as its small consumption causing the concentration change can be ignorable. Hence, Equation (8.5) simplifies to

$$r = k'\theta_B \tag{8.8}$$

where $K' = K\theta_A$

K': pseudo true rate constant

$$r = \frac{k'K_BX_B}{1+K_BX_B} \tag{8.9}$$

Therefore, if $X = X_{max}$, thus $\theta_B = 1$

Hence, Equation (8.9) simplified to $r = k'$

or if $X \ll X_{max}$, thus $\theta_B = \frac{K_BX_B}{1+K_BX_B} \approx k'$

Hence, Equation (8.9) simplified to

$$r = k'K_BX_B = K_{App}X_B \tag{8.10}$$

k_{App}: Apparent first-order constant.

8.3.1.2 Direct–Indirect Model (D-I Model)

Satoca et al. brought in a replaced kinetic approach, the "Direct–Indirect" (D-I) model (Monllor-Satoca et al. 2007). The model proposes a few significant ideas on an immediate, roundabout, adiabatic, and inelastic interfacial exchange of charge, which further sets up physical importance to the motor boundaries included. The L-H model examines the energy of the response in an equilibrated adsorption/desorption of reactants on the outside of the semiconductor material under a consistent brightening. Besides, it additionally neglects to set up an important connection between the approaching brilliant transition (Φ) and the reactant focus (X). Nonetheless, the current model outfits a practical reliance on the photooxidation rate (r) on the test boundaries (photon transition and toxin concentration). The rate of photooxidation relies upon all the interfacial charge transfer and the reaction between the conduction band electrons with the liquified O2 to form the expected receptive species (Gerischer 1995). The conversation unearths the discussion on the interest of photogenerated valence band openings (h_f+) in a direct reaction (DT) with the natural substrate or moves by means of an indirect transfer (IT) component by using hydroxyl revolutionaries (surface cupped orifice h_s+). The short number of potential courses of immediate and circuitous exchange is:

$$hf^{+} + [X]_{aq} \rightarrow [X]_{aq}^{*} + H_{aq}^{+} \tag{8.11}$$

Photogenerated gaps responding with reactants in the watery solution (not ingested on the impetus surface).

$$hf^{+} + [X]_{s} \rightarrow [X]_{s}^{*} + H_{aq}^{+} \tag{8.12}$$

Photogenerated gaps reacting with reactants absorbed on the catalyst surface.

$$hs^{+} + [X]_{aq} \rightarrow [X]_{aq}^{*} + H_{aq}^{+} \tag{8.13}$$

Surface cupped orifice reacting with reactants in the aqueous solution (not absorbed on the catalyst surface).

$$hs^{+} + [X]_{s} \rightarrow [X]_{s}^{*} + H_{aq}^{+} \tag{8.14}$$

Surface cupped orifice reacting with reactants absorbed on the catalyst surface. At the point when the natural particles are not adsorbed on the impetus surface, the gap move between the reactants in the fluid arrangement happens adiabatically. Marcus built up a Fluctuating Energy Level Model which characterizes this exchange system and later Gerischer interpreted it on the semiconductor electrolyte interface (Marcus 1956). This model helps to predict an orifice transfers rate constant (k_{ox}^{adb}).

$$k_{ox}^{adb} \propto \exp\left[\frac{-(E_{red} - E_s)^2}{4\lambda kT}\right] \tag{8.15}$$

where E_s: Energy of the surface cupped orifice, E_{red}: The most probable energy of the occupied energy levels of reactants in the aqueous solution, and λ: The reorganization energy (between 0.5 and 1.0 eV).

By the way, solid collaboration expands the serious adsorption between the reactant particles and the water atoms. In such a case, the orifice moves' mechanism isn't adiabatic yet inelastic and isn't administered by this model any longer. In this manner, the changed rate is consistent (k_{oxins})

$$k_{ox}^{ins} \propto \sigma\hat{c} \tag{8.16}$$

where σ: orifice capture cross-section of filled surface states, ĉ: thermal velocity of a free hole.

The L-H model relies on the equilibrated adsorption of the reactants. Therefore, the relationship between the concentration of the dissolved substance and the adsorbent is given in the isotherm of Lamir type adsorption.

$$[X]_s = \frac{ab[X]_{aq}}{1+a[X]_{aq}} \tag{8.17}$$

where a: adsorption constant and b: desorption constant.

8.3.2 Disinfection Kinetic Models of PCO

The PCO purification measure is an extremely intricate marvel. It includes numerous variables for instance pH of the example, distinctive impetus stacking, illumination force, turbidity, the temperature of the response blend, and in particular the intricate structure of the microorganism (Marugán et al. 2011). The kinetic models depict endurance bends, a graphical outline connoting the semilog plot of inactivation on contact time. The variety looking like the endurance bends plays an urgent factor in the evaluation of the component of microbial inactivation. The kinetic inactivation models are inferred depending on the accompanying assumptions: (a) uniform appropriation of microorganisms and the disinfectant particles, (b) Steady pH, temperature, and the centralization of the impetus (disinfectant), and (c) Sufficient blending to dodge fluid dissemination as a possible, restricting operator in the synthetic response. There exists four basic motor models: (1) Chick's model, (2) Chick–Watson model, (3) Delayed Chick–Watson model, and (4) Hom model. The overall articulation of the differential rate law utilized by these dynamic models is as in Equation (8.18) (Dalrymple et al. 2010, Rachmadi et al. 2020)

$$\frac{dN}{dt} = -KmN^xC^nt^{m-1} \tag{8.18}$$

where: $\frac{dN}{dt}$ Rate of inactivation, N: Number of survivors at contact time t, K: Reaction rate constant, C: Concentration of the disinfectant, and m, n, and x: Empirical constants.

8.3.2.1 Chick's Model

Chick published paper "The laws of disinfection" in 1908, and he compared the bacterial inactivation to a chemical reaction, where individual bacterium was treated as molecules (Watson 1908):

$$xN + nC \xrightarrow{K1} S \quad (\text{deal microorganisms}) \tag{8.19}$$

The rate of reaction for the following reaction is given as

$$\frac{dN}{dt} = -K_1N^xC^n \quad (\text{deal microorganisms}) \tag{8.20}$$

Here, x and $n = 1$ and considering the reaction is irreversible, thus $S = 0$. Hence, Equation (8.20) results in

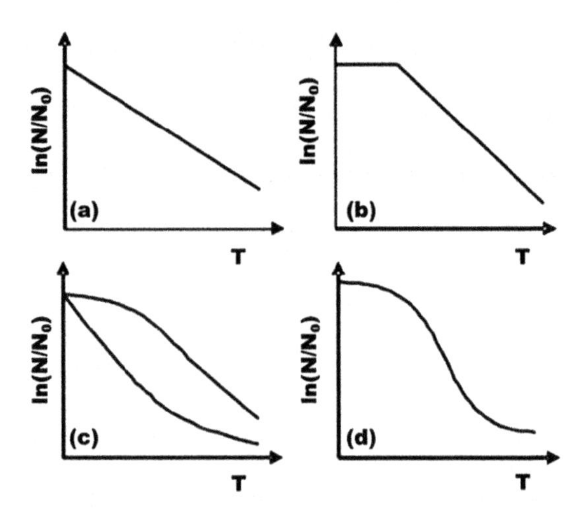

FIGURE 8.3 Survival curves were observed for different kinetic models used in the PCO disinfection. Curve (a) first-order kinetics of inactivation, illustrates the exponential death of microbes with time, curve (b) exponential disinfection but with a shoulder initially, curve (c) curve with a shoulder and a tailing off, and curve (d) with a tailing off sequence at the end. Reprinted with permission of Marugan et al. Full details are given in the respective publication (Marugán et al. 2008; Ganguly et al. 2018).

$$\frac{dN}{dt} = -K * N \tag{8.21}$$

where $K*$ is considered as the pseudo-first-order rate constant where $K* = K_1C$. Thus Equation (8.21) is the generalized rate law expression of Chick's law. It demonstrates that the rate of inactivation is proportional to the number of surviving microorganisms at a stable disinfectant concentration (Figure 8.3a). Integration of Equation (8.21) results in

$$\ln\frac{N}{N_0} = -K * T \tag{8.22}$$

where N_0: Initial microbial population.

In recent researches, the importance of the true order of this kinetic model was explained by Danae and Dionissios (Venieri and Mantzavinos 2017). Authors interpreted that the reaction parameters including catalyst concentration and temperature are accounted as a constant termed as pseudo-first-order constant. Thus, at various N_0 values, however, a consistent exploratory arrangement should deliver steady K* esteem. The ROS in the response blend is a prevalent factor overseeing the pace of cleansing; therefore, the "N" value can't be the only one to be considered as a selective factor. The abundance of ROS will possibly cause snappier sanitization and show first request energy, in the interim, a low ROS fixation will transform into an expected restricting specialist in the cleaning cycle. Accordingly, ROS in the cleansing cycle is absolutely a chief factor overseeing the energy of the cycle. Besides, the ROS created in the response relies upon the steady test factors Thus the exploratory boundaries are the genuine regulator of the cleansing energy.

8.3.2.2 Chick–Watson Model

The Chick's law hypothesis first requests energy and in quite a while, the microorganic inactivation doesn't follow this suspicion. Watson in the identical year altered the Chick's law by fusing the concentration-time item which considers the impact of shifting concentration of disinfectant (Equation 8.23)

$$K = C^n T \tag{8.23}$$

where K: a constant for a microorganism and a series of conditions, n: a constant, and T: the time needed to reach a certain activation point. An n value of less than 1 indicates the increased importance of the contact time, more than the disinfectant concentration. The addition of Watson's function in Chick's law leads to a Chick–Watson pseudo first-order rate law as (Haas and Karra 1984):

$$\frac{dN}{dt} = -KNC^n \tag{8.24}$$

where K: the first order rate constant. Integration of the above expression results in (Figure 8.3a) (Haas and Karra 1984)

$$\ln\frac{N}{N_0} = -KC^n T \tag{8.25}$$

$$\frac{N}{N_0} = e^{-KC^n T} \tag{8.26}$$

8.3.2.3 Delayed Chick–Watson Model

The rearranged idea of the Chick–Watson model makes it impartial in many investigations. Nonetheless, the Chick–Watson model has its entanglements, it expects that the organisms are of single strain and the inactivation happens on a solitary hit and at a solitary spot. It shows neither the underlying shoulder (slack stage) nor the following off nature. The postponed Chick–Watson model introduces a time lag parameter to conquer the shoulder observed in the cleansing cycle (Figure 8.3b) (del Carmen et al. 2017; Cho et al. 2006).

$$\frac{N}{N_0} = \begin{cases} 1, T \le T_{lag} \\ e^{-KC^n T}, T > T_{lag} \end{cases} \tag{8.27}$$

8.3.2.4 Hom's and Modified Hom's Model

The postponed Chick–Watson model shows the shoulder or off at a time, but the simultaneous presence of both can't be communicated utilizing this model. Hom's cleansing model proposed in 1972 gives a summed-up articulation of the time-fixation item. It is fundamentally the same as the Chick–Watson model, an expansion of a force factor m for time is watched. When $m = 1$, the condition lessens to the Chick–Watson model $m > 1$ outcomes in the presence of the underlying shoulder in the bends, while $m < 1$ leads to the presence of the following off. Thus, Hom's model additionally doesn't permit the event of both the underlying slack and the following off in the bend simultaneously (Haas and Joffe 1994, Chong, Jin, and Saint 2011).

$$\ln \frac{N}{N_0} = -KC^n T^m \tag{8.28}$$

From that point forward, a change is presented in Hom's model which permits the presence of the underlying slack, the log-straight stage, and the following off toward the finish of the bend. The changed Hom's model is as (Haas and Joffe 1994, Chong, Jin, and Saint 2011):

$$\ln \frac{N}{N_0} = -K_1 [\ ] 1 - \exp(-K_2 T)^{K3} \tag{8.29}$$

where K_1, K_2, and K_3 are the empirical constants in this model.

8.4 FACTORS INFLUENCING THE DISINFECTION MECHANISM

8.4.1 Irradiation Length and Intensity

It was seen that a ceaseless illumination with no between eruption permitted the total cleansing yet intruded-on irradiation decreased the bactericidal movement. It is perceived that the self-protection instruments of the microbes help them to recuperate. The bacterial cell creates a superoxide dismutase (SOD) compound as a cautious system to adapt to the lopsided measure of ROS inside the cell and lessens the oxidative pressure. The ROS at the end produced within the sight of the impetus on the due illumination of the light is altogether high which eventually prompts total bacterial sanitization. In a PCO reaction, the irradiation intensity (ϕ) is undoubtedly a major parameter of concern, higher irradiation results in increased ROS production (Murcia et al. 2017).

8.4.2 pH and Catalyst Loading

The surface charge of the catalyst and its microorganic cell wall is reliant on the general pH of the framework. Additionally, the course of action of the band structure and the size of the impetus totals rely upon this working boundary. The catalyst or the disinfectant's surface charge and the pH has a characterized connection. The pH level at which the surface charge of the catalyst surface is zero (purpose of zero charges) is referred to as the isoelectric point. Catalyst loading is another significant working factor of concern. It has been seen in a few examinations that the purification cycle improves with an expansion in the impetus focus. Nonetheless, it quickly arrives at an immersion limit, where the pace of the response stays steady even with the expansion in the impetus fixation. The expansion in the impetus focus, in the long run, prompts turbidity and blocks the ingestion of the approaching radiation. Studies have been performed to advance the mass of the impetus required for the particular sanitization measure, to maintain a strategic distance from abundance utilization (Murdoch et al. 2011, Akyol, Yatmaz, and Bayramoglu 2004).

8.4.3 Temperature and Turbidity

In a photocatalytic purification framework, the temperature isn't a basic factor. In any case, by and large, the photocatalytic sanitization rate diminishes with the expansion in operational temperature (Gaya and Abdullah 2008). The thermal energy granted to the framework is not enough to defeat the band hole hindrance. The inconsequential amount of warmth energy present in the sterilization blend is utilized in the enactment cycle of the reactant on the impetus surface. The enactment energy required by the reactants increments at a low operating temperature and remains modestly successful in a medium temperature between 20°C and 80°C. Conversely, when the temperature of the framework begins to increment past 80°C, the adsorption of the reactants on the impetus surface becomes conceded and expands the recombination of the charged transporters (Malato et al. 2009).

Turbidity is another key factor that influences the cleaning cycle. The presence of little insoluble sub-particles for instance earth, microscopic fishes, and microorganisms is the reason for the turbidity. The turbid water brings about the weakening of approaching light and the ingestion level differs altogether. It was seen that an expanded degree of turbidity in the water test diminishes the photocatalytic inactivation rate. The expanded particulate level reduces the capacity of the approaching illumination to penetrate through the turbid water. Besides, the impetus added leads to conglomeration, which decreases the cleansing movement and the warmth enamored in the framework animates the bacterial development (Rincon and Pulgarin 2004, Tang and Chen 2004).

ACKNOWLEDGEMENT

This work was supported by the research project of Student Grant Competition of the Technical University of Liberec no. 21406/2020 granted by the Ministry of Education, Youth, and Sports of Czech Republic and the Ministry of Education, Youth and Sports of the Czech Republic, European Union – European Structural and Investment Funds in the Frames of Operational Programme Research, Development, and Education – project Hybrid Materials for Hierarchical Structures (HyHi, Reg. No. CZ.02.1.01/0.0/0.0/16_019/0000843), the Ministry of Education, Youth, and Sports in the frames of support for researcher mobility (VES19China-mobility, Czech–Chinese cooperation) Design of multilayer micro-/nanofibrous structures for air filters applications, Reg. No. 8JCH1064., and project "Intelligent thermoregulatory fibers and functional textile coatings based on temperature resistant encapsulated PCM" SMARTTHERM (Project No. TF06000048).

REFERENCES

Akyol, A, HC Yatmaz, and M Bayramoglu. 2004. "Photocatalytic decolorization of Remazol Red RR in aqueous ZnO suspensions." *Applied Catalysis B: Environmental* 54 (1):19-24.

Ashikaga, Takao, Masayoshi Wada, Hiroshi Kobayashi, Masaaki Mori, Yoshio Katsumura, Hiroshi Fukui, Shinobu Kato, Michihiro Yamaguchi, and Tasuku Takamatsu. 2000. "Effect of the photocatalytic activity of TiO2 on plasmid DNA." *Mutation Research/Genetic Toxicology and Environmental Mutagenesis* 466 (1):1–7.

Boeniger, Mark F. 1995. "Use of ozone generating devices to improve indoor air quality." *American Industrial Hygiene Association Journal* 56 (6):590–598.

del Carmen Huesca, Espitia, Luz Luz, Veronica Aurioles-López, Irwing Ramírez, Jose Luis Sánchez-Salas, and Erick R Bandala. 2017. "Photocatalytic inactivation of highly resistant microorganisms in water: a kinetic approach" *Journal of Photochemistry and Photobiology A: Chemistry* 337:132–139.

Chen, Haihan, Charith E Nanayakkara, and Vicki H Grassian. 2012. "Titanium dioxide photocatalysis in atmospheric chemistry." *Chemical Reviews* 112 (11):5919–5948.

Cho, Min, Yunho Lee, Wonyong Choi, Hyenmi Chung, and Jeyong Yoon. 2006. "Study on Fe (VI) species as a disinfectant: Quantitative evaluation and modeling for inactivating Escherichia coli." "*Water Research* 40 (19):3580–3586.

Chong, Meng Nan, Bo Jin, and Christopher P Saint. 2011. "Bacterial inactivation kinetics of a photo-disinfection system using novel titania-impregnated kaolinite photocatalyst." *Chemical Engineering Journal* 171 (1):16–23.

Dalrymple, Omatoyo K, Elias Stefanakos, Maya A Trotz, and D Yogi Goswami. 2010. "A review of the mechanisms and modeling of photocatalytic disinfection." *Applied Catalysis B: Environmental* 98 (1-2):27–38.

Donaldson, Kenneth, Paul H Beswick, and Peter S Gilmour. 1996. "Free radical activity associated with the surface of particles: a unifying factor in determining biological activity?" *Toxicology letters* 88 (1–3):293–298.

Doyle, Jeff J, and Jan L Doyle. 1987. A rapid DNA isolation procedure for small quantities of fresh leaf tissue.

Dunford, Rosemary, Angela Salinaro, Lezhen Cai, Nick Serpone, Satoshi Horikoshi, Hisao Hidaka, and John Knowland. 1997. "Chemical oxidation and DNA damage catalysed by inorganic sunscreen ingredients." *FEBS Letters* 418 (1-2):87–90.

Fujishima, Akira, and Kenichi Honda. 1972. "Electrochemical photolysis of water at a semiconductor electrode." *Nature* 238 (5358):37–38.

Ganguly, Priyanka, Ciara Byrne, Ailish Breen, and Suresh C Pillai. 2018. "Antimicrobial activity of photocatalysts: fundamentals, mechanisms, kinetics and recent advances." *Applied Catalysis B: Environmental* 225:51–75.

Garrido, Carmen, and Guido Kroemer. 2004. "Life's smile, death's grin: vital functions of apoptosis-executing proteins." *Current Opinion in Cell Biology* 16 (6):639–646.

Gaya, Umar Ibrahim, and Abdul Halim Abdullah. 2008. "Heterogeneous photocatalytic degradation of organic contaminants over titanium dioxide: a review of fundamentals, progress and problems." *Journal of Photochemistry and Photobiology C: Photochemistry Reviews* 9 (1):1–12.

Gerischer, Heinz. 1995. "Photocatalysis in aqueous solution with small TiO2 particles and the dependence of the quantum yield on particle size and light intensity." *Electrochimica Acta* 40 (10):1277–1281.

Goswami, DY, DM Trivedi, and SS Block. 1997. "Photocatalytic disinfection of indoor air."

Haas, Charles N, and Josh Joffe. 1994. "Disinfection under dynamic conditions: modification of Hom's model for decay." *Environmental Science & Technology* 28 (7):1367–1369.

Haas, Charles N, and Sankaram B Karra. 1984. "Kinetics of microbial inactivation by chlorine—I Review of results in demand-free systems." *Water Research* 18 (11):1443–1449.

Helbock, Harold J, Kenneth B Beckman, and Bruce N Ames. 1999. "8-Hydroxydeoxyguanosine and 8-hydroxyguanine as biomarkers of oxidative DNA damage." *Methods in Enzymology*, 156–166. Elsevier.

Herrmann, Jean-Marie. 2010. "Photocatalysis fundamentals revisited to avoid several misconceptions." *Applied Catalysis B: Environmental* 99 (3-4):461–468.

Hoffmann, Michael R, Scot T Martin, Wonyong Choi, and Detlef W Bahnemann. 1995. "Environmental applications of semiconductor photocatalysis." *Chemical Reviews* 95 (1):69–96.

Huang, Zheng, Pin-Ching Maness, Daniel M Blake, Edward J Wolfrum, Sharon L Smolinski, and William A Jacoby. 2000. "Bactericidal mode of titanium dioxide photocatalysis". *Journal of Photochemistry and Photobiology A: Chemistry* 130 (2-3):163–170.

Kim, Soohyun, Kashif Ghafoor, Jooyoung Lee, Mei Feng, Jungyeon Hong, Dong-Un Lee, and Jiyong Park. 2013. "Bacterial inactivation in water, DNA strand breaking, and membrane damage induced by ultraviolet-assisted titanium dioxide photocatalysis". *Water Research* 47 (13):4403–4411.

Lammel, Tobias, Paul Boisseaux, Maria-Luisa Fernández-Cruz, and José M Navas. 2013. "Internalization and cytotoxicity of graphene oxide and carboxyl graphene nanoplatelets in the human hepatocellular carcinoma cell line Hep G2." *Particle and Fibre Toxicology* 10 (1):27.

Malato, Sixto, Pilar Fernández-Ibáñez, Manuel I Maldonado, Julián Blanco, and Wolfgang Gernjak. 2009. "Decontamination and disinfection of water by solar photocatalysis: recent overview and trends." *Catalysis Today* 147 (1):1–59.

Marcus, Rudolph A. 1956. "On the theory of oxidation-reduction reactions involving electron transfer. I." *The Journal of Chemical Physics* 24 (5):966–978.

Marugán, Javier, Rafael van Grieken, Carlos Sordo, and Cristina Cruz. 2008. "Kinetics of the photocatalytic disinfection of Escherichia coli suspensions." *Applied Catalysis B: Environmental* 82 (1-2):27–36.

Marugán, Javier, Rafael Van Grieken, Pablos Cristina, M Lucila Satuf, Alberto E Cassano, and Orlando M Alfano. 2011. "Rigorous kinetic modelling with explicit radiation absorption effects of the photocatalytic inactivation of bacteria in water using suspended titanium dioxide." *Applied Catalysis B: Environmental* 102 (3–4):404–416.

Matsunaga, Tadashi, Ryozo Tomoda, Toshiaki Nakajima, and Hitoshi Wake. 1985. "Photoelectrochemical sterilization of microbial cells by semiconductor powders." *FEMS Microbiology Letters* 29 (1-2):211–214.

Minero, Claudio, Giuseppe Mariella, Valter Maurino, and Ezio Pelizzetti. 2000. "Photocatalytic transformation of organic compounds in the presence of inorganic anions. 1. Hydroxyl-mediated and direct electron-transfer reactions of phenol on a titanium dioxide– fluoride system." *Langmuir* 16 (6):2632–2641.

Monllor-Satoca, Damián, Roberto Gómez, Manuel González-Hidalgo, and Pedro Salvador. 2007. "The "Direct–Indirect" model: An alternative kinetic approach in heterogeneous photocatalysis based on the degree of interaction of dissolved pollutant species with the semiconductor surface." *Catalysis Today* 129 (1-2):247–255.

Murcia, JJ, EG Ávila-Martínez, H Rojas, José Antonio Navío, and MC Hidalgo. 2017. "Study of the E. coli elimination from urban wastewater over photocatalysts based on metal-lized TiO2." *Applied Catalysis B: Environmental* 200:469–476.

Murdoch, M, GIN Waterhouse, MA Nadeem, JB Metson, MA Keane, RF Howe, J Llorca, and H Idriss. 2011. "The effect of gold loading and particle size on photocatalytic hydrogen production from ethanol over Au/TiO 2 nanoparticles". *Nature Chemistry* 3 (6):489–492.

Ogawa, Wakano, Motoyasu Onishi, Ruiting Ni, Tomofusa Tsuchiya, and Teruo Kuroda. 2012. "Functional study of the novel multidrug efflux pump KexD from Klebsiella pneumoniae". *Gene* 498 (2):177–182.

Ollis, David F. 2005. "Kinetics of liquid phase photocatalyzed reactions: an illuminating approach." *The Journal of Physical Chemistry B* 109 (6):2439–2444.

Rachmadi, Andri Taruna, Masaaki Kitajima, Tsuyoshi Kato, Hiroyuki Kato, Satoshi Okabe, and Daisuke Sano. 2020. "Required chlorination doses to fulfill the credit value for disinfection of enteric viruses in water: A critical review." *Environmental Science & Technology* 54 (4):2068–2077.

Rincón, AG, and C Pulgarin. 2003. "Photocatalytical inactivation of E. coli: effect of (continuous–intermittent) light intensity and of (suspended–fixed) TiO2 concentration." *Applied Catalysis B: Environmental* 44 (3):263–284.

Rincon, Angela-Guiovana, and Cesar Pulgarin. 2004. "Effect of pH, inorganic ions, organic matter and H_2O_2 on E. coli K12 photocatalytic inactivation by TiO_2: implications in solar water disinfection."*Applied Catalysis B: Environmental* 51 (4):283–302.

Safapour, Negar, and Robert H Metcalf. 1999. "Enhancement of solar water pasteurization with reflectors." *Applied and Environmental Microbiology* 65 (2):859–861.

Satuf, María L, María J Pierrestegui, Lorena Rossini, Rodolfo J Brandi, and Orlando M Alfano. 2011. "Kinetic modeling of azo dyes photocatalytic degradation in aqueous TiO2 suspensions. Toxicity and biodegradability evaluation." *Catalysis Today* 161 (1):121–126.

Serpone, Nick, Angela Salinaro, Satoshi Horikoshi, and Hisao Hidaka. 2006. "Beneficial effects of photo-inactive titanium dioxide specimens on plasmid DNA, human cells and yeast cells exposed to UVA/UVB simulated sunlight." *Journal of Photochemistry and Photobiology A: Chemistry* 179 (1-2):200–212.

Tang, C, and V Chen. 2004. "The photocatalytic degradation of reactive black 5 using TiO2/UV in an annular photoreactor." *Water Research* 38 (11):2775–2781.

Teta, Monica, Matthew M Rankin, Simon Y Long, Geneva M Stein, and Jake A Kushner. 2007. "Growth and regeneration of adult β cells does not involve specialized progenitors." *Developmental Cell* 12 (5):817–826.

Venieri, Danae, and Dionissios Mantzavinos. 2017. "Disinfection of waters/wastewaters by solar Photocatalysis." In *Advances in Photocatalytic Disinfection*, 177–198. Springer.

Watson, Herbert Edmeston. 1908. "A note on the variation of the rate of disinfection with change in the concentration of the disinfectant." *Epidemiology & Infection* 8 (4):536–542.

Ying, Wang, Yang Xudong, Wang Yunqiu, Wang Yongbao, and Han Zhiyong. 2005. "Disinfection and bactericidal effect using photocatalytic oxidation." *HKIE Transactions* 12 (1):39–43.

Yoshihara, Toshitada, Ryuzi Katoh, Akihiro Furube, Yoshiaki Tamaki, Miki Murai, Kohjiro Hara, Shigeo Murata, Hironori Arakawa, and M Tachiya. 2004. "Identification of reactive species in photoexcited nanocrystalline TiO2 films by wide-wavelength-range (400–2500 nm) transient absorption spectroscopy." *The Journal of Physical Chemistry B* 108 (12):3817–3823.

Zhao, Juan, and Xudong Yang. 2003. "Photocatalytic oxidation for indoor air purification: a literature review." *Building and Environment* 38 (5):645–654.

Zhao, Feng, Ying Zhao, Ying Liu, Xueling Chang, Chunying Chen, and Yuliang Zhao. 2011. "Cellular uptake, intracellular trafficking, and cytotoxicity of nanomaterials." *Small* 7 (10):1322–1337.

9 Titanium Dioxide
Enhanced Disinfection

Muhammad Zaman Khan, Jiri Militký, Jakub Wiener and Azam Ali
Technical University of Liberec, Czech Republic

CONTENTS

9.1 INTRODUCTION

Fabrics are often considered our second skin. The textiles bring comfort as well as protect our bodies. High-performance functional fabrics enjoy a high level of growth in the market. The modification of the textiles is carried out to impart the functional properties for domestic and industrial applications. Textiles with functional properties such as self-cleaning, super-hydrophobicity, antimicrobial, antistatic, and antipollution have been developed rapidly in recent years. Textile performance can often be achieved not only with emerging production technologies and materials (e.g., 3D structures) but also with new advances in textile modification. The functional properties can be incorporated into the textiles by functionalization with functional materials, such as organic and inorganic nanomaterials. The photocatalyst is mostly used among these functional materials. The photocatalyst material has emerged as a promising functional material due to its ability to catalyze reaction under the light. The semiconductor photocatalyst materials are becoming promising technology for environmental decontamination and also use energy from the sunlight and artificial light to eradicate and eliminate pollution by using the energy of natural sunlight or indoor artificial light that is available in the world. The mechanism of these functions is based on the in situ generated highly reactive oxygen species (ROS; e.g., OH and O_2) for the mineralization of organic compounds. Photocatalysts, such as TiO_2, ZnO, Fe_2O_3, CdS, WO_3, SnO_2, and ZnS, are commonly used to degrade a wide range of organic materials into readily biodegradable compounds and subsequently convert them to carbon dioxide (CO_2) and water (H_2O). Among the reported research articles, titanium dioxide and its composites are the most common photocatalyst used in textiles materials. Most functional textiles synthesized with titanium dioxide as the catalyst is activated by ultraviolet (UV) light. The important features for photocatalysts are as follows:

- Low toxicity and eco-friendly.
- Ambient operation temperature and pressure.
- Complete mineralization of organics without secondary pollution.
- Low operational cost.
- Wide range of activity toward a variety of contaminants.
- Photocatalytic activity in both indoor and outdoor light (Wang et al. 2015).

Some metal oxides powdered to nanoparticles (NPs) ($1 < \varphi \leq 100$ nm) have been of great interest to scientists representing various fields of science. The photocatalytic property of these materials is the major reason for the growing interest. Photocatalytic properties of TiO_2 were first reported in the 1970s (Fujishima and Honda 1972) and later confirmed in several studies. At the end of the twentieth century, studies conducted at various research institutes found that other metal oxides, such as zinc oxide (ZnO), after they have been powdered to the NP form, exhibit formerly undisplayed photocatalytic properties. Due to those newly discovered characteristics, TiO_2 and ZnO in the NP form have found many new applications, such as ingredients of photocatalytic layers covering various work surfaces. These coated surfaces attained self-disinfecting and self-cleaning properties (Costa et al. 2013). It is due to the advanced oxidation processes (AOPs) initiated by the UV radiations. The photocatalytic surfaces are responsible for the inactivation of infectious substances and the degradation of organic compounds. Due to the antiviral, antibacterial, and antifungal properties, these semiconductor photocatalysts are coated onto the surfaces in miscellaneous applications, such as hospitals, abattoirs, farms, and laboratories. The effectiveness of photocatalysis has also been proven in many areas, such as water treatment, purification of drinking water, and air disinfection (Zyoud et al. 2017; Yadavalli and Shukla 2017; Scorb et al. 2009). The researchers presume that the introduction of photocatalytic surfaces in the food industry, animal production, and health care facilities will be helpful and it will help to prevent food poisonings and food contaminations and improve the efficiency of pathogen eradication (Bogdan, Zarzyńska, and Pławińska-Czarnak 2015). The metallic and metal oxide nanoparticles, including titanium dioxide nanoparticles, among polymeric nanoparticles, liposomes, micelles, quantum dots, dendrimers, or fullerenes are considered important due to their potential applications in novel medical therapies.

The titanium dioxide (titanium (IV) oxide, titania, TiO_2), is an inorganic photocatalytic material. The bulk production of TiO_2 was started in the early twentieth century as a substituent material for white dye in paints due to its nontoxicity. Currently, the annual production of titanium dioxide exceeds four million tons per year and titanium dioxide has many applications in everyday products (see Figure 9.1), as an excipient in the pharmaceutical industry for sun cream production in the cosmetics industry, as a colorant in white plastics, and as a relatively cheap and

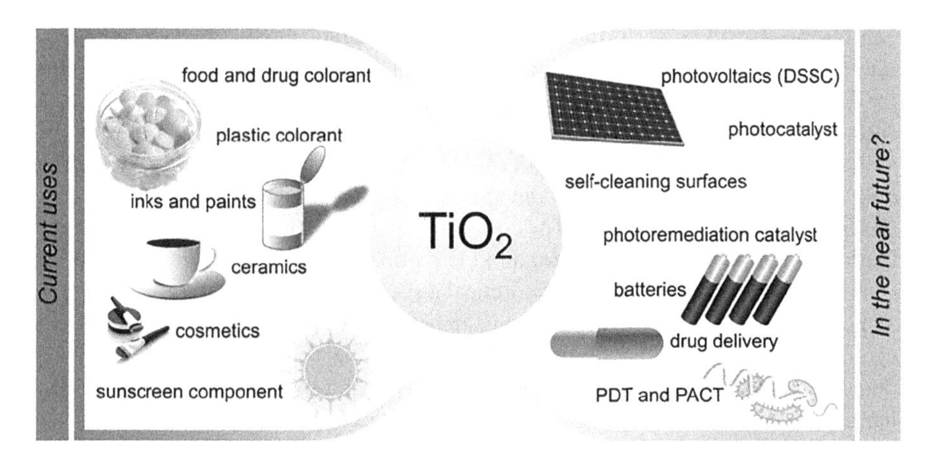

FIGURE 9.1 Current applications and potential future use of TiO_2; PDT, photodynamic therapy; PACT, antimicrobial photodynamic therapy; DSSC, dye-sensitized solar cell (Reprinted under the terms of the Creative Commons Attribution license from Nanomaterials, MDPI Publications (Ziental et al. 2020)).

nontoxic food pigment approved by the relevant European Union authorities for the safety of food additives. The research on the possible applications of TiO_2 nanoparticles started in 1985, when one of the first works on the subject of photocatalytic disinfection was published. Since that time the use of TiO_2 nanoparticles in photodynamic therapy studies has been constantly increasing. It concerns TiO_2 nanoparticle applications as photosensitizing agents in the treatment of cancer as well as in photodynamic inactivation of antibiotic-resistant bacteria. The TiO_2 nanoparticles themselves as well as their composites and combinations or hybrids with other molecules were successfully tested as photosensitizers in photodynamic therapy. Titanium dioxide nanoparticles were applied inter alia in the synthesis of bioconjugates with cell-specific monoclonal antibodies for the treatment of malignant tumors or the preparation of black TiO_2 nanoparticles for antimicrobial therapy of antibiotic-resistant bacteria (Ziental et al. 2020).

9.2 TIO_2 PHOTOCATALYSTS

The nanostructured titanium dioxide is one of the most widely studied photocatalysts and extensively explored for photocatalytic viral disinfection over the past many years. Among these studies, Degussa P25 was the most popular photocatalyst due to its high photoactivity, long-term stability, nontoxicity, and low cost.

9.2.1 OD – 3D TiO_2 STRUCTURES

The photocatalytic performance of titanium dioxide depends strongly on its structure, size, and morphology. The titanium dioxide is mainly used as nanoparticles for any study of its photoactivity. Besides, some other morphologies have exhibited outstanding performances for the photocatalytic annihilation of pathogenic microorganisms. For instance, the nanotube architecture exhibits an excellent behavior in the eradication of microorganisms due to its intrinsic features such as large surface-to-volume ratio and improved light harvesting. In recent years, the nanotubes of titanium dioxide have been utilized for the eradication of the *Staphylococcus aureus*, *Escherichia coli*, *Pseudomonas fluorescence*, and *Sphaerotilus natans* (Podporska-Carroll et al. 2015). The conventional anodization method is used to synthesize nanotube arrays of titanium dioxide on titanium foil as substrate (Cantarella et al. 2016; Baram et al. 2011). A theoretical and experimental work performed by Yu and coworkers has reported the relationship between the photocatalytic performance and the morphology of several titanium dioxide nanostructures (Yu et al. 2014). The photoactive titanium dioxide nanostructures may be arranged in the following order: nanosheets (2D) > nanotubes (1D) > nanoparticles (0D). The band structure and density of states of these titanium dioxide nanostructures were studied using density functional theory calculations. The morphological transition of the nanoparticles to nanotubes/nanosheets can contribute to the widening of the bandgap in these structures. Hence, the bottom edge of the conduction band of the titanium dioxide nanotubes is greater than that for nanoparticles. Therefore, photoexcited electrons in nanotube can perform superior reduction activity. The 3D morphology-based hierarchical titanium dioxide structures have a greater ability of reflection and scattering of light inside nanostructures and also enhance the light-harvesting. The hierarchical structures, such as nanorods spheres, have been reported for the annihilation of *E. coli and S. aureus* (Zhang et al. 2015; Bai et al. 2013). The 3D morphological structures of dendritic microspheres based on rutile titanium dioxide nanoribbons have also been studied for antimicrobial applications. Moreover, nanostructures like titanate nanotubes are also utilized for disinfection. Generally, the growth of these nanotubes is carried out by the hydrothermal method under alkaline conditions using titanium dioxide precursor.

9.2.2 Metal and Nonmetal Doped TiO_2

In many previously reported studies, the modification of TiO_2 is carried out by single doping, co-doping, and impregnation with different metal and nonmetal ions, to improve the photocatalytic performance in the UV light and visible-light region. Hence, the doping with cations and anions in the crystal structure of TiO_2 is used to create intra-band gap states near the edges of the conduction band (CB) and valence band (VB), causing absorption in the visible light region. Many years ago, Asahi et al. reported that the anionic doping of TiO_2 with nitrogen ($TiO_{2-x}N_x$), could be considered as the most effective method to favor the shift of the absorption edge toward the visible region ($\lambda < 500$ nm) due to the relatively small ionic radius of nitrogen, only ~6% greater than the ionic radius of the oxygen atom (Nithya et al. 2018; Schlur et al. 2014; Veréb et al. 2013). Moreover, the photocatalytic performance of the titanium dioxide-based materials may be enhanced using metal particles doping on the TiO_2 surface. The interfacial contact between semiconductor surface and metal nanoparticles can create an electric field facilitating an interfacial process of electron transfer from the photoexcited semiconductor to the deposited metal

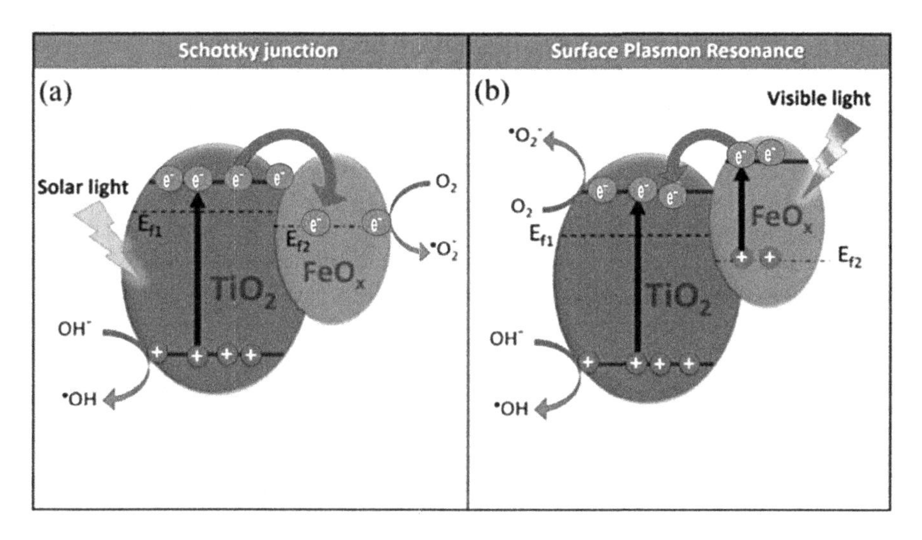

FIGURE 9.2 Schematic representation of electron transfer via Schottky barrier formation in a metal–semiconductor interface junction under (a) solar light irradiation (UV-component) and (b) under visible-light irradiation (Reprinted under the terms of the Creative Commons Attribution license from Coatings, MDPI Publications (Kiwi and Rtimi 2018)).

(Gholipour et al. 2015). The formed Schottky barrier acts as an efficient electron trap by decreasing the probability of recombination of the photogenerated charge carriers increasing the photocatalytic behavior of the system. The presence of the metal in the semiconductor also shifts the absorption into the visible region, due to its localized surface plasmon resonance (LSPR) properties. Figure 9.2a shows that the TiO_2 nanocomposites irradiated with solar energy photons, having energy higher than the TiO_2 band gap, photoexcite electrons from the valance band to the conduction band, leaving holes in the valance band. Figure 9.2b shows the interfacial charge transfer (IFCT) at the TiO_2-FeO_x heterojunction under visible light. The close contact between FeO_x NPs and TiO_2 in the sputtered films acts as an electron sink to promote the reduction of oxygen on their surfaces. Subsequently, the holes in the valence band of TiO_2 migrated, inducing bacterial oxidation. A local electric field is developed by the SPR of FeO_x in contact with TiO_2. The increased charge separation, due to the FeO_x NPs sputtered on TiO_2, increased the lifetime of the TiO_2 charge carriers. This was due to FeO_x partly substituting the lattice Ti^{4+} sites in TiO_2, which modifies the visible light absorbance of TiO_2 (Kiwi and Rtimi 2018).

Silver is the most interesting among all the metals that are deposited on the titanium dioxide. Silver–titanium dioxide is commonly synthesized by photoreduction under UV light, sol-method, and incipient wet impregnation method. Since the Fermi level of titanium dioxide is very high compared to silver, the electron transfer from the conduction band of the semiconductor to the silver nanoparticles is thermodynamically possible. The Schottky barrier formed in the physical junction of semiconductor and metal built in the physical junction of both substances halts the electrons transfer from silver to TiO_2. Although, silver exhibits LSPR under visible light, where the collective oscillation of its electrons can yield an interband excitation, it is able to provide enough energy to electrons that move to the interface to overcome the Schottky barrier (Lee, Obregón, and Rodríguez-González 2015; Tahir et al. 2016).

Copper is also widely used in conjunction with TiO_2 for antimicrobial purposes. This metal itself can perform antibacterial and antiviral activities because the copper ions can infiltrate across their cell membrane. Moreover, Cu ions can change the charge balance of the microorganism that can provide its deformation until cell lysis. According to theoretical studies, which states the incorporation of copper ions in the titanium dioxide should be less than 0.3% to cause a substitution in the crystal lattice. Therefore, the Cu^{2+}/Cu^{+} ions may replace the Ti^{4+} from the TiO_2 crystal lattice resulting in the formation of single and double oxygen vacancies. As a consequence, the copper-doped TiO_2 increases its charge transfer resistance and decreases the capacitance. Moreover, exhibiting a shift in the optical absorption edge to the visible region indicates a narrowing of the bandgap in the semiconductor. In previous studies, the use of a Cu/TiO_2 system for the removal of several microorganisms like *E. coli* has been reported (Rtimi et al. 2016). Other metal-TiO_2 systems using Au, Pt, and Pd have also been studied for the photocatalytic removal of microorganisms (Rezaeian-Delouei, Ghorbani, and Mohsenzadeh 2011; Tang et al. 2019; Tseng et al. 2013; Rodríguez-González et al. 2020).

9.2.3 TiO_2 Heterojunctions Systems

To enhance the photocatalytic activity of the semiconductor, the heterojunction systems are developed to improve the spatial separation of the photogenerated charge carriers in the semiconductor. Generally, two semiconductors show a close contact forming heterostructures based on the physical junction of their particles (Rodríguez-González et al. 2020). The heterojunction is a contact interface that is formed as a result of hybridization between two semiconductors. The semiconductor must show different bandgaps and also the narrow bandgap must lie in the visible region, before being used for the heterojunction. The combination of titanium dioxide with other semiconductors can enhance the photocatalytic efficiency of the TiO_2. This technique is not only able to improve the effective utilization rate of the electrons by promoting the photogenerated electrons and holes to transfer in the opposite direction, but it can also expand the spectral response range of the composite to visible light and even the near-infrared region. Generally, the TiO_2-based heterojunctions are mostly classified into two different types depending on the charge carrier separation mechanism, which are conventional type and direct Z-scheme. Therefore, according to the valence band and conduction band potentials of the semiconductors, there are three types of heterojunction systems which are shown in Figure 9.3 (Li, Li, and Zhou 2020).

According to the different bands and electronic structures, the conventional type can be divided into three main types: type-I (straddling gap), type-II (staggered gap), and type-III (broken gap) heterojunctions. For type-I heterojunctions, the level of the conduction band (CB) of semiconductor-I is higher than that of semiconductor-II, while the valence band (VB) of semiconductor-I is lower than that of semiconductor-II. The recombination can occur due to the difference between the band gaps and the

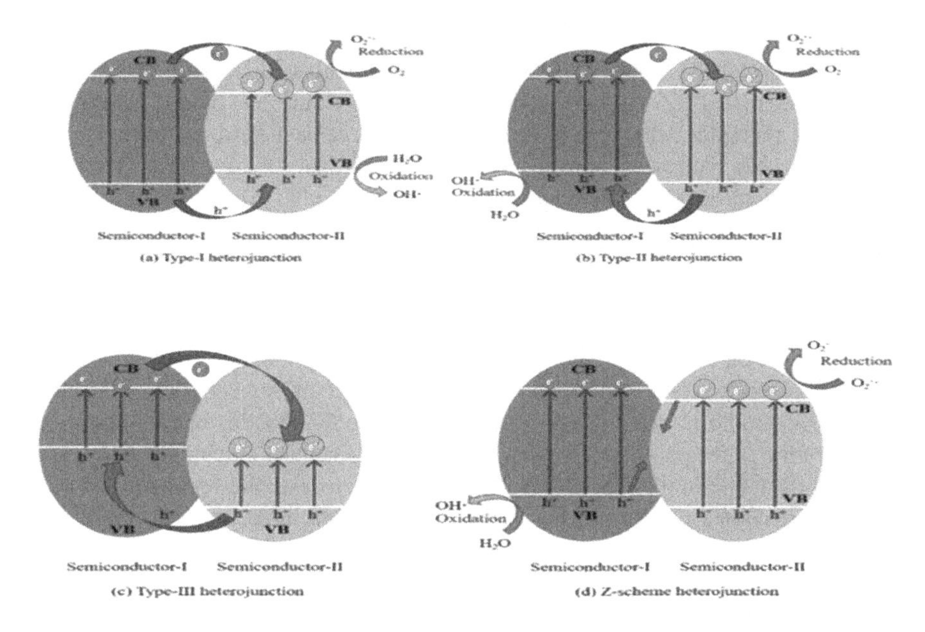

FIGURE 9.3 Types of heterojunction systems, based on two semiconductors (Reprinted under the terms of the Creative Commons Attribution license from Catalysts, MDPI Publications (Li, Li, and Zhou 2020)).

photoinduced charges accumulated on smaller bandgap semiconductors. A type-I heterojunction is composed of two semiconductors, where semiconductor-I has a conduction band edge higher than that of semiconductor-II (Figure 9.3a). Moreover, the top of the valence band (VB) of the semiconductor-I displays a lower value than the top edge of the valence band (VB) of the semiconductor-II. Under this configuration, the hole electron pairs photogenerated in the semiconductor-II migrate to the semiconductor-I, which acts as a recombination center of these charge carriers. Therefore, the heterojunction type-I commonly exhibits poor photocatalytic performance. In type-II heterojunctions (Figure 9.3b), the level of the conduction band and the valence band of semiconductor-II are higher than those of semiconductor-I. Additionally, the migration of charge carriers to the opposite directions can be promoted, because the difference between the chemical potentials causes a phenomenon called band bending. For a type-II heterojunction, the bottom edge of the conduction band of the semiconductor-I is more negative than the bottom of the CB of the semiconductor-II. In contrast, the top edge of the VB of the semiconductor-II has a more positive potential than the valence band of the semiconductor-I. These differences in the edge potentials are the driving force to provide the efficient transfer of the photogenerated charge carriers between both semiconductors, thus, reducing their recombination and increasing the photocatalytic performance of the coupled system. The band structure of the type-III heterojunctions (Figure 9.3c) is similar to that of type-II except that the staggering gap becomes so wide that the bandgaps do not overlap. Among these conventional heterojunctions, the type-II heterojunction attracts the attention of more researchers. A type-III heterojunction exhibits a band configuration similar to type-II, although, the difference in the potentials of the VB and CBs is more pronounced. This type of configuration is commonly called a Z-scheme heterojunction in which a "Z" shaped transport path is carried out. The type-II heterojunctions have the same band arrangement as the Z-scheme heterojunctions, but the electron transfer path between semiconductors is different (Figure 9.3d). The electron transfer path between semiconductors is similar to the English letter "Z". During the photocatalytic reaction, the photogenerated electrons with lower reduction ability in semiconductor-II recombine with the photogenerated holes in semiconductor-I with lower oxidation ability. Thus, the photogenerated electrons with high reduction ability in semiconductor-I and the photogenerated holes with high oxidation ability in semiconductor-II can be maintained. Moreover, the electrostatic attraction between the photogenerated electron in the conduction band (CB) of semiconductor-II and the photogenerated holes in the valence band (VB) of the semiconductor-I will promote the migration of the photogenerated electron from the semiconductor-II to the semiconductor-I, while in the type-II heterojunction, the electrostatic repulsion between the photogenerated electron of the semiconductor-I and the semiconductor-II will hinder the migration of electrons from semiconductor-I to semiconductor-II. (Rodríguez-González et al. 2020; Kang et al. 2019; Li, Li, and Zhou 2020; Zhou, Yu, and Jaroniec 2014; Low et al. 2017).

Furthermore, a photocatalytic Z-scheme system can be classified as direct or indirect depending on whether an electron mediator is necessary to achieve the transfer mechanism. For an indirect system, the Fe^{3+}/Fe^{2+} and IO_3^-/I^- redox couples are electron mediators frequently used in the liquid phase, while noble-metal nanoparticles are reported as electron mediators in all-solid-state photocatalytic systems. Besides, an S-scheme heterojunction has also been proposed from the Z-scheme basis. Therefore, the S-scheme system is composed of two n-type photocatalytic semiconductors representing an oxidation and a reduction photocatalyst. In this transfer mechanism, the driving force mainly comes from the internal electric field of the system (Fu et al. 2019). TiO_2 has been widely reported in the formation of heterojunction systems with other semiconductors for the degradation of organic pollutants, hydrogen production from water splitting, and CO_2 photoreduction. Although, only a few TiO_2 coupled systems have been studied for photocatalytic eradication of pathogenic microorganisms, some previous works on TiO_2-based heterojunction systems have been reported for the photocatalytic inactivation of several microorganisms (Chen, Tsai, and Chen 2008; Ma et al. 2009; Zhu et al. 2017).

9.2.4 Modification of TiO_2 with Graphene/Carbonaceous Materials

The graphene was first reported by Novoselov in the year (2004) and gained a lot of attention since then. The graphene material is composed of one atom thick layer of sp^2 hybridized carbon atoms forming six-member rings arranged in a two-dimensional hexagonal lattice. Many methods are used for the production of graphene by carrying out the exfoliation of the π-stacked carbon layers, such as the chemical vapor deposition (CVD) and the micromechanical cleavage of graphite. The Hummers method is most commonly used due to its low-cost and simplicity. The Hummers method consists of the strong oxidation and exfoliation of the bulk graphite and subsequent thermal/chemical reduction. Therefore, the complete removal of the oxygen functional groups due to the oxidation process may not be complete. So, the reduced form is commonly referred to as reduced graphene oxide. The coupling of the graphene with TiO_2 may increase the photocatalytic performance of the semiconductor by reducing the recombination rate of the photogenerated charge carriers. Therefore, the graphene/TiO_2 and reduced graphene oxide/TiO_2 have been used in the disinfection process of water contaminated with pathogenic microorganisms (Liu et al. 2011; Rahimi et al. 2015; Cao et al. 2013). In previously reported studies, the photoexcited electrons in TiO_2 can be transferred to the π–π conjugated network of the graphene, therefore increasing the efficiency of the photocatalytic process. Single-walled carbon nanotubes (SWCNTs) and multiwalled carbon nanotubes (MWCNTs) have also been used in the formation of composites with titanium dioxide for the disinfection process (Krishna et al. 2005; Czech and Buda 2015). Moreover, the photogenerated electrons in TiO_2 can be transferred and stored in the carbon nanotubes (CNTs). The CNTs can delocalize these charge carriers, due to their high electron-accepting properties and thus can increase the photocatalytic performance of TiO_2. According to the previous studies, the composite exhibited the formation of the Ti–C and Ti–O–C carbonaceous bonds at the heterojunction and it contributed to the charge transfer between TiO_2 and MWCNTs. Some other carbon-based materials, like carbon quantum dots (CQDs) have also been coupled with TiO_2 for the elimination of microorganisms. The CQDs exhibit interesting properties, such as photoinduced electron transfer, up and down conversion photoluminescence, and electron storage. Many photoinduced mechanisms for these materials have been explained, such as the transfer of photogenerated electrons of the irradiated TiO_2 to the CQDs, acting as electron acceptors, as well as the direct injection of electrons into the conduction band of TiO_2 coming from up-conversion and down-conversion processes of the CQDs. The activated carbon (AC) supported TiO_2 nanoparticles (TiO_2/AC) have been studied for the photocatalytic inhibition of the *E. coli* bacteria (Lee et al. 2014). It has been reported that carbon in AC can reduce the TiO_2 to form some Ti^{3+} ions (LI et al. 2008). The Ti^{3+} ions as active centers can trap the photoexcited electrons and can reduce the recombination rate of the charge carrier pairs. The other carbonaceous material such as the chitosan has also been used in the preparation of TiO_2 nanocomposites for the inactivation of *E. coli* and *S. aureus* (Raut et al. 2016). The chitosan is a linear polysaccharide composed of β-(1-4) D-glucosamine and *N*-acetyl-D-glucosamine units, which has shown antimicrobial properties toward bacteria, viruses, and fungi. Moreover, it is worth noting that the carbon doping of TiO_2 or incorporation of carbon atoms within the crystal structure of TiO_2 provides an extended absorption to the visible light range and efficient separation of the photogenerated charge carrier. The simplest procedure for carrying out the preparation of carbon-doped TiO_2 is the use of carbohydrates, such as glucose and sucrose, as carbon precursors. In this way, the incorporation of the carbonaceous species in TiO_2 occurs during the calcination process of the organic precursors (Rodríguez-González et al. 2020).

9.2.5 TiO_2 Functionalization Methods and Applications

Nanotechnology is mostly used for functionalization and deposition of nanoparticles onto textiles, which deals with the development and utilization of nanostructures. Nanoparticle manufacturing is an essential part of nanotechnology because many important properties are obtained from a nanoparticle, nanocrystal, nanolayer level, and assembling of nanoparticles. Nanotechnology is recognized as an emerging technology for the twenty-first century (Roco 1999). Many methods are used for synthesis and coating of nanomaterials on textile surfaces including hydrothermal technique (Li et al. 2015; Ashraf et al. 2013), sol-gel method (Xiong et al. 2011), dip coating (Shirgholami et al. 2011; Khan et al. 2018), electrochemical method (Liu et al. 2016), layer-by-layer method (Gustafsson, Larsson, and Wågberg 2012), CVD (Boscher et al. 2014; Aminayi and Abidi 2013), and spray coating (Zhang et al. 2018; Lu et al. 2015; Latthe and Rao 2012). In recent years, the use of TiO_2 for photocatalytic disinfection processes has been extended to commercial applications. The most recurrent use is the removal of bacteria in aqueous systems. Even the purification of storm water, which is usually of better quality than wastewater, has also been submitted to photocatalytic disinfection processes. Moreover, the use of TiO_2 as photocatalytic materials for air purification has been increased recently. Every year many studies, industrial equipment, and patents are reported regarding the concrete, pavement, or air conditioning filters. The heating ventilation air conditioning (HVAC) systems use photocatalytic TiO_2 filters for the elimination of airborne bacterial consortia. These HVAC systems are used on daily basis in offices, hospitals, malls, houses, aircrafts, factories, and buildings. The indoor air quality is improved due to the disinfection photocatalytic action of the TiO_2-coated filters in the air conditioning system. In this regard, TiO_2 thin films have been used in several everyday commodities from industries such as food, construction, environmental, medical, among others.

9.3 CONCLUSION

Titanium dioxide is considered one of the most promising photocatalysts materials, due to its unique electronic configuration, availability, low cost, chemical stability, photostability, inertness, and nontoxicity. The TiO_2 is well-known to exhibit excellent photocatalytic antimicrobial activity over a broad spectrum of microorganisms, which is due to the high redox potential of ROS generated by the photoexcitation. Nanomaterials of TiO_2 were extensively reported to kill bacteria and viruses. Nowadays, titanium dioxide is widely used as an additive in various consumer goods, in particular foods, cosmetics, pharmaceuticals, and paints as a white pigment. Despite the great progress in synthesis and modification of TiO_2, still there are many challenges and opportunities to improve the efficiency of TiO_2-based disinfection. This chapter comprehensively discussed several strategies of TiO_2 synthesis and doping techniques. Moreover, developing an antimicrobial material with UV/visible light-responsive photocatalysts is very important and necessary. Through modification of TiO_2-based materials, the photocatalytic performance of TiO_2 can be greatly improved. The disinfection of the TiO_2 effect is not specific for pathogenic microorganisms, but it can also damage beneficial microorganisms in real conditions.

The general mechanism of disinfection described in this chapter deals with the complete and effective eradication of microorganisms, because of ROS, physical damage, and the biocide properties of the TiO_2-based materials. Besides, the heterojunction systems are also very useful for antimicrobial applications. Metals or oxides for nanocomposites must have biocompatible features without photocorrosion effect. Due to the described mechanism, TiO_2-based nanomaterials can be a good alternative to decrease antimicrobial resistance. Therefore, researchers must understand the mechanism of dealing with actual microorganisms. Moreover, the efficiency and stability of the modified TiO_2 must be improved. The performance and efficiency of modified TiO_2 are currently limited by the physicochemical properties of these materials.

REFERENCES

Aminayi, Payam, and Noureddine Abidi. 2013. "Imparting Super Hydro/oleophobic Properties to Cotton Fabric by Means of Molecular and Nanoparticles Vapor Deposition Methods." *Applied Surface Science* 287. Elsevier B.V.: 223–231. doi:10.1016/j.apsusc.2013.09.132.

Ashraf, Munir, Christine Campagne, Anne Perwuelz, Philippe Champagne, Anne Leriche, and Christian Courtois. 2013. "Development of Superhydrophilic and Superhydrophobic Polyester Fabric by Growing Zinc Oxide Nanorods." *Journal of Colloid and Interface Science* 394 (1): 545–553. doi:10.1016/j.jcis.2012.11.020.

Bai, Hongwei, Zhaoyang Liu, Lei Liu, and Darren Delai Sun. 2013. "Large-Scale Production of Hierarchical TiO2 Nanorod Spheres for Photocatalytic Elimination of Contaminants and Killing Bacteria." *Chemistry – A European Journal* 19 (9): 3061–3070. doi:10.1002/chem.201204013.

Baram, Nir, David Starosvetsky, Jeana Starosvetsky, Marina Epshtein, Robert Armon, and Yair Ein-Eli. 2011. "Photocatalytic Inactivation of Microorganisms Using Nanotubular TiO2." *Applied Catalysis B: Environmental* 101 (3-4). Elsevier B.V.: 212–219. doi:10.1016/j.apcatb.2010.09.024.

Boscher, Nicolas D., Véronique Vaché, Paul Carminati, Patrick Grysan, and Patrick Choquet. 2014. "A Simple and Scalable Approach towards the Preparation of Superhydrophobic Surfaces-Importance of the Surface Roughness Skewness." *Journal of Materials Chemistry A* 2 (16): 5744–5750. doi:10.1039/c4ta00366g.

Cantarella, Maria, Ruy Sanz, Maria Antonietta Buccheri, Lucia Romano, and Vittorio Privitera. 2016. "PMMA/TiO2 Nanotubes Composites for Photocatalytic Removal of Organic Compounds and Bacteria from Water." *Materials Science in Semiconductor Processing* 42. Elsevier: 58–61. doi:10.1016/j.mssp.2015.07.053.

Cao, Baocheng, Shuai Cao, Pengyu Dong, Jing Gao, and Jing Wang. 2013. "High Antibacterial Activity of Ultrafine TiO2/graphene Sheets Nanocomposites under Visible Light Irradiation." *Materials Letters* 93. Elsevier: 349–352. doi:10.1016/j.matlet.2012.11.136.

Chen, Wei Jen, Pei Jane Tsai, and Yu Chie Chen. 2008. "Functional Fe3O4/TiO2 Core/shell Magnetic Nanoparticles as Photokilling Agents for Pathogenic Bacteria." *Small* 4 (4): 485–491. doi:10.1002/smll.200701164.

Costa, Anna Luisa, Simona Ortelli, Magda Blosi, Stefania Albonetti, Angelo Vaccari, and Michele Dondi. 2013. "TiO2 Based Photocatalytic Coatings: From Nanostructure to Functional Properties." *Chemical Engineering Journal* 225. Elsevier B.V.: 880–886. doi:10.1016/j.cej.2013.04.037.

Czech, Bozena, and Waldemar Buda. 2015. "Photocatalytic Treatment of Pharmaceutical Wastewater Using New Multiwall-Carbon nanotubes/TiO2/SiO2 Nanocomposites." *Environmental Research* 137: 176–184. doi:10.1016/j.envres.2014.12.006.

Fu, Junwei, Quanlong Xu, Jingxiang Low, Chuanjia Jiang, and Jiaguo Yu. 2019. "Ultrathin 2D/2D WO3/g-C3N4 Step-Scheme H2-Production Photocatalyst." *Applied Catalysis B: Environmental* 243 (October 2018). Elsevier: 556–565. doi:10.1016/j.apcatb.2018.11.011.

Fujishima, Akira, and Kenichi Honda. 1972. "Electrochemical Photolysis of Water at a Semiconductor Electrode." *Nature* 238 (5358): 37–38. doi:10.1038/238038a0.

Gholipour, Reza, Mohammad, Cao Thang Dinh, François Béland, and Trong On Do. 2015. "Nanocomposite Heterojunctions as Sunlight-Driven Photocatalysts for Hydrogen Production from Water Splitting." *Nanoscale* 7 (18). Royal Society of Chemistry: 8187–8208. doi:10.1039/c4nr07224c.

Gustafsson, Emil, Per A. Larsson, and Lars Wågberg. 2012. "Treatment of Cellulose Fibres with Polyelectrolytes and Wax Colloids to Create Tailored Highly Hydrophobic Fibrous Networks." *Colloids and Surfaces A: Physicochemical and Engineering Aspects* 414. Elsevier B.V.: 415–421. doi:10.1016/j.colsurfa.2012.08.042.

Kang, Xiaolan, Sihang Liu, Zideng Dai, Yunping He, Xuezhi Song, and Zhenquan Tan. 2019. "Titanium Dioxide: From Engineering to Applications." *Catalysts* 9 (2). doi:10.3390/catal9020191.

Khan, Muhammad Zaman, Vijay Baheti, Jiri Militky, Azam Ali, and Martina Vikova. 2018. "Superhydrophobicity, UV Protection and Oil/water Separation Properties of Fly ash/Trimethoxy(octadecyl)silane Coated Cotton Fabrics." *Carbohydrate Polymers* 202 (August). Elsevier: 571–580. doi:10.1016/j.carbpol.2018.08.145.

Kiwi, John, and Sami Rtimi. 2018. "Mechanisms of the Antibacterial Effects of TiO2-FeOx under Solar or Visible Light: Schottky Barriers versus Surface Plasmon Resonance." *Coatings* 8 (11). doi:10.3390/coatings8110391.

Krishna, V., S. Pumprueg, S. H. Lee, J. Zhao, W. Sigmund, B. Koopman, and B. M. Moudgil. 2005. "Photocatalytic Disinfection with Titanium Dioxide Coated Multi-Wall Carbon Nanotubes." *Process Safety and Environmental Protection* 83 (4 B): 393–397. doi:10.1205/psep.04387.

Latthe, Sanjay S., and A. Venkateswara Rao. 2012. "Superhydrophobic SiO 2 Micro-Particle Coatings by Spray Method." *Surface and Coatings Technology* 207. Elsevier B.V.: 489–492. doi:10.1016/j.surfcoat.2012.07.055.

Lee, Hyun Uk, Gaehang Lee, Ji Chan Park, Young Chul Lee, Sang Moon Lee, Byoungchul Son, So Young Park, et al. 2014. "Efficient Visible-Light Responsive TiO2 Nanoparticles Incorporated Magnetic Carbon Photocatalysts." *Chemical Engineering Journal* 240. Elsevier B.V.: 91–98. doi:10.1016/j.cej.2013.11.054.

Lee, Soo Wohn, S. Obregón, and V. Rodríguez-González. 2015. "The Role of Silver Nanoparticles Functionalized on TiO2 for Photocatalytic Disinfection of Harmful Algae." *RSC Advances* 5 (55): 44470–44475. doi:10.1039/c5ra08313c.

Li, Youji, Mingyuan MA, Xiaohu Wang, and Xiaohua Wang. 2008. "Inactivated Properties of Activated Carbon-Supported TiO2 Nanoparticles for Bacteria and Kinetic Study." *Journal of Environmental Sciences* 20 (12). The Research Centre for Eco-Environmental Sciences, Chinese Academy of Sciences: 1527–1533. doi:10.1016/S1001-0742(08)62561-9.

Li, Shuhui, Jiangying Huang, Mingzheng Ge, Chunyan Cao, Shu Deng, Songnan Zhang, Guoqiang Chen, Keqin Zhang, Salem S. Al-Deyab, and Yuekun Lai. 2015. "Robust Flower-Like TiO2@Cotton Fabrics with Special Wettability for Effective Self-Cleaning and Versatile Oil/Water Separation." *Advanced Materials Interfaces* 2 (14): 1500220. doi:10.1002/admi.201500220.

Li, Ruixiang, Tian Li, and Qixing Zhou. 2020. "Impact of Titanium Dioxide (TiO2) Modification on Its Application to Pollution Treatment—a Review." *Catalysts* 10 (7): 1–33. doi:10.3390/catal10070804.

Liu, Jincheng, Lei Liu, Hongwei Bai, Yinjie Wang, and Darren D. Sun. 2011. "Gram-Scale Production of Graphene Oxide-TiO2 Nanorod Composites: Towards High-Activity Photocatalytic Materials." *Applied Catalysis B: Environmental* 106 (1–2). Elsevier B.V.: 76–82. doi:10.1016/j.apcatb.2011.05.007.

Liu, Hui, Shou Wei Gao, Jing Sheng Cai, Cheng Lin He, Jia Jun Mao, Tian Xue Zhu, Zhong Chen, et al. 2016. "Recent Progress in Fabrication and Applications of Superhydrophobic Coating on Cellulose-Based Substrates." *Materials* 9 (3): 124. doi:10.3390/ma9030124.

Low, Jingxiang, Jiaguo Yu, Mietek Jaroniec, Swelm Wageh, and Ahmed A. Al-Ghamdi. 2017. "Heterojunction Photocatalysts." *Advanced Materials* 29 (20). doi:10.1002/adma.201601694.

Lu, Yao, Sanjayan Sathasivam, Jinlong Song, Colin R. Crick, Claire J. Carmalt, and Ivan P. Parkin. 2015. "Robust Self-Cleaning Surfaces That Function When Exposed to Either Air or Oil." *Science* 347 (6226): 1132–1135. doi:10.1126/science.aaa0946.

Ma, Ning, Xinfei Fan, Xie Quan, and Yaobin Zhang. 2009. "Ag-TiO2/HAP/Al2O3 Bioceramic Composite Membrane: Fabrication, Characterization and Bactericidal Activity." *Journal of Membrane Science* 336 (1–2): 109–117. doi:10.1016/j.memsci.2009.03.018.

Nithya, N., G. Bhoopathi, G. Magesh, and C. Daniel Nesa Kumar. 2018. "Neodymium Doped TiO2 Nanoparticles by Sol-Gel Method for Antibacterial and Photocatalytic Activity." *Materials Science in Semiconductor Processing* 83 (April). Elsevier Ltd: 70–82. doi:10.1016/j.mssp.2018.04.011.

Podporska-Carroll, Joanna, Eugen Panaitescu, Brid Quilty, Lili Wang, Latika Menon, and Suresh C. Pillai. 2015. "Antimicrobial Properties of Highly Efficient Photocatalytic TiO2 Nanotubes." *Applied Catalysis B: Environmental* 176–177. Elsevier B.V.: 70–75. doi:10.1016/j.apcatb.2015.03.029.

Rahimi, Rahmatollah, Solmaz Zargari, Azam Yousefi, Marzieh Yaghoubi Berijani, Ali Ghaffarinejad, and Ali Morsali. 2015. "Visible Light Photocatalytic Disinfection of E. Coli with TiO_2 - Graphene Nanocomposite Sensitized with tetrakis(4-Carboxyphenyl) porphyrin." *Applied Surface Science* 355. Elsevier B.V.: 1098–1106. doi:10.1016/j.apsusc.2015.07.115.

Raut, A. V., H. M. Yadav, A. Gnanamani, S. Pushpavanam, and S. H. Pawar. 2016. "Synthesis and Characterization of Chitosan-TiO2:Cu Nanocomposite and Their Enhanced Antimicrobial Activity with Visible Light." *Colloids and Surfaces B: Biointerfaces* 148. Elsevier B.V.: 566–575. doi:10.1016/j.colsurfb.2016.09.028.

Rezaeian-Delouei, M., M. Ghorbani, and M. Mohsenzadeh. 2011. "An Enhancement in the Photocatalytic Activity of TiO2 by the Use of Pd: The Question of Layer Sequence in the Resulting Hierarchical Structure." *Journal of Coatings Technology and Research* 8 (1): 75–81. doi:10.1007/s11998-010-9274-1.

Roco, M. C. 1999. "Nanoparticles and Nanotechnology Research." *Journal of Nanoparticle Research*, 1–6.

Rodríguez-González, Vicente, Sergio Obregón, Olga A Patrón-Soberano, Chiaki Terashima, and Akira Fujishima. 2020. "An Approach to the Photocatalytic Mechanism in the TiO2-Nanomaterials Microorganism Interface for the Control of Infectious Processes." *Applied Catalysis B: Environmental* 270 (March). Elsevier: 118853. doi: 10.1016/j.apcatb.2020.118853.

Rtimi, S., S. Giannakis, R. Sanjines, C. Pulgarin, M. Bensimon, and J. Kiwi. 2016. "Insight on the Photocatalytic Bacterial Inactivation by Co-Sputtered TiO2-Cu in Aerobic and Anaerobic Conditions." *Applied Catalysis B: Environmental* 182. Elsevier B.V.: 277–285. doi:10.1016/j.apcatb.2015.09.041.

Schlur, Laurent, Sylvie Begin-Colin, Pierre Gilliot, Mathieu Gallart, Gaëlle Carré, Spiros Zafeiratos, Nicolas Keller, et al. 2014. "Effect of Ball-Milling and Fe-/Al-Doping on the Structural Aspect and Visible Light Photocatalytic Activity of TiO2 towards Escherichia Coli Bacteria Abatement." *Materials Science and Engineering C* 38 (1). Elsevier B.V.: 11–19. doi:10.1016/j.msec.2014.01.026.

Scorb, E. V., L. I. Antonovskaya, N. A. Belyasova, and D. V. Sviridov. 2009. "Photocatalysts for Reagentless Disinfection on the Basis of Titanium Dioxide Films Modified by Silver Nanoparticles." *Catalysis in Industry* 1 (2): 165–170. doi:10.1134/s2070050409020135.

Shirgholami, Mohammad A., Mohammad Shateri Khalil-Abad, Ramin Khajavi, and Mohammad E. Yazdanshenas. 2011. "Fabrication of Superhydrophobic Polymethylsilsesquioxane Nanostructures on Cotton Textiles by a Solution-Immersion Process." *Journal of Colloid and Interface Science* 359 (2). Elsevier Inc.: 530–535. doi:10.1016/j.jcis.2011.04.031.

Tahir, Kamran, Aftab Ahmad, Baoshan Li, Sadia Nazir, Arif Ullah Khan, Tabassum Nasir, Zia Ul Haq Khan, Rubina Naz, and Muslim Raza. 2016. "Visible Light Photo Catalytic Inactivation of Bacteria and Photo Degradation of Methylene Blue with Ag/TiO2 Nanocomposite Prepared by a Novel Method." *Journal of Photochemistry and Photobiology B: Biology* 162. Elsevier B.V.: 189–198. doi:10.1016/j.jphotobiol.2016.06.039.

Tang, Yanan, Hang Sun, Yinxing Shang, Shan Zeng, Zhen Qin, Shengyan Yin, Jiayi Li, Song Liang, Guolong Lu, and Zhenning Liu. 2019. "Spiky Nanohybrids of Titanium Dioxide/gold Nanoparticles for Enhanced Photocatalytic Degradation and Anti-Bacterial Property." *Journal of Colloid and Interface Science* 535. Elsevier Inc.: 516–523. doi:10.1016/j.jcis.2018.10.020.

Tseng, Yao Hsuan, Der Shan Sun, Wen Shiang Wu, Hao Chan, Ming Syuan Syue, Han Chen Ho, and Hsin Hou Chang. 2013. "Antibacterial Performance of Nanoscaled Visible-Light Responsive Platinum-Containing Titania Photocatalyst in Vitro and in Vivo." *Biochimica et Biophysica Acta – General Subjects* 1830 (6). Elsevier B.V.: 3787–3795. doi:10.1016/j.bbagen.2013.03.022.

Veréb, G., L. Manczinger, A. Oszkó, A. Sienkiewicz, L. Forró, K. Mogyorósi, A. Dombi, and K. Hernádi. 2013. "Highly Efficient Bacteria Inactivation and Phenol Degradation by Visible Light Irradiated Iodine Doped TiO2." *Applied Catalysis B: Environmental* 129. Elsevier B.V.: 194–201. doi:10.1016/j.apcatb.2012.08.037.

Wang, Jinfeng, Jian Zhao, Lu Sun, and Xungai Wang. 2015. "A Review on the Application of Photocatalytic Materials on Textiles." *Textile Research Journal* 85 (10): 1104–1118. doi:10.1177/0040517514559583.

Xiong, Dean, Guojun Liu, Liangzhi Hong, and E. J. Scott Duncan. 2011. "Superamphiphobic Diblock Copolymer Coatings." *Chemistry of Materials* 23 (19): 4357–4366. doi:10.1021/cm201797e.

Yadavalli, Tejabhiram, and Deepak Shukla. 2017. "Role of Metal and Metal Oxide Nanoparticles as Diagnostic and Therapeutic Tools for Highly Prevalent Viral Infections." *Nanomedicine: Nanotechnology, Biology, and Medicine* 13 (1). Elsevier Inc.: 219–230. doi:10.1016/j.nano.2016.08.016.

Yu, Yanlong, Peng Zhang, Limei Guo, Zhandong Chen, Qiang Wu, Yihong Ding, Wenjun Zheng, and Yaan Cao. 2014. "The Design of TiO2 Nanostructures (nanoparticle, Nanotube, and Nanosheet) and Their Photocatalytic Activity." *Journal of Physical Chemistry C* 118 (24): 12727–12733. doi:10.1021/jp500252g.

Zhang, Weijuan, Wenkai Chang, Baozhen Cheng, Zenghe Li, Junhui Ji, Yang Zhao, and Jun Nie. 2015. "Formation of Rod-like Nanostructure by Aggregation of TiO2 Nanoparticles with Improved Performances." *Bulletin of Materials Science* 38 (6): 1617–1623. doi:10.1007/s12034-015-0969-x.

Zhang, Zhi Hui, Hu Jun Wang, Yun Hong Liang, Xiu Juan Li, Lu Quan Ren, Zhen Quan Cui, and Cheng Luo. 2018. "One-Step Fabrication of Robust Superhydrophobic and Superoleophilic Surfaces with Self-Cleaning and Oil/water Separation Function." *Scientific Reports* 8 (1). Springer US: 1–12. doi:10.1038/s41598-018-22241-9.

Zhou, Peng, Jiaguo Yu, and Mietek Jaroniec. 2014. "All-Solid-State Z-Scheme Photocatalytic Systems." *Advanced Materials* 26 (29): 4920–4935. doi:10.1002/adma.201400288.

Zhu, Qi, Xiaohong Hu, Mishma S. Stanislaus, Nan Zhang, Ruida Xiao, Na Liu, and Yingnan Yang. 2017. "A Novel P/Ag/Ag2O/Ag3PO4/TiO2 Composite Film for Water Purification and Antibacterial Application under Solar Light Irradiation." *Science of the Total Environment* 577: 236–244. doi:10.1016/j.scitotenv.2016.10.170.

Ziental, Daniel, Beata Czarczynska-Goslinska, Dariusz T. Mlynarczyk, Arleta Glowacka-Sobotta, Beata Stanisz, Tomasz Goslinski, and Lukasz Sobotta. 2020. "Titanium Dioxide Nanoparticles: Prospects and Applications in Medicine." *Nanomaterials* 10 (2). doi:10.3390/nano10020387.

Zyoud, Ahed H., Majdi Dwikat, Samar Al-Shakhshir, Sondos Ateeq, Jumana Ishtaiwa, Muath H.S. Helal, Maher Kharoof, et al. 2017. "ZnO Nanoparticles in Complete Photo-Mineralization of Aqueous Gram Negative Bacteria and Their Organic Content with Direct Solar Light." *Solar Energy Materials and Solar Cells* 168 (April). Elsevier B.V.: 30–37. doi:10.1016/j.solmat.2017.04.006.

10 Impact of Copper and Ions against Coronavirus
Impact of Copper and Ions

Azam Ali
Technical University of Liberec, Czech Republic
Saeed Ahmad
University of Peshawar, Pakistan
Jiri Militký
Technical University of Liberec, Czech Republic
Hira Khaleeq
Government College University, Pakistan
Muhammad Zaman Khan
Technical University of Liberec, Czech Republic

CONTENTS

10.1 INTRODUCTION

Viruses are small obligate intracellular parasites, which by definition contain either an RNA or DNA genome surrounded by a protective, virus-coded protein coat. In 2019, a new strain of coronavirus called SARS-CoV-2 started circulating and causing the coronavirus disease 2019 or disease COVID-19. The tragedy is that the virus does not show its impact on the human body in the first few days about 4–7 days (incubation period of the virus inside the human body). This infected person (during the time of incubation of the virus) has been visiting the places and remain in touch with surfaces and people. It becomes too late when a person realizes the severe symptoms, and till that time an infected person has delivered the viruses to many places, communities, or even to family members. The other people (secondarily infected person) receive the virus from a contaminated environment and infectious viruses have been transferred to facial mucosa. Hence, it is the need of the hour and it becomes necessary to clean the contaminated surfaces to overcome the spread of coronavirus. To achieve the cleaning goal many surface cleaning materials based on bleaching agents has been introduced. The main drawback of using surface cleaning materials is their limited life. As they remain effective only for a few minutes or hours, the cleaned surfaces turn to neutral again. Besides, the surface cleaning materials and agents leave residual particles and colonies of viruses that are enough to incubate again and initiate infection. In this situation, surfaces should be developed or coated with permanent biocidal materials to control the viability and infection of viruses. Copper, its alloys, and ions have proved to have excellent antiviral, antifungal, and antibacterial activity against a wide range of pathogens.

10.2 MECHANISM OF ACTION ON THE HUMAN BODY

When a virus enters the body, it moves freely. It uses its outermost shell protein to intact with the cells. In the first few days, its effects on the upper respiratory system are commonly called sinusitis. Then it goes down in the throat and causes pain in the throat. After that, it further moves down to the larynx and then bronchi (lower respiratory system). It uses its outer crown-like protein spikes as a key to intact or unlocks the receptors and get into cells. Most probably, the H2 receptors in the lungs are activated. The virus starts to release the RNA (genetic material) after entering into the cells. The cell starts to replicate the virus and makes thousands of copies. During the time of replication, human feelings are quite normal and have no symptoms. This is called the incubation time of the virus. It takes about 4–7 days for the symptoms to appear. In the severe condition of acute respiratory distress syndrome or ARDS, lungs develop too much inflammation damage. The circulation of oxygen through the lungs and blood cells becomes thicker. Hence, the painful cough and shortness of breath occurrs.

10.3 TREATMENT OF COVID-19

Currently, there is no vaccine available for the treatment of COVID-19. However, scientists have been successful in replicating the virus, which will help in early detection and treatment of the virus in the affected people (who are not yet showing symptoms but have the virus in their bodies). Antibiotics cannot be used for the treatment of SARS as it is a viral disease, whereas supplemental oxygen, antipyretics, and ventilation are much supportive in the treatment of SARS. In this case, if someone is SARS positive, then he/she must be kept in complete isolation with complete precautions taken by medical staff in contact with the patients. Initially, there was unreliable support for the use of some steroids and the antiviral drug (ribavirin). However, there has been no published evidence for the support of this treatment. It

should be taken into account that no extra efforts were put into serious research for the treatment of coronavirus by some antivirus drug before the emergence of SARS. In the current scenario, the high death rate and rapid transmission of COVID-19 is a global threat. This disastrous situation is created due to the lack of research and effective therapy against SARS patients. However, just some empirical strategies have been subjected to treat the patients of coronavirus previously. The only way to avoid life-threatening coronavirus is to kill that virus before transmitting inside any human body. That is why the selection of antiviral common materials used in daily life is necessary. Copper is the only metal that is most effective and reduces the viability of the coronavirus within 4 h.
Approved Tests

a. Chest X-ray
b. ELISA (Enzyme-Linked Immunosorbent Assay)
c. Immunofluorescence
d. PCR (*polymerase chain reaction)*

10.4 COMMONLY USED MATERIALS FOR INACTIVATION OF CORONAVIRUS

Some active and speedier agents (mostly disinfectants) that have proved to deactivate the coronavirus within a minute are chlorine-based bleaches, quaternary compounds, sodium hypochlorite-based bleach, and hydrogen peroxide-based bleach. Ultraviolet rays especially UV-C (shortwave ultraviolet light), ozone gas, higher temperatures, and humidity also tend to result in other coronaviruses dying quicker. Although research has an emphasis that a related coronavirus family that causes SARS could be inactivated by temperatures above 56°C at a rate of about 10,000 viral particles every 15 min, it has been concluded that coronavirus could be inactivated by disinfecting surfaces with sodium hypochlorite 0.25 %, or chlorine dioxide (99% purity) diluted at 1/2.5 relation, 62–71% alcohol (ethanol), or 0.5% hydrogen peroxide bleach.

10.5 COPPER AND OXIDE

Copper is an inorganic metal, and it belongs to the first row of transition series of elements, which consists of Sc, Ti, V, Cr, Mn, Fe, Ni, Cu, Ni, Co, and Zn. Copper resides in the middle of the periodic table in group 11, along with Ag and Au. The element has the symbol Cu, an atomic number of 29, and an atomic mass of 63. Copper has two main oxidation states (+1 and +2) and two naturally occurring isotopes (^{63}Cu and ^{65}Cu), with abundances of 69.17% and 30.83%, respectively. The oxidation states of copper make it more favorable against antimicrobial behavior. Despite a similarity in electronic structure, there are few resemblances between the chemistry of the three elements in group 11, although certain complexes of Cu^{2+} and Ag^{2+} are isomorphous. There are 29 isotopes of copper. ^{63}Cu and ^{65}Cu are stable, with ^{63}Cu comprising approximately 69% of naturally occurring copper (Audi, Bersillon, Blachot, Wapstra 2003).

10.6 USE OF COPPER AGAINST PATHOGENS

Copper is the most effective tool/matter against bacteria and viruses. The pathogens of influenza *Escherichia coli*, superbugs MRSA (methicillin-resistant *Staphylococcus aureus*), and coronavirus (SARS-COV-2) become inactive when they fall on the copper surface. Their viability is completely inactive and remain harmless (Warnes, Little, and Keevil 2015) (Wu, Fernandez-lima, and Russell 2010) (Ali, Baheti, Javaid, Militky 2018b) (Ali et al. 2020). While the other solid surfaces are made up of plastic, aluminum, steel, cardboard, or any other polymer or metal, these pathogens remain active for 4 days to even more than a week (Warnes, Little, and Keevil 2015). Previous studies have shown that murine norovirus (MNV) and human norovirus, highly infectious nonenveloped viruses that are resistant to environmental stress and impervious to many cleaning agents, are destroyed on copper and copper alloy surfaces (Ruuskanen, Lahti, Jennings, and Murdoch 2011). One more study was conducted within the intensive care units (ICUs) of three separate US hospitals. The study was approved by the institutional review boards for all sites as well as by the Office of Risk Protection of the United States Army. The Medical University of South Carolina (hospital 1) located in Charleston, South Carolina, is a 660-bed academic facility with 17 medical ICU beds. The Memorial Sloan Kettering Cancer Centre (hospital 2) is located in New York. The microbial burden (MB) associated with commonly touched surfaces in ICUs was determined by sampling six objects in 16 rooms in ICUs in three hospitals over 43 months. At month 23, copper-alloy surfaces, with inherent antimicrobial properties, were installed onto six monitored objects in eight of 16 rooms. Copper was found to cause a significant (83%) reduction in the average MB found on the objects (465 CFU/100 cm^2) compared to the controls (2,674 CFU/100 cm^2).

The introduction of copper surfaces to objects formerly covered with plastic, wood, stainless steel, and other materials found in the patient care environment significantly reduced the overall MB, thereby providing a potentially safer environment for hospital patients, health care workers (HCWs), and visitors (Schmidt, Attaway, Sharpe, John, Sepkowitz, Morgan, Fairey, Singh, Steed, Cantey, and Freeman, 2012) (Warnes, Little, and Keevil 2015). Researchers have extensively used copper and its ions against a variety of pathogens. In a recent study, cotton fabric was first pretreated with copper, silver, and stannous nanoparticles and subsequently, copper plating was performed to enhance the antimicrobial properties (see Figure 10.1). Figures 10.2 and 10.3 are showing the clear zone of inhibition around the copper-coated fabric samples against Gram-positive and Gram-negative bacteria (Ali et al. 2020). The elemental composition and deposition of copper particles are analyzed with the help of EDS and scanning electron microscope and the results are shown in (Figure 10.1).

In another study, copper and cuprous particles were deposited on the nylon stretchable surface. The activated surface was further deposited with a silver fine layer. The prepared substrate was used as conductive electrodes in the transcutaneous electrical nerve stimulation device and triboelectric device. The effectiveness of electrodes against *S. aureus* and *E. coli* was determined by the zone of inhibition method

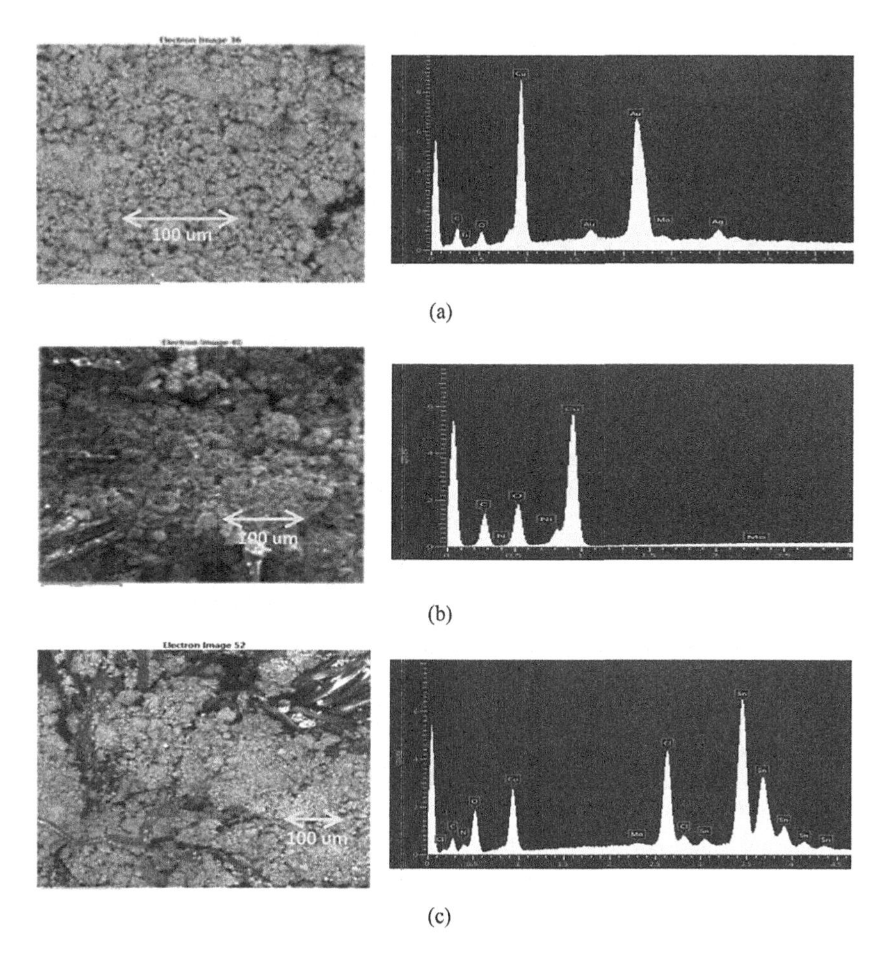

FIGURE 10.1 SEM images of copper coating over (a) copper particles, (b) silver, and (c) stannous particles and their EDS analysis (Ali et al. 2020).

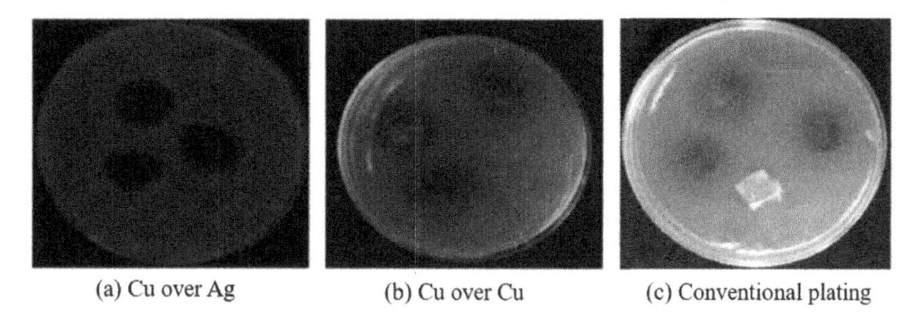

(a) Cu over Ag (b) Cu over Cu (c) Conventional plating

FIGURE 10.2 Antimicrobial properties of copper-plated fabric against *E. coli* (Ali et al. 2020).

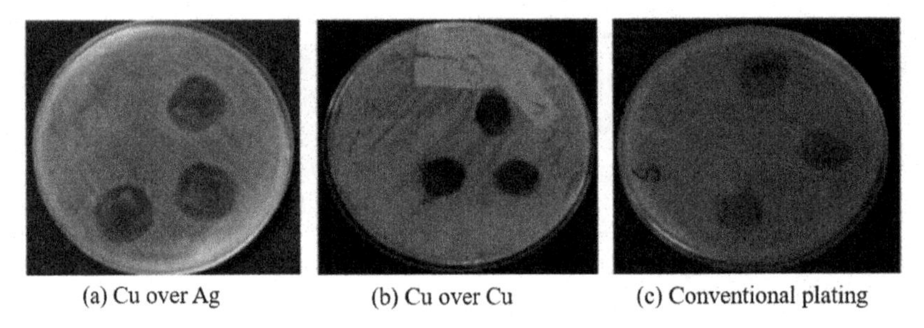

(a) Cu over Ag (b) Cu over Cu (c) Conventional plating

FIGURE 10.3 Antimicrobial properties of the copper-plated fabric against *S. aureus* (Ali et al. 2020).

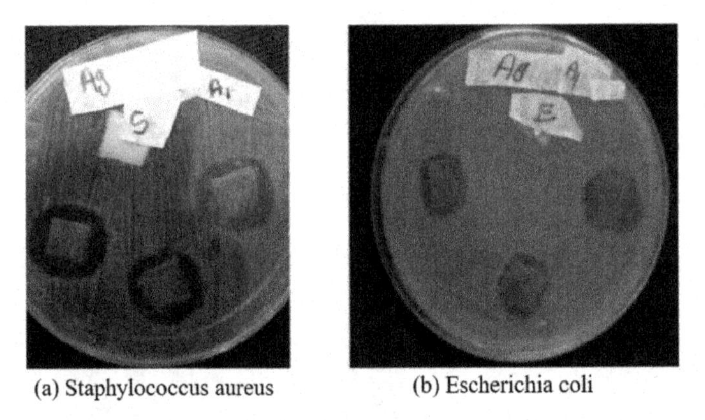

(a) Staphylococcus aureus (b) Escherichia coli

FIGURE 10.4 Antimicrobial properties of silver copper complex deposition (Ali, Baheti, Javaid, Militky 2018b; Ali, Baheti, Militky, Khan 2018c).

(Ali et al. 2018a) (Ali, Baheti, Javaid, Militky 2018b, 2018c, Ali, Baheti, Militky, Khan 2018c). The results are shown in (Figure 10.4).

10.7 EFFECT OF COPPER ON CORONAVIRUS

Copper, ions, and copper alloy surfaces degrade the viral genome and ensure that the fragmentation of coronavirus becomes irreversible. A combination of genetic mixing with segmented genomes and point of mutations (mostly evident in viruses that cause disease in the respiratory tract such as coronavirus and influenza virus) results in constantly changing antigenicity and host immune response envision. Hence, it is attached to the host cell and gets transmitted from animals to humans. It can occur only if the mutation results in an increased ability of the virus to bind to human cells. If this is accompanied by a decrease in binding to the original host, then the human-to-human transmission can occur, presenting a substantial threat of the rapid spread of a novel virus throughout the community. The copper ionic structures are most effective in killing bacteria and inactivating norovirus. The ions of copper and complex have direct impact against microbial pathogens to inhibit their growth.

The procedure of bacterial inhibition on copper and ion surfaces is very complex and involves not only direct action of copper ions on multiple sites, but also the generation of destructive oxygen radicals, resulting in "metabolic suicide." This shows that the hydroxyl radicals and superoxide generation may be important in the inactivation and degradation of coronaviruses on copper alloys but that inactivation on 100% copper surfaces is primarily due to the direct effect of copper ions. Coronavirus genome RNA to copper and alloys is very specific in fragmentation of entire genome that can also be observed at the gene level by the reduction in copy number of a small fragment of proteins, and this extent of damage increases with contact time. The capsid integrity of norovirus allows access of copper ions to the genome inactivating the virus. For coronavirus, the envelope and nucleoprotein are likewise compromised, and the process occurs more rapidly than with nonenveloped norovirus, which has a resistant capsid, to allow copper ion and/or ROS to destroy the genome. Excitingly, there was a 10-min delay in inhibition of simulated wet-droplet contamination which may show the time taken to breach the envelope and disrupt the nucleoprotein which allows access of copper ions to the coronavirus genome.

According to the theory of Mis-metalation, the metal (copper) will replace some of the other metals (iron–sulfur cluster proteins are changed to the copper–sulfur ones) that are present in specific proteins of bacteria and viruses and results in the fragmentation of their proteins (Imlay 2014). The major structure of coronavirus is composed up of proteins. Proteins have a bond with polypeptides, coenzymes, cofactors, or also with another proteins and macromolecules like DNA and RNA. The S protein of coronavirus (responsible for the attachment of the virus to host cell) is composed of oligosaccharides which are covalently attached to the side groups of the polypeptide chains (oligomers, polymers, nucleic acids, polysaccharides, oligosaccharides) (Nelson and Cox 2005). In a previous work, peptides of Cu^+ and Cu^{2+} showed the relative binding with specific amino acids and their fragmentation (Wen, Yalcin, Harrison 1995). Lim et al. found the Cu_2^+ binding sites in copper metalloproteins for oxidizing the amino acids by using mass spectrometry (Lim, Vachet 2003). If the protein is translated from messenger RNA, then the C-terminal carboxyl group (also known as the carboxyl-terminus (COOH-terminus)) is an important Cu^+ ligand, while the coordination of Cu^+ is less with N-terminal (α-amino acid). The difference is due to deprotonation of the acidic C-terminal carboxyl group and charge solvation of the copper ions (Wu, Fernandez-Lima, Perez, Russell 2009). Due to *charge solvation* (CS), the metal (Cu) *charge* is stabilized by interaction with one or more electron-rich sites.

10.8 MECHANISM OF COPPER ACTION ON PATHOGENIC BACTERIA

Hygiene has acquired importance in the field of protective textiles. In the field of medical and high-tech applications, textile consumers are looking for antibacterial fabrics. Microorganism growth is dangerous for both living and nonliving matters. Microorganisms have a severed effect on textile raw materials, wet processing chemicals, rolls, or bulk materials in the storage room. Some critical effects of the microbial attack are unpleasant smells from socks and undergarments, the spread of stains, degradation of textile fibers, and even some allergic diseases. Famous species of microorganisms are yeast, mold, fungus, and mildew. Both good and bad types of microorganisms are part of our everyday lives. Some natural fibers are better food and living place for microbes due to their inherent natural properties of hydrophilicity and their porous structure. Microorganisms like the moist and warm environment in socks and developed rapid colonies. The pathogenic microbes have an adverse effect on human skin and cause severe infection. Besides, the staining, bad odor, and loss of the functional properties of textile substrates are the results of microbial attack. Antimicrobial agents are used for killing or inhibiting the growth of bacteria. Agents that kill bacteria are called bactericidal, while the agents that inhibit their growth are known as bacteriostatic. Usually, antimicrobial agents are classified into two groups: the first is natural antimicrobial agents, and the second is synthetic antimicrobial agents. Effective natural antimicrobial agents are chitosan, clove, turmeric, tulsi, neem, pomegranate, and aloe vera, while among synthetic agents are antimicrobial dyes, quaternary ammonium compounds, Triclosan (2, 4, 4′-trichloro-2′ hydroxy diphenyl ether), regenerable N-halamine polyhexamethylene biguanides (PHMB), peroxyacids, and metals and metal salts such as copper, silver, zinc, and nanoparticles of noble metals and metal oxides. Microbes are always present on the human body even on clean skin. The normal range of microorganisms on the human hand is between 100 and 1,000 microbes/cm^2. Textiles, having features approximate to the human body provide an excellent medium for growth, adherence, propagation, and transfer of infection caused by microbial species. The antibacterial property of coated fabrics can be attributed to the combination of chemical and physical interactions of bacteria with particles. Nanoparticles can incorporate into the cell via endocytosis mechanisms. Afterward, the cellular uptake of ions increased as ionic species were subsequently released within the cells by nanoparticle dissolution (Studer, Limbach, van Duc, Krumeich, Athanassiou, Gerber, Moch, and Stark, 2010). This resulted in high intracellular concentration gained within the cell for further massive oxidative stress. Micrometric metal does not cause cell damage as compared to highly biocidal nanoparticles of the same mass (Karlsson, Cronholm, Hedberg, Tornberg, de Battice, Svedhem, and Wallinder, 2013). The unique properties of the antimicrobial effect of nanoparticles arise from a variety of aspects, including the similar size of nanoparticles and biomolecules such as proteins and polynucleic acids (Zheng, Davidson, and Huang, 2003). The antibacterial behavior of copper nanoparticles could be a result due to the physical, chemical, and combination of physio-chemical interactions and behaviors.

a. **Chemical interactions can occur between:**
   (i) Cu^{++} ions and components in the interior of cell structure, (ii) Transportation of Cu^{++} ions into the cell, (iii) Cell membrane components and hydrogen peroxide (H_2O_2) generated due to the presence of copper nanoparticles, (iv) The generation of chemical species generated due to Cu nanoparticles and components in the interior of the cell, and (v) Copper nanoparticles can form organic complexes with sulfur-, nitrogen-, or oxygen-containing functional groups present in the microorganism (Ahmad et al. 2017). This may result in defects

in the conformational structure of nucleic acids and proteins, besides changes in oxidative phosphorylation and osmotic balance. Finally, microorganisms exposed to toxic doses of some metal particles upregulate genes involved in the elimination of ROS generating oxidative stress (Lemire, Harrison, and Turner, 2013; Chattopadhyay, Patel 2010).

b. **Physical interactions can be as follows:**
(i) Physical blockage of the transport channels of cell membranes by Cu nanoparticles, (ii) Physical damage to the membrane components by Cu-NPs due to abrasion, (iii) Penetration of Cu-NPs particles through the cell membrane to interact with the interior of the cell, and (iv) direct interaction between Cu-NPs and bacterial cell membrane components through electrostatic effect.

c. **A combination of the physical and chemical interaction as described above**:
The combination of physical and chemical interaction is a very complex behavior. These types of behavior result in either the participation of antimicrobial copper nanoparticles itself or their ions can participate in the biocide mechanisms. Anyway, independent of that, in the end, metal nanoparticles are the active biocidal agents. Copper is most essential for life; therefore, all cells possess a sufficient amount of inherent copper. Hence, to keep intracellular copper at safe levels all cells possess copper homeostatic mechanisms. However, under different environmental conditions, external copper causes an imbalance of the homeostatic system and leads to overload the intracellular copper. The condition becomes too severe and approaches the toxic level. The toxic effect of copper has a wide range of mechanisms and till now it is unknown to describe with certainty which mechanism is active against a particular bacterium. The redox property of copper is well-known and is the most occurring mechanism resulting in lethal oxidative damage to cells. All different mechanisms are still under research. In fact, under different environmental and growth conditions the behavior of copper is different on cells. There are numerous unknown pathways which support to enter the copper in cells. The cytoplasm of the cell can reduce the copper to Cu^{+}, which in turn participate in Fenton type reactions and produce highly reactive hydroxyl radicals. This leads to a reaction nonspecifically with proteins, lipids, nucleic acid, and free amino acids. The anaerobic conditions make the copper-glutathione complexes, which behave as copper-donors for metalloenzymes. The dominant toxicity mechanism is referred to as the displacement of iron from iron–sulfur cluster proteins by Cu^{+} (Solioz 2018).

10.9 CONCLUSION

This chapter focused on the discussion about the evolution, transmission of a novel coronavirus SARS-CoV-2, and the effectiveness of copper against their specific proteins. The spread and outbreak of the novel coronavirus SARS-CoV-2 are still uncontrolled and going on continuously spreading all over the world. The death toll and economic disruption are increasing day by day. The spread is not only increasing leaps and bounds by touching but also transmitting human-to-human (by contact, droplets, and fomites). The viability of coronavirus is from several hours to several days on different communal areas and surfaces. In this regard, the copper, copper ions, copper alloys, and copper coating exploit the full potential to degrade the pathogenic bacteria and viruses. Copper is very important in the different biological and biochemical processes due to its ions Cu^{2+} and Cu^{+} having functions including oxidation, dioxygen transport, and electron transfer. Copper has the peculiar nature of oxidation and reduction due to its unique position in the periodic table among oxidation metals. Copper is a redox-active metal usually shifts between Cu^{+} (reduction) and Cu^{2+} (oxidized) states. The peculiar nature of copper to donate and accept electrons generate Cu^{+} and Cu^{2+} ions. The function of the redox reaction ($Cu^{2+} \leftrightarrow Cu^{+}$) is expediated by reactive oxygen species. Reactive oxygen species are normally generated on alloy surfaces and responsible for the inactivation of pathogens. The denaturing of specific proteins of coronaviruses by the interaction of copper and its ions is well elaborated.

ACKNOWLEDGEMENT

This work was supported by the Ministry of Education, Youth and Sports of the Czech Republic and the European Union – European Structural and Investment Funds in the frames of Operational Programme Research, Development and Education under project Hybrid Materials for Hierarchical Structures [HyHi, Reg. No. CZ.02.1.01/0.0/0.0/16_019/0000843].

REFERENCES

Ahmad, Sheraz, Munir Ashraf, Azam Ali, Khubab Shaker, Muhammad Umair, Ali Afzal, Yasir Nawab, and Abher Rasheed. 2017. "Preparation of Conductive Polyethylene Terephthalate Yarns by Deposition of Silver & Copper Nanoparticles." *Fibres and Textiles in Eastern Europe* 25 (5): 25–30. doi: 10.5604/01.3001.0010.4623.

Ali, Azam, Vijay Baheti, Jiri Militky, Zaman Khan, and Syed Qummer Zia Gilani. 2018a. "Comparative Performance of Copper and Silver Coated Stretchable Fabrics." *Fibers and Polymers* 19 (3). doi: 10.11007/s12221-018-7917-5.

Ali, Azam, Vijay Baheti, Muhammad Usman Javaid, Jiri Militky. 2018b. "Enhancement in Ageing and Functional Properties of Copper-Coated Fabrics by Subsequent Electroplating." *Applied Physics A* 124 (9): 651.

Ali, Azam, Vijay Baheti, Jiri Militky, and Zaman Khan. 2018c. "Utility of Silver-Coated Fabrics as Electrodes in Electrotherapy Applications." *Journal of Applied Polymer Science* 135: 46357. doi: 10.1002/app.46357.

Ali, Azam, Vijay Baheti, Michal Vik, and Jiri Militky. 2020. "Copper Electroless Plating of Cotton Fabrics after Surface Activation with Deposition of Silver and Copper Nanoparticles." *Journal of Physics and Chemistry of Solids* 137 (October 2018): 109181. doi: 10.1016/j.jpcs.2019.109181.

Audi, Georges, Bersillon, Olivier, Blachot, Jean, Wapstra. 2003. "The NUBASE Evaluation of Nuclear and Decay Properties." *Nuclear Physics A* 729: 3–128.

Chattopadhyay, DP, Patel, BH. 2010. "Effect of Nanosized Colloidal Copper on Cotton Fabric." *Journal of Engineered Fabrics & Fibers* 5 (3): 1–6.

Imlay, James A. 2014. "The Mismetallation of Enzymes During Oxidative Stress." *Journal of Biological Chemistry* 289: 41. 21–28.

Karlsson, Hanna L., Pontus Cronholm, Yolanda Hedberg, Malin Tornberg, Laura De Battice, Sofia Svedhem, and Inger Odnevall Wallinder. 2013. "Cell Membrane Damage and Protein Interaction Induced by Copper Containing Nanoparticles—Importance of the Metal Release Process." *Toxicology* 313: 59–69.

Lemire, Joseph A., Joe J. Harrison, and Raymond J. Turner. 2013. "Antimicrobial Activity of Metals: Mechanisms, Molecular Targets and Applications." *Naure Reviews Microbiology* 11: 371–384.

Lim, Jihyeon, and Richard W. Vachet. 2003. "Development of a Methodology Based on Metal-Catalyzed Oxidation Reactions and Mass Spectrometry to Determine the Metal Binding Sites in Copper Metalloproteins." *Analytical Chemistry* 75: 1164–1172.

Nelson, David L., Cox, Michael M. 2005. *Principles of Biochemistry*. 4th ed. New York: W. H. Freeman.

Ruuskanen O, Lahti E, Jennings LC, Murdoch DR. 2011. "Viral Pneumonia." *Lancet* 10: 61459-6.

Schmidt, M.G., Attaway, H.H., Sharpe, P.A., John, J., Sepkowitz, K.A., Morgan, A., Fairey, S.E., Singh, S., Steed, L.L., Cantey, J.R. and Freeman, K.D., 2012. "Sustained Reduction of Microbial Burden on Common Hospital Surfaces through Introduction of Copper." *J Clin Microbiol* 50: 2217–2223.

Solioz, Marc. 2018. *Copper and Bacteria : Evolution, Homeostasis and Toxicity*. Bern: Springer.

Studer, A.M., Limbach, L.K., van Duc, L., Krumeich, F., Athanassiou, E.K., Gerber, L.C., Moch, H.; Stark, W.J. 2010. "Nanoparticle Cytotoxicity Depends on Intracellular Solubility: Comparison of Stabilized Copper Metal and Degradable Copper Oxide Nanoparticles." *Toxicol. Lett* 197: 169–174.

Thiel V., Ed. 2007. *Coronaviruses: Molecular and Cellular Biology*. 1st ed. Caister Academic Press, Norfolk, United Kingdom

Warnes, Sarah L., Zoë R. Little, and C. William Keevil. 2015. "Human Coronavirus 229E Remains Infectious on Common Touch Surface Materials." *MBio* 6 (6). doi: 10.1128/mBio.01697-15.

Wen, D., Yalcin, T., Harrison, A. G. 1995. "Fragmentation Reactions of Cu_- Cationated _-Amino-Acids." *Rapid Commun. Mass Spectrom* 9: 1155–1157.

Wu, Z., Fernandez-Lima, F. A., Perez, L. M., Russell, D. H. 2009. "A New Copper Containing MALDI Matrix That Yields High Abundances of [Peptide _ Cu]_ Ions." *Journal of the American Society for Mass Spectrometry* 20: 1263–1271.

Wu, Zhaoxiang, Francisco A Fernandez-lima, and David H Russell. 2010. "Amino Acid Influence on Copper Binding to Peptides: Cysteine Versus Arginine." *JAM* 21 (4): 522–533. doi: 10.1016/j.jasms.2009.12.020.

Zheng M, Davidson F and Huang X. 2003. "Ethylene Glycol Monolayer Protected Nanoparticles for Eliminating Nonspecific Binding with Biological Molecules." *Journal of the American Chemical Society* 125: 7790–7791.

Section III

Textile Applications

11 Copper-Coated Textiles for Viruses Dodging

Shi Hu, Dana Kremenakova, Jiri Militký and Aravin Prince Periyasamy
Technical University of Liberec, Czech Republic

CONTENTS

11.1 INTRODUCTION

As early as more than 2,000 years ago, the ancient Chinese people had used coating glue on the surfaces of fabrics (Wei, Ma, and Schreiner 2012). At that time, they were mostly natural compounds such as lacquer and tung oil, which were mainly used in the production of waterproof cloth ("The Virtues of Ancient Glues – Gilles Perrault" 2020). Modern textile coating technology enables textiles to combine multifunction, especially in the medical field.

Copper and copper alloy have been proved for certain bacteria or viruses. The metallization methods for textiles, including dry process and wet process are described in this chapter as well. With the developments in textile technology, advanced copper-coated products are available on the market and applied to protect people against this pandemic. Besides, several discussions regarding developing safe and effective cooper-coated materials which prevent COVID-19 are described.

11.2 ANTIBACTERIAL AND ANTIVIRAL PROPERTIES OF COPPER AND COPPER ALLOY

The antibacterial and antiviral properties of copper and copper alloy were discovered almost 2,000 years ago (Vincent, Hartemann, and Engels-Deutsch 2016). To sterilize water and wounds, the ancient Egyptians, Greeks, Romans, and Aztecs used copper compounds for good hygiene. Hippocrates treated open wounds and skin irritations with copper (Grass, Rensing, and Solioz 2011). The Romans cataloged numerous medicinal uses of copper for various diseases. The Aztecs treated sore throats with copper, while people from Persia and India applied copper to treat boils, eye infections, and venereal ulcers (Michels 2006). Since 1976, people found that after saturating the grains seed with cupric sulfate, the treated seeds can have anti-seed-borne fungi property, and the experience of the following few decades has fully proved that the strategy of infiltrating seeds with cupric sulfate is safe and reliable (Dias, Cicero, and Novembre 2015). Thanks to its antimicrobial properties, copper can be utilized in a hospital or any high-hygienic demand places. Depending on the reports from Ben Ando, in Trafford General Hospital, copper door handles and work surfaces are replacing stainless steel in the fight against the superbug. The new fittings have contributed to the hospital's Methicillin-resistant *Staphylococcus aureus* (MRSA)-free status for the past 2 years (Ando 2011). According to another report by Shayla Love, by replacing 10% of the items in the patient room in the hospital with copper materials reduces the presence of microbes by 83%, and Health care-Associated Infections (HAIS) by 58% (Love 2020). There are many studies to prove that the utilization of antibacterial copper surface treatment facilities can maintain a level below the terminal cleaning standard of 5 cfu/cm^2. For items with a high frequency of contact, after using copper for surface antibacterial treatment, the danger of HAIS rate of patients living in it will be significantly reduced (Schmidt 2011).

COVID-19 spread during December 2019; it widely splurges all over the world. This virus is highly infectious and causes huge economic and personnel losses. On October 1, 2020, globally, it is accounted that there are 33,722,075 confirmed cases and 1,009,270 death cases (World Health Organization 2019). To protect their people from the COVID-19, every country startes to apply positive policies and measures. To protect the people, the simplest method is to wear masks which are already promoted by many countries as an effective measure. According to the Jeremy Howard research, the conservative estimate of R0 for COVID-19 is 2.4 (Howard et al. 2020). For wearing masks, they find that R0 is reduced by a factor $(1 - ep_m)^2$, where e is the efficacy of trapping viral particles inside the mask, and p_m is the percentage of the population (Howard et al. 2020). Assuming that the use rate and action rate of masks are both 50%, R0 will be reduced to 1.35 after the use of masks. For infectious diseases, this is an order for magnitude reduction (Ferguson et al. 2020). In this case, assuming 100 cases (COVID-19 positive) during the beginning of the month, it will become 31,280 cases by end of the month as R0 = 2.4. However, with R0 = 1.35, it will be only 584 cases. Fewer cases can ensure the local medical capacity to deal with more cases, and at the same time, there will be conditions for tracking infected cases, thereby fundamentally inhibiting the spread of the virus in the local area (Howard et al. 2020). But no matter N95 masks or medical masks, for most of the masks which are available in the market, the main purpose is to stop the virus from passing through the protective materials. The active material masks to inhibit the spread of novel coronavirus are still not widely used to fight against COVID-19 during the start period of the pandemic (Figure 11.1). Due to increase of virus spread, governments all over the world are searching for more effective proactive textiles to inhibit the virus. Active copper-coated textile plays a significant role in this part. Hong Kong SAR China government offered free masks for local

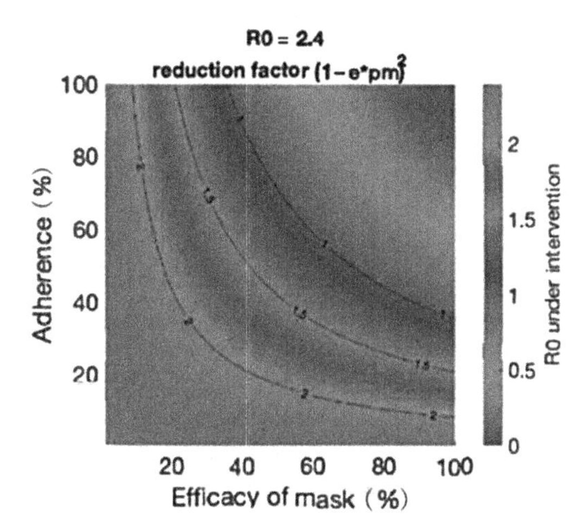

FIGURE 11.1 Impact of public mask-wearing under the full range of mask adherence and efficacy scenarios. The shade indicates the resulting reproduction number R0 from an initial R0 of 2.4 (Reprinted from (Ferguson et al. 2020), with the kind permission of Creative Commons Attribution-Non-Commercial-No Derivatives 4.0 International License).

citizens for protecting themselves. This mask is constructed from six layers of textiles, two layers among them are copper-coated fabrics, which have been used as an active layer (Parry 2020). The UK manufacturer Promethean Particles reports it is collaborating with textile companies and research institutes to explore the copper nanoparticles with antiviral properties, which are designed for use in fabrics and Personal Protective Equipment (PPE) for the health care sector. The large-scale use of copper-plated textiles to combat the new coronavirus is impossible under current conditions. There are many reasons, such as the effectiveness of copper-plated textiles; safety remains to be studied, and the rate of mass production cannot be guaranteed. But, it is undeniable that copper-plated textiles have significant effects as an antivirus.

Researcher Gadi Borkow and Jeffrey Gabbay found that the survival rate of *Candida albicans* on the surface of copper-plated fabrics changed with time. After 60 min, in the control group, the untreated fabric did not show an inhibitory effect on fungi; however, the fungus exposed to the copper surface fabric can be completely suppressed (Borkow and Gabbay 2004). On the surface of different materials, the survival time of the virus is different, and the same is true for the COVID-19 viruses. Regarding the research on surface stability of COVID-19, it was most stable on plastic and stainless steel and the viable virus could be detected up to 72 h post-application, though the virus titer was reduced (plastic, from 103.7 to 100.6 TCID 50/mL after 72 h; stainless steel, from 103.7 to 100.6 TCID 50/mL after 48 h). SARS-CoV-1 had similar stability kinetics (polypropylene, from 103.4 to 100.7 TCID 50/mL after 72 h; stainless steel from 103.6 to 100.6 TCID 50/mL after 48 h). After 4 h, there is no viable HCoV-19 virus that could be measured on copper for; same as the results for SARS-CoV-1 after 8 h (Van Doremalen et al. 2020).

Copper and copper alloy surfaces have significant properties that kill bacteria. As reported by researcher Harold T. Michels, the bacteria counts taken on an alloy containing 99% copper show a rapid 7-log falloff to zero within75 min: 99.9% reduction in live MRSA bacteria at 20°C (Michels et al. n.d.). Not only as antibacterial, copper and copper alloys also have properties that kill the virus. Researcher J. O. Noyce had tested the Influenza A Virus on the copper surface, at 22°C at 50–60% relative humidity, after incubation for 24 h on stainless steel, and 500,000 virus particles were still infectious, but after incubation for 6 h on copper, only 500 particles were active (Noyce, Michels, and Keevil 2007). After coating copper on the textile surface, the antibacterial property of the coated fabrics is increased. As per the research of Galani Irene, copper-coated para-aramid fabrics showed a significant reduction of *Klebsiella pneumoniae* and *Candida parapsilosis* bacterial inoculum in 15 min. For *Enterococcus faecium*, *Pseudomonas aeruginosa*, and *C. parapsilosis*, the effect was seen at 1 h. In the case of *Acinetobacter baumannii*, this reduction fulfilled the criteria for bactericidal action even at 15 min. Copper/polyester significantly reduced the bacterial inoculum and showed bactericidal effect after 1 h of contact except for *A. baumannii*- for which cidal activity was exhibited even at 15 min (Irene et al. 2016).

In another research about copper-coated textile antimicrobial properties, researcher A. Ali and his colleagues researched on copper-coated cotton fabrics; the zone of inhibitions was evidenced against both types of bacteria *S. aureus* and *Escherichia coli* after the copper coating. From the research, we know that compared to *E. coli*, *S. aureus* depicted the highest sensitivity. The zone of inhibitions for *S. aureus* increased from 9.5 to 15.5 mm, while for *E. coli*, it increased from 7.5 to 12 mm with an increasing number of dips (Ali et al. 2018). In another work, comparison was made on the inhibition zone on agar medium and the antibacterial activity of different samples (Suryaprabha and Sethuraman 2017a). Normal cotton fabric, which was used as a control, did not show any antibacterial activity. The distinct zone of the Cu-coated samples was 27 mm in the case of Gram-positive bacteria and 25 mm in the case of Gram-negative bacteria. The results of the study clearly showed that the Cu-coated specimens are more sensitive toward microorganisms and have high bacterial inhibition efficiency in both cases (Suryaprabha and Sethuraman 2017b). By bimetallic deposition of copper and silver on cotton fabric, Suryaprabha found out that normal cotton did not show any antibacterial activity against Gram-negative and Gram-positive bacteria, while the metal-coated cotton in the case of *S. aureus*, showed an inhibition zone of 14 mm and for *E. coli*, the inhibition zone was 17 mm. This result suggested that the bimetallic deposited cotton fabric performs an effective inhibitor for bacterial growth on cotton fabric against both Gram-positive and Gram-negative organisms (Suryaprabha and Sethuraman 2017a).

Regarding the principle of copper-oxide antimicrobial properties, filters containing copper oxide-impregnated polypropylene fibers can reduce infectious titers of a panel of viruses spiked into culture media. Enveloped; nonenveloped; RNA and DNA viruses were affected, suggesting the possibility of using copper oxide-containing devices to deactivate a wide spectrum of infectious viruses found in filterable suspensions. Prolongation of the exposure of these microorganisms to the copper oxide-containing fibers further reduced their viable titers (Borkow, Felix, and Gabbay 2010).

11.3 MECHANISMS OF COPPER AND COPPER OXIDE DISINFECTION FOR VIRUS

From the last chapter, we know that the antibacterial property of copper was already approved by many researches or articles. However, the structures of bacteria and virus are different; the virus has a more simplified structure compared to bacteria, for example, size. Generally, the size of a bacteria is 10–35 times bigger than that of a virus. Bacteria are single-celled, living organisms (Verity 2000). They have a cell wall and all the components necessary to survive and reproduce, although some may derive energy from other sources. Bacteria are single-celled organisms that live all around us – including inside our gut. However, a small percentage of bacteria can be harmful and cause illness. Viruses are parasitic, which means they need a live host to survive. Viruses infect the cells of living beings, including humans, and can cause diseases like the common cold or AIDS (Burch 2020). Although most viruses are potentially harmful, some can be beneficial. Also, the infection mechanisms to humans or other animals are different. As a result, we can't simply reach the conclusion that the effectiveness of copper antibacterial property will also directly apply to antivirus property.

The outcome of copper and copper oxide disinfection for the virus is different. Researcher Teguh Hari Sucipto tested the antiviral property of copper chloride dihydrate against dengue virus type two. Results of this study suggest that copper(II) chloride dihydrate demonstrated significant anti-DENV-2 inhibitory activities and is not toxic within the Vero cells (Sucipto et al. 2017). Another research by Sarah L. Warnes and Zoe R. Little, titled *Human Coronavirus 229E Remains Infectious on Common Touch Surface Materials*, demonstrated that human coronavirus 229E was rapidly inactivated at various time points and assayed for infectivity on a range of copper alloys (within a few minutes for simulated fingertip contamination) and Cu/Zn brasses were very effective for antivirus use at lower copper concentration. Approximately, 103 PFU HuCoV-229E (20 μL infected-cell lysate) was applied to 1 cm^2 coupons of a range of brasses, and control metal surfaces that did not contain copper (stainless steel, zinc, and nickel). The results show that the human coronavirus occuring on brass and copper–nickel surfaces were inactivated: inactivated in <40 min on brasses and in 120 min on copper nickels containing less than 70% copper. Analysis of the initial 30 min of contact between virus and brasses reveals an initial lag followed by rapid inactivation. Stainless steel and nickel did not demonstrate any antiviral activity, although mild antiviral activity was observed on zinc (this was significant only at 60 min [P = 0.046]). The same inoculum was applied as 1μL/cm^2, was dried immediately to simulate fingertip touch contamination, and was found to have inactivated the virus approximately eight times faster (Warnes, Little, and Keevil 2015). It was demonstrated that the different ratios of copper in the alloy can also have different properties on anti-Human Coronavirus 229E. From this finding, we can clearly understand that the antivirus properties of Cu(I) and Cu(II) are different, and the different components of copper can also contribute to different antivirus results. The viral genomes and virus morphology, like the disintegration of envelope and dispersal of surface spikes, can be destroyed and irreversibly affected when exposed to the copper surface. Cu(I) and Cu(II) were responsible for the inactivation of the human coronavirus, which was enhanced by reactive oxygen species generation on alloy surfaces, resulting in even faster inactivation than was seen with nonenveloped viruses on copper.

Viruses don't show any metabolic activity when it stays outside of a host cell. Thus, the interaction between metals and a virus must involve a mechanism that does not require a metabolic process. Disinfectants must remove or destroy viruses to such an extent that successful reproduction in a susceptible cell is prevented. This could be accomplished by permanently immobilizing viruses on a surface, blocking or destroying host-cell receptors on the virus, or inactivating the nucleic acid in the viral capsid. If proteins are complexed or altered, such that they cannot perform their normal function, cell death, or viral inactivation may be imminent. Heavy metals, such as copper, may disrupt enzyme structure and function by binding thiol or other groups on protein molecules. Replacement of naturally occurring metals in enzyme prosthetic groups may alter enzyme function (Sterritt and Lester 1980; Thurman and Gerba 1989).

As mentioned before, the molecular mechanisms of copper antivirus are various. The following theories support this property of copper:

- Radicals can be formed by copper complexes which inactivate viruses (Kuwahara et al. 1986).
- Copper may disrupt enzyme structures, and functions by binding to sulfur- or carboxylate-containing groups and amino groups of proteins (Sterritt and Lester 1980).
- In inactivation experiments on the flu strain, H1N1, which is nearly identical to the H5N1 avian strain and the 2009 H1N1 (swine flu) strain, researchers hypothesized that copper's antimicrobial action probably attacks the general structure of the virus and so includes a broad-spectrum effect (Michels 2006).

Depending on the research by June Kuwahara, the Cu(II)-CPT (camptothecin) system significantly caused single- and double-strand breaks of DNA to form circular and linear duplexes by the irradiation of 365-nm light (Kuwahara et al. 1986). Researcher Martin stated that it is the substitution of copper ions for other metal ions which provides the basis for the toxicity of copper for antivirus property (Crisponi and Nurchi 2015). He suggested that the higher the charge density or the ratio of charge to the radius of the metal ion, the more likely the metal ion is to undergo hydrolysis in an aqueous solution, yielding hydroxy complexes which may then form polynuclear complexes or precipitates. Metal ions commonly substitute for other metal ions of a similar size. Positive metal ions, such as copper(II), favor binding sites with an increased number of N donors (Thurman and Gerba 1989).

Copper nanoparticles are quite effective against both bacteria and fungi (Kanhed et al. 2014, Usman et al. 2013) but are not extensively used in the antimicrobial coating of textiles due to their relatively high toxicity than other metal nanoparticles and also their strong inherent color that may affect the color of the textile substrate. In some researches, the antibacterial activity of copper nanoparticles is not as effective as that of silver nanoparticles (Ruparelia et al. 2008). As mentioned in the previous part, viruses do not show metabolic activity outside a host cell. Thus, the virus inactivation process between metals and a virus does not require a metabolic process. Due to the interaction of molecules, the stability or instability of the molecule will affect its function, thereby affecting the survival of the virus (Šponer et al. 2006). Martin and Mariam noted that copper ions in low ionic strength solutions compete with hydrogen bonds in macromolecules, which can denature DNA. Rifkind et al. (1976) demonstrated that as the concentration of copper(II) increased, the disordering of poly (A) and poly (C) from double- to single-stranded increased (see Fig 2.8). Ueda et al. reported that copper(II) has a specific affinity for DNA and can bind and disorder helical structures by crosslinking within and between strands. Since metal ions can simultaneously coordinate with several ligand atoms, a single metal ion may link either two phosphates, two bases, or a phosphate and a base. Each crosslink may interrupt several bases inter- or intramolecularly (Thurman, Gerba, and Bitton 1989). Martin and Mariam stated that as a constituent of DNA, metal ions have many potential binding sites on nucleic acids. Virtually, all nitrogen and oxygen donors have been proposed to be involved in coordination to metal ions, including OH groups on the ribose, N and O groups on the bases, and negatively charged O atoms in the phosphate residues. Potential metal-ion binding sites have a proton affinity. There are many factors that may influence the ability of metal ions to interact with various ligands, including basicity, pH, competition with hydroxide ions for binding sites, the nature of the donor and metal atoms, the ability to form chelation complexes or hydrogen bonds, kinetic limitations, the degree of base stacking of nucleic acids, and the availability of N(1) and N(7) in purines (Šponer et al. 2006).

Hutchinson suggested that copper(II) may complex mRNA and therefore, play an antiviral role in disinfection (Hutchinson 1985). Buffle stated that copper(II) has a strong affinity for either 0, N, or S sites. Reduction to the Cu(I) state may occur. Cu(I) and Ag(I) are generally found in very low concentrations and display a robust affinity for N or S sites (Turner et al. 1986). Researcher Zimmer reported that using the addition of Cu(II) as an electrolyte in high concentration may allow renaturation, which depends upon the temperature, Cu(II) concentration, and base composition of the DNA (Zimmer et al. 1971). Eichhorn and Clark (Eichhorn and Clark 1965) suggested that the copper(II)–nucleoside coordination bonds could also be of the identical order of stability because of the DNA helix. At room temperature, the copper(II)–nucleoside bond doesn't overcome the attractive forces of the DNA strands, yet, at higher temperatures, the copper complex becomes stronger relative to the DNA forces, causing denaturation. An outsized increase in ionic strength increases the stability of the double helix relative to the copper complex since competition between sodium and copper decreases the local concentration of copper(II). As a result, the intercalated copper ions are released (Hay and Morris 1976).

11.4 METHOD OF METALLIZATION OF TEXTILE

To combine metal particles with textile, the widely used method is coating. As this chapter mentioned in the previous part, one type of metal has a different antimicrobial effect compared with other metals. However, when using pigments or metal particles for coating the textile, besides the composition, the particle size, shape, and surface composition are also important parameters influencing the antimicrobial action. Due to the fact that the particles are applied as part of a coating composition , coating matrix, the interaction between the particles and matrix, and the localization of the particle in the coating can also be factors influencing the antibacterial properties (Mahltig 2017). As mentioned before, copper is one optimized method as antivirus, and to apply copper on textile products, the metallization process is necessary. The metallization process for fabrics is a textile finishing process that adds value and improves functions. With the advent of synthetic fibers, less expensive ways have been found to prepare metallic fabrics in the twentieth century. Either physical or chemical reaction can be involved via wet or dry processes to produce metalized textiles. Wet processing include pretreatment, coloration including dyeing, printing, and finishing. In that, we use different techniques, technologies, and materials. It is mostly supposed that the antimicrobial activity of a particle increases with a decrease in particle size. The reason for this is probably the increasing surface-to-volume ratio with decreasing particle diameter. Usually, particles with diameters of 1 or only some nanometers should be considered as nanoparticles. However, because of the popularity of "nanomaterials," even particles with diameters shortly below 100 nm are claimed to be nanoparticles (Mathiazhagan and Joseph 2011). In line with the definition by researcher Mahltig, particles with diameters of 0.5–5 μm are often described as microsized and named metal pigments. Particles with diameters of 1–50 nm are nanoscales and named nanoparticles. Particles within the range between 50 and 500 nm can be best described as submicron pigments or submicron particles (Mahltig 2017).

The utilization of such metallic-effect pigments as a part of the antimicrobial coating has been intensively reported by Topp et al. In this study, metallic-effect pigments made of silver, copper, silver-coated copper, and steel have been investigated. The gained antibacterial effect increases as a function of the pigment concentration in the coating and the thickness of the coating. However, as mentioned before, it's still not clear that if this rule is also valid for COVID-19 virus (Topp et al. 2014). Wet processing technology has its advantages, such as low production cost, simple equipment operation, and universal application for large-scale production. However, the shortcomings of this technology are particularly obvious, because the process involves the use of too many chemical reagents, which makes the sewage treatment and waste treatment after production more difficult, and the cost increases. Secondly, the adhesion of dyeing and finishing agents to fabrics is relatively weak. To improve reliability, additional processing is required, which increases the production cost and complexity (El-Naggar et al. 2003).

One wet process of cotton fiber treatment was applied by Tencel Ltd., London, UK, via electroless plating process which includes the following steps: 1) cotton fibers having a diameter of ~11–13 μm are soaked for 5 s in 1% $SnCl_2$, pH 3.5, at room temperature; 2) the fibers are then soaked for 5 s in $PdCl_2$, pH 4, at room temperature, producing activated fibers; and 3) the activated cotton fibers are then exposed to formaldehyde, $CuSO_4$, and polyethylene glycol at pH 9. After 5 min, the cotton fibers are plated with cationic copper [Cu(II) and Cu(I)]. Finally, the fibers are dried and run through a textile carding machine, which separates and aligns them. Hereafter, these fibers will be referred to as copper fibers. Hereafter, fabrics containing

TABLE 11.1
Antimicrobial Performance of Woven Fabrics Prepared by the Dry Process (Rtimi, Pulgarin, and Kiwi 2017)

	Method	Fabric	Biocide	Antimicrobial Performance
Dry Process	DCMS	Cotton	Cu	*E. coli* Inactivation time: 2 h
	DCPMS	Cotton	Cu	*E. coli* Inactivation time: 0.2 h
	DCMS	Polyester	Cu	*E. coli* Inactivation time: 140 min
	HIPIMS	Polyester	Cu	*E. coli* Inactivation time: 90 min
	HIPIMS	Polyester	Brass	*S. aureus*, *E. coli* Reduction rate: 99.99%

20% (w/w) copper fibers and 80% noncopper-treated cotton fibers will be referred to as copper fabrics (Tal 2020). Another method for metallization of textiles, compared to the wet process, is generally named as a dry process. In recent years, there are quite a few innovations that were developed through the use of dry processes, which are used for the fabrics to obtain antimicrobial function. Table 11.1 lists some dry process methods and relevant information. The biggest advantage of the dry process is that no water pollution is generated in the process; after coating, the metal has a good adhesive property to the fabrics, which means no additional process is needed compared to the wet process. However, the production costs of most of the dry process methods are relatively higher than the wet process. Fortunately, for antimicrobial use, the coated textile requires only a very thin antimicrobial reagent (e.g., silver or copper) over the fabric surface, which presents a very effective antimicrobial function. Due to less quantity of antimicrobial reagent, high throughput of vacuum-coated antimicrobial fabric and affordability is enabled in many cases.

In a variety of dry deposition techniques, the sputter deposition techniques are the most mature and widely used. The evolution of technologies of this kind persists today. The advanced high-power impulse magnetron sputter (HIPIMS) technique is one of the best dry deposition techniques, which wipes out the drawbacks of conventional magnetron sputter techniques by employing a repeated megawatt electrical impulse onto the cathode material to instantaneously generate a significant portion of high-energy ions state of very high-density plasma. The results show that the arrival ions on the material surface give a perfect growth condition for obtaining very strongly adhered film with very dense microstructure. Notably, the film is often obtained at a comparatively low deposition temperature in comparison with those many other conventional magnetron sputter deposition techniques. Rifkind's works demonstrated that a 1-min batch-type deposition run using HIPIMS to deposit silver and brass coating on polyester fabric can fulfill the antimicrobial requirement by a regular test for antimicrobial efficacy (ISO 105-X12) (Rifkind et al. 1976).

Except for HIPIMS, according to the research by Karol Bula et al., ultra-thin metallic-coated textile can be manufactured by using the cathode sputtering method, and the best screening efficiency of all the samples tested was stated for the copper-sputtered (Cu-PP) polypropylene nonwoven; the efficiency differed with the change in the frequency (Bula, Koprowska, and Janukiewicz 2006). Research Ying-Hung

Chen (Chen, Wu, and He 2015) did a research on antimicrobial brass coatings prepared on polyethylene terephthalate textile by HIPIMS; in this research, the results show that brass adheres to the surface of the fabric in a stable form, and the ratio of copper to zinc in the plated brass can maintain a relatively stable level. Brass plating after oxygen plasma pretreatment can significantly improve the adhesion of the coating and the fabric. For *S. aureus* and *E. coli*, after 1 min of copper plating treatment, good antibacterial effects can be obtained. Through similar research, we can see that HPIMS coating technology is of great significance to the research and production of antibacterial-coated textiles.

As mentioned in the previous section, textiles coated with nanocopper particles also have better antibacterial and antiviral properties. Many people believe that elemental copper coatings have advantages in antibacterial properties compared to copper oxide coatings. According to this theory, coating nanocopper particles directly on textiles faces many challenges in antibacterial and antiviral applications. Due to the large surface area, nanocopper ions are easily oxidized into monovalent copper ions and divalent copper ions during production and storage. To solve this problem, a nanocopper coating was synthesized on the surface of the fabric by using some polymers (e.g., polyvinylpyrrolidone, polyethylene glycol, and chitosan) and surfactants as stabilizers (Usman et al. 2013).

11.5 DISCUSSION

One Japanese company, NBC, developed monovalent copper nanoparticles, Cufitec™, Cufitec™ is a trademark of NBC Meshtec, Inc., which can be applied to textile material. The test result of this material is shown in Figure 11.2. After 60 min of applying the virus on Cufitec™, a 99% reduction of the virus can be observed. According to this product's information, they already successfully used masks and other medical textiles (NBCM Meshtec, Inc 2018).

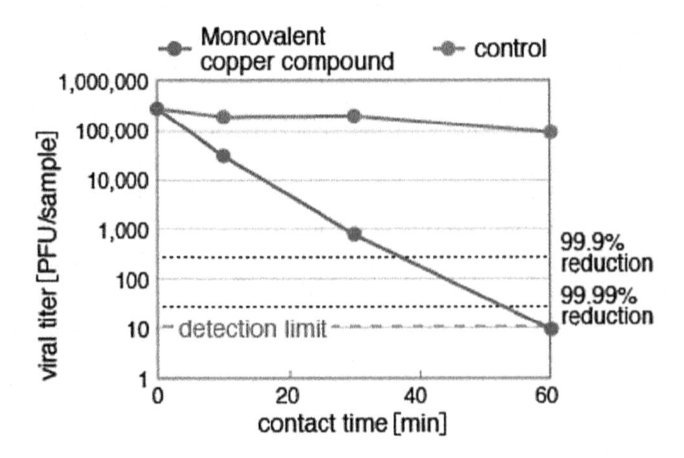

FIGURE 11.2 Effects of Cufitec® in research results (Reproduced from (NBCM Meshtec, Inc 2018), with the permission of NBCM Inc).

Inspired by this product, copper metallization on textile can be promising for the anti-COVID-19 virus. There are still some questions that are not yet answered by literature research:

- COVID-19 is a new virus, and its characteristics are still not fully clear. Copper has a significant antiviral action on anti-Human Coronavirus 229E, but not on COVID-19. A relevant experiment must be done to prove that copper has significant anti-COVID-19 virus property.
- As mentioned before, copper metalized textile has already been developed, but which form of copper (such as copper, brass, copper alloy, nanoparticle) is the most effective for the COVID-19 virus is still not clear and needs to be proofed by experiments.

Most researches recommend textile metallization by using nanocopper particles. But as scientists already proved, the nanoparticle is harmful to the human body. Regarding the research of Huan Meng, the ultrahigh chemical reactivity of nanocopper results in the specific nanotoxicity which is fully proved by in vitro and in vivo experiments. Using chemical kinetics study (in vitro) and blood gas and plasma electrolytes analysis (in vivo), they found that high reactivity causes the big toxicological difference between small size (23.5 nm) and big size (17 m). For viruses like COVID-19, protective textile material will be used for air filtration equipment, such as masks or air conditioner filtration net. If the nanocopper particle is coated on masks, there is the risk that few particles will be breathed into the body via breathing, which can be risky for human health. Depending on different situations, "people wearing masks" time is varies from a few minutes to more than 4 h. Researcher Maria Helena Barbosa researched the influence of wearing time on the efficacy of disposable surgical masks as a microbial barrier. Regarding his research, disposable surgical masks with 95% BFE are efficient microbial barriers for up to 2–3 hours of wearing time, therefore, they are indicated for every critical invasive procedure. However, another conclusion is that their bacterial filtration efficacy decreases significantly after 4 h (Figure 11.3; Barbosa and Graziano 2006).

As we mentioned above, no matter if it is nanoparticle/solid copper, brass, or copper alloy, the inactive time of the virus is still more than 30 min. For a short-time user, with a mask-wearing time less than 30 min, the time range is not good enough for copper-coated masks to inactivate the virus. For users wearing masks for more than 4 h, the filtering efficiency of the mask itself would be significantly weakened, and even if coated with copper, the antivirus property cannot be proved to have any influence. On the other hand, the longer the time wearing copper-coated masks, the higher the risk of breathing in the metal particles. Considering the production costs and protective effectiveness, most of the currently used face masks are made from nonwoven textiles or cotton fiber and are disposable. It would be a big waste after one-time use of the face masks as there is no chance to sterilize the used face masks. After metallization, whether a user has the chance to reuse the mask after UV-light treatment or other methods is yet to be investigated.

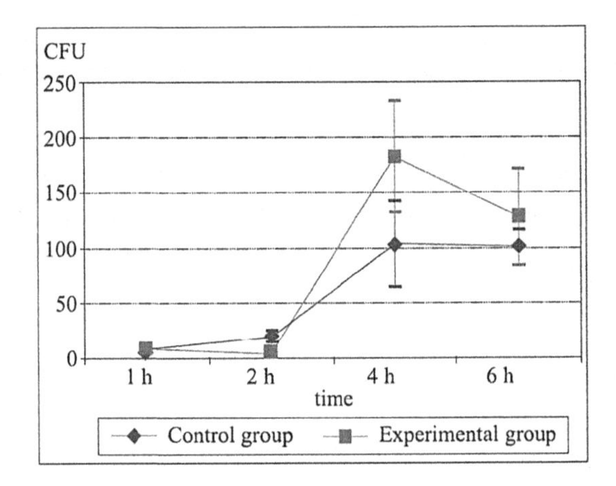

FIGURE 11.3 Counts of CFU on operation tables, after 1 h, 2 h, 4 h, and 6 h of exposition (Reprinted from (Barbosa and Graziano 2006) with the kind permission of Creative Commons Attribution-Non-Commercial-No Derivatives 4.0 International License).

11.6 CONCLUSION AND FUTURE PROSPECTIVE

Copper has been proofed as an effective material for antivirus; even Human-Coronavirus 229E will be disinfected <40 min on brasses and in 120 min on copper nickels containing less than 70% copper. However, it's still not clear that if it's also effective for the COVID-19 virus. At present, textiles coated by the copper nanoparticles, copper, and copper alloys perform as effective antiviruses; for metallization of textile, there are several methods; theoretically, nanoparticles should be the most effective disinfectant compared to others since due to their increased surface, relevant research can be done. From a production point of view, the metallization method and costs should be considered. However, the impact of using copper nanoparticles on human health must also be proved.

ACKNOWLEDGEMENT

This work was supported by the Ministry of Education, Youth and Sports of the Czech Republic, European Union – European Structural and Investment Funds in the Frames of Operational Programme Research, Development and Education – project Hybrid Materials for Hierarchical Structures (HyHi, Reg. No. CZ.02.1.01/0.0/0.0/16_019/0000843).

REFERENCES

Ali, Azam, Vijay Baheti, Jiri Militky, Zaman Khan, Veronika Tunakova, and Salman Naeem. 2018. "Copper Coated Multifunctional Cotton Fabrics." *Journal of Industrial Textiles* 48 (2). SAGE Publications Ltd: 448–464. doi:10.1177/1528083717732076.

Ando, B. 2011. "*Copper Fittings Help Stop MRSA*." BBC. https://www.bbc.com/news/av/health-15134986/copper-fittings-help-stop-mrsa-scientists-say.

Barbosa, Maria Helena, and Kazuko Uchikawa Graziano. 2006. "Influence of Wearing Time on Efficacy of Disposable Surgical Masks as Microbial Barrier." *Brazilian Journal of Microbiology* 37 (3): 216–217. doi:10.1590/S1517-83822006000300003.

Beesoon, Sanjay, Nemeshwaree Behary, and Anne Perwuelz. 2020. "Universal Masking during COVID-19 Pandemic: Can Textile Engineering Help Public Health? Narrative Review of the Evidence." *Preventive Medicine* 139:106236. Academic Press Inc. doi:10.1016/j.ypmed.2020.106236.

Borkow, Gadi, and Jeffrey Gabbay. 2004. "Putting Copper into Action: Copper-impregnated Products with Potent Biocidal Activities." *The FASEB Journal* 18 (14). Wiley: 1728–1730. doi:10.1096/fj.04-2029fje.

Borkow, Gadi, Anthony Felix, and Jeffrey Gabbay. 2010. "Copper-Impregnated Antimicrobial Textiles; an Innovative Weapon to Fight Infection." In *Medical and Healthcare Textiles*, 14–22. Elsevier Ltd. doi:10.1533/9780857090348.14.

Bula, Karol, Joanna Koprowska, and Jarosław Janukiewicz. 2006. "Application of Cathode Sputtering for Obtaining Ultra-Thin Metallic Coatings on Textile Products." *Fibres and Textiles in Eastern Europe* 14 (5): 75–79.

Burch, Kelly. 2020. "What Is the Difference between Bacteria and Viruses – Insider." Accessed March 12. https://www.insider.com/what-is-the-difference-between-bacteria-and-virus.

Chen, Ying-Hung, Guo-Wei Wu, and Ju-Liang He. 2015. "Antimicrobial Brass Coatings Prepared on Poly(Ethylene Terephthalate) Textile by High Power Impulse Magnetron Sputtering." *Materials Science and Engineering: C* 48 (March). Elsevier Ltd: 41–47. doi:10.1016/j.msec.2014.11.017.

Crisponi, Guido, and Valeria M. Nurchi. 2015. "Metal Ion Toxicity." In *Encyclopedia of Inorganic and Bioinorganic Chemistry*, 1–14. Chichester, UK: John Wiley & Sons, Ltd. doi:10.1002/9781119951438.eibc0126.pub2.

Davydova, S. L. 1979. "Macromolecules of Biological Interest in Complex Formation." *Metal Ions in Biological Systems* 8 (Sigel, H., Ed): 183. Marcel Dekker, Inc, New York and Basel.

Dias, Marcos Altomani Neves, Silvio Moure Cicero, and Ana Dionísia Luz Coelho Novembre. 2015. "Uptake of Seed-Applied Copper by Maize and the Effects on Seed Vigor." *Bragantia* 74 (3). Instituto Agronomico: 241–246. doi:10.1590/1678-4499.0044.

Eichhorn, G. L., and P. Clark. 1965. "Interactions of metal ions with polynucleotides and related compounds." *Proceedings of the National Academy of Sciences of the United States of National Academy of Sciences* 53 (3): 586–593. doi:10.1073/pnas.53.3.586.

El-Naggar, A. M., M. H. Zohdy, M. S. Hassan, and E. M. Khalil. 2003. "Antimicrobial Protection of Cotton and Cotton/Polyester Fabrics by Radiation and Thermal Treatments. I. Effect of ZnO Formulation on the Mechanical and Dyeing Properties." *Journal of Applied Polymer Science* 88 (5). John Wiley & Sons, Ltd: 1129–1137. doi:10.1002/app.11722.

Ferguson, Neil M, Daniel Laydon, Gemma Nedjati-Gilani, Natsuko Imai, Kylie Ainslie, Marc Baguelin, Sangeeta Bhatia, et al. 2020. "Of Non-Pharmaceutical Interventions (NPIs) to Reduce COVID-19 Mortality and Healthcare Demand." Accessed August 25. doi:10.25561/77482.

Grass, Gregor, Christopher Rensing, and Marc Solioz. 2011. "Metallic Copper as an Antimicrobial Surface." *Applied and Environmental Microbiology*. American Society for Microbiology (ASM). doi:10.1128/AEM.02766-10.

Hay, R. W. and Morris, P. J., Metal ion promoted hydrolysis, in Metal Ions in Biological Systems, Vol. 5, Sigel, H., Ed., Marcel Dekker, New York, 1976. "*Metal Ion Promoted Hydrolysis*." Biological Systems, Vol. 5: 173.

Howard, Jeremy, Austin Huang, Zhiyuan Li, Zeynep Tufekci, Vladimir Zdimal, Helene-Mari van der Westhuizen, Arne von Delft, et al. 2020. "Face Mask Against COVID-19: An Evidence Review." *British Medical Journal*, no. April (April). Preprints: 1–8. doi:10.20944/preprints202004.0203.v1.

Hutchinson, D. W. 1985. "Metal Chelators as Potential Antiviral Agents." *Antiviral Research*. Elsevier. doi:10.1016/0166-3542(85)90024-5.

NBC Meshtec, Inc. 2018. "Technical Overview I Anti-Viral and Anti-Bacterial Technology Cufitec®IProduct&service I NBC Meshtec Inc. –The Most Advanced Mesh Technology in the World." Accessed January 1. http://www.nbc-jp.com/eng/product/cufitec/.

Irene, Galani, Priniotakis Georgios, Chronis Ioannis, Tzerachoglou Anastasios, Plachouras Diamantis, Chatzikonstantinou Marianthi, Westbroek Philippe, and Souli Maria. 2016. "Copper-Coated Textiles: Armor against MDR Nosocomial Pathogens." *Diagnostic Microbiology and Infectious Disease* 85 (2). Elsevier Inc.: 205–209. doi:10.1016/j.diagmicrobio.2016.02.015.

Kanhed, Prachi, Sonal Birla, Swapnil Gaikwad, Aniket Gade, Amedea B. Seabra, Olga Rubilar, Nelson Duran, and Mahendra Rai. 2014. "In Vitro Antifungal Efficacy of Copper Nanoparticles against Selected Crop Pathogenic Fungi." *Materials Letters* 115: 13–17. doi:10.1016/j.matlet.2013.10.011.

Kuwahara, June, Tadashi Suzuki, Kyoko Funakoshi, and Yukio Sugiura. 1986. "Photosensitive DNA Cleavage and Phage Inactivation by Copper(II)-Camptothecin." *Biochemistry* 25 (6). American Chemical Society: 1216–1221. doi:10.1021/bi00354a004.

Love, S. 2020. "*Copper Destroys Viruses and Bacteria. Why Isn't It Everywhere*." Vice Media Group.https://www.vice.com/en_us/article/xgqkyw/copper-destroys-viruses-and-bacteria-why-isnt-it-everywhere.

Mahltig, Boris. 2017. "Metal Pigments as Antimicrobial Agent and Coating Additives." In *Handbook of Antimicrobial Coatings*, 283–299. Elsevier. doi:10.1016/B978-0-12-811982-2.00014-7.

Mathiazhagan, A., and Rani Joseph. 2011. "Nanotechnology – A New Prospective in Organic Coating – Review." *International Journal of Chemical Engineering and Applications*, 225–237. doi:10.7763/IJCEA.2011.V2.108.

Michels, H. T. 2006. "Anti-Microbial Characteristics of Copper."

Michels, H. T., S. A. Wilks, J. O. Noyce, and C. W. Keevil. 2005 "Copper Alloys for Human Infectious Disease Control." *Stainless Steel*, 77000(55.0), 27.0.

Mitchell, A., M. Spencer, and C. Edmiston. 2015. "Role of Healthcare Apparel and Other Healthcare Textiles in the Transmission of Pathogens: A Review of the Literature." *Journal of Hospital Infection*. 90(4) : 285–292. W.B. Saunders Ltd. doi:10.1016/j.jhin.2015.02.017.

Noyce, J. O., H. Michels, and C. W. Keevil. 2007. "Inactivation of Influenza A Virus on Copper versus Stainless Steel Surfaces." *Applied and Environmental Microbiology* 73 (8). American Society for Microbiology (ASM): 2748–2750. doi:10.1128/AEM.01139-06.

Parry, Jane. 2020. "Covid-19: Hong Kong Government Supplies Reusable Face Masks to All Residents." *BMJ (Clinical Research Ed.)* 369 (May). NLM (Medline): m1880. doi:10.1136/bmj.m1880.

Petukh, Marharyta, and Emil Alexov. 2014. "Ion Binding to Biological Macromolecules." *Asian Journal of Physics: An International Quarterly Research Journal* 23 (5). NIH Public Access: 735–744. http://www.ncbi.nlm.nih.gov/pubmed/25774076.

Rajendran, Subbiyan, and Subhash C. Anand. 2016. "Smart Textiles for Infection Control Management." In *Advances in Smart Medical Textiles: Treatments and Health Monitoring*, 93–117. Elsevier Inc. doi:10.1016/B978-1-78242-379-9.00005-0.

Rifkind, Joseph M., Yong A. Shin, Jane M. Heim, and Gunther L. Eichhorn. 1976. "Cooperative Disordering of Single-stranded Polynucleotides through Copper Crosslinking." *Biopolymers* 15 (10). 1879–1902. doi:10.1002/bip.1976.360151002.

Rtimi, Sami, Cesar Pulgarin, and John Kiwi. 2017. "Recent Developments in Accelerated Antibacterial Inactivation on 2D Cu-Titania Surfaces under Indoor Visible Light." *Coatings*. doi:10.3390/coatings7020020.

Ruparelia, Jayesh P., Arup Kumar Chatterjee, Siddhartha P. Duttagupta, and Suparna Mukherji. 2008. "Strain Specificity in Antimicrobial Activity of Silver and Copper Nanoparticles." *Acta Biomaterialia* 4 (3). 707–716. doi:10.1016/j.actbio.2007.11.006.

Schmidt, MG. 2011. "Copper Surfaces in the ICU Reduced the Relative Risk of Acquiring an Infection While Hospitalized." *BMC Proceedings* 5 (S6): O53. doi:10.1186/1753-6561-5-S6-O53.

Šponer, Judit E, Jaroslav V Burda, Jerzy Leszczynki, and Jií Šponer. 2006. "Interaction of metal cations with nucleic acids and their building units a Comprehensive View from Quantum Chemical Calculations." In *Computational Studies of RNA and DNA*, 389-410. Springer, Dordrecht.

Sterritt, R. M., and J. N. Lester. 1980. "Interactions of Heavy Metals with Bacteria." *Science of the Total Environment*, 14 (1). : 5–17. doi:10.1016/0048-9697(80)90122-9.

Sucipto, Teguh Hari, Siti Churrotin, Harsasi Setyawati, Tomohiro Kotaki, Fahimah Martak, and Soegeng Soegijanto. 2017. "Antiviral activity of copper(ii)chloride dihydrate against dengue virus type-2 in vero cell." *Indonesian Journal of Tropical and Infectious Disease* 6 (4). Universitas Airlangga: 84. doi:10.20473/ijtid.v6i4.3806.

Suryaprabha, Thirumalaisamy, and Mathur Gopalakrishnan Sethuraman. 2017a. "Design of Electrically Conductive Superhydrophobic Antibacterial Cotton Fabric through Hierarchical Architecture Using Bimetallic Deposition." *Journal of Alloys and Compounds* 724. Elsevier Ltd: 240–248. doi:10.1016/j.jallcom.2017.07.009.

Suryaprabha, Thirumalaisamy, and Mathur Gopalakrishnan Sethuraman. 2017b. "Fabrication of Copper-Based Superhydrophobic Self-Cleaning Antibacterial Coating over Cotton Fabric." *Cellulose* 24 (1). Springer Netherlands: 395–407. doi:10.1007/s10570-016-1110-z.

Tal, Meirav. 2020. "US Patent for Metallized Textile Patent (Patent # 5,871,816 Issued February 16, 1999) – Justia Patents Search." Accessed August 25. https://patents.justia.com/patent/5871816.

"The Virtues of Ancient Glues – Gilles Perrault." 2020. Accessed October 1. https://www.gillesperrault.com/the-virtues-of-ancient-glues/.

Thurman, Robert B., and Charles P. Gerba. 1989. "The Molecular Mechanisms of Copper and Silver Ion Disinfection of Bacteria and Viruses." *Critical Reviews in Environmental Control* 18 (4). Taylor & Francis Group: 295–315. doi:10.1080/10643388909388351.

Topp, Kristin, Hajo Haase, Christoph Degen, Gerhard Illing, and Boris Mahltig. 2014. "Coatings with Metallic Effect Pigments for Antimicrobial and Conductive Coating of Textiles with Electromagnetic Shielding Properties." *Journal of Coatings Technology and Research* 11 (6). Springer New York LLC: 943–957. doi:10.1007/s11998-014-9605-8.

Turner, D. R., M. S. Varney, M. Whitfield, R. F.C. Mantoura, and J. P. Riley. 1986. "Electrochemical Studies of Copper and Lead Complexation by Fulvic Acid. I. Potentiometric Measurements and a Critical Comparison of Metal Binding Models." *Geochimica et Cosmochimica Acta* 50 (2). Pergamon: 289–297. doi:10.1016/0016-7037(86)90177-8.

Usman, Muhammad Sani, Mohamed Ezzat El Zowalaty, Kamyar Shameli, Norhazlin Zainuddin, Mohamed Salama, and Nor Azowa Ibrahim. 2013. "Synthesis, Characterization, and Antimicrobial Properties of Copper Nanoparticles." *International Journal of Nanomedicine* 8 (November). Dove Medical Press Ltd.: 4467–4479. doi:10.2147/IJN.S50837.

Van Doremalen, Neeltje, Trenton Bushmaker, Dylan H. Morris, Myndi G. Holbrook, Amandine Gamble, Brandi N. Williamson, Azaibi Tamin, et al. 2020. "Aerosol and Surface Stability of SARS-CoV-2 as Compared with SARS-CoV-1." *New England Journal of Medicine*. Massachussetts Medical Society. doi:10.1056/NEJMc2004973.

Verity, P. G. 2000. "Grazing Experiments and Model Simulations of the Role of Zooplankton in Phaeocystis Food Webs." In *Journal of Sea Research*, 43:317–343. Elsevier. doi:10.1016/S1385-1101(00)00025-3.

Vincent, Marin, Philippe Hartemann, and Marc Engels-Deutsch. 2016. "Antimicrobial Applications of Copper." *International Journal of Hygiene and Environmental Health* 219 (7). Elsevier GmbH: 585–591. doi:10.1016/j.ijheh.2016.06.003.

Warnes, Sarah L., Zoë R. Little, and C. William Keevil. 2015. "Human Coronavirus 229E Remains Infectious on Common Touch Surface Materials." *MBio* 6 (6). American Society for Microbiology. doi:10.1128/mBio.01697-15.

Wei, Shuya, Qinglin Ma, and Manfred Schreiner. 2012. "Scientific Investigation of the Paint and Adhesive Materials Used in the Western Han Dynasty Polychromy Terracotta Army, Qingzhou, China." *Journal of Archaeological Science* 39 (5). Academic Press: 1628–1633. doi:10.1016/j.jas.2012.01.011.

World Health Organization. 2019. "Coronavirus Disease 2019." https://www.who.int/emergencies/diseases/novel-coronavirus-2019.

Zimmer, Ch., G. Luck, H. Fritzsche, and H. Triebel. 1971. "DNA-Copper(II) Complex and the DNA Conformation." *Biopolymers* 10 (3). John Wiley & Sons, Ltd: 441–463. doi: 10.1002/bip.360100303.

12 Eradicating Spread of Virus by Using Activated Carbon

Daniel Karthik, Jiri Militký and Mohanapriya Venkataraman
Technical University of Liberec, Czech Republic

CONTENTS

12.1 INTRODUCTION

The rapid spread of pandemic viral diseases calls for more effective protective materials. Coronavirus disease 2019 has recently caught much attention after its lethal outbreak representing a serious issue to public health. Protective measures are currently being undertaken with most of what is at our disposal. However, with ever-advancing and mutating novel viruses that increasingly surround us, the need for extensive research to develop more suitable, effective protective materials and systems, keeping in mind the costs. To reduce exposure to toxic gases, viral and bacterial microorganisms in the air, several types of respiratory protective items such as cotton cloth, surgical masks, and N95 respirator masks are being employed. While the N95 respirator mask is designed to protect the wearer from the environment, surgical masks and N95 masks are designed to filter viruses, bacteria, and dust (Khayan et al., 2019).

There are various treatment technologies for the removal of chemical compounds which include biological/chemical reduction, ion exchange (IX), adsorption by granular activated carbon, tailored granular activated carbon, granular ferric hydroxide, and membrane filtration. Among these treatment methods, the adsorption process is more advantageous due to its low cost and high treatment efficiency. An adsorbent containing a positive charge on the surface provides an adsorption site for the perchlorate because it is an anionic pollutant (Xu et al., 2011). Porous carbon materials, due to their extensive specific surface area, high adsorption capacity, microstructure, and special surface reactivity have been widely used in separation, purification, and catalytic processes. The existence of activated carbons is believed to contribute a synergetic effect for providing an attachment surface for bioregeneration (microorganisms) and serving as a nucleus for the occurrence of floc formation (Byeon et al., 2008; Ellerie et al., 2013). A high degree of porosity and an extended inter particulate surface area is revealed by activated carbon, resulting in a superior level of adsorption capacity. The capacity to adsorb is greatly influenced by the chemical structure of the carbon surface (Bansal and Goyal, 2005). Virus removal by activated carbon may be influenced by many factors such as raw material, specific surface area, element content, surface functional group, pore size distribution, and surface charge (Matsushita et al., 2013). Despite the sizeable market for activated carbon, the particular adsorption mechanisms of various compounds that take place on this adsorbent material are still uncertain. The complex, heterogeneous nature of the activated carbon surface has led to contradicting mechanisms being proposed in the literature. As a result, the prediction of adsorption capacity is limited to idealized cases, and the design of most practical systems is semiempirical in nature (Franz et al., 2000).

12.2 ACTIVATED CARBON-BASED SYSTEMS AND ITS APPLICATIONS

Activated carbons are excellent and versatile adsorbents. Their important applications are the adsorptive removal of color, odor, taste, and other undesirable organic and inorganic pollutants from drinking water; in the treatment of industrial wastewater; for the purification of air in inhabited areas, such as in food processing, chemical industries, and restaurants; for the purification of various chemical, food, and pharmaceutical products; in respirators for workers under harsh environments; and in a variety of gas-phase applications. The most important application of activated carbon adsorption is in the purification of air and water, where large quantities of activated carbons are being consumed and the consumption is constantly increasing. There are mainly two types of adsorption systems for the purification of air. One is the purification of air for instantaneous use in inhabited areas, where free and clean air is a necessity, and the other prevents air pollution in the industrial exhaust streams. The former functions at pollutant concentrations below 10 ppm, generally around 2–3 ppm. As the concentration of the pollutant is low, the adsorption filters can work for a longer period and the consumed carbon material can be disposed of because regeneration could be expensive. The control of air pollution requires special adsorption setup to cope with larger concentrations of the pollutants. The saturated carbon needs to be regenerated by the use of steam, air, or nontoxic gaseous treatments. These two

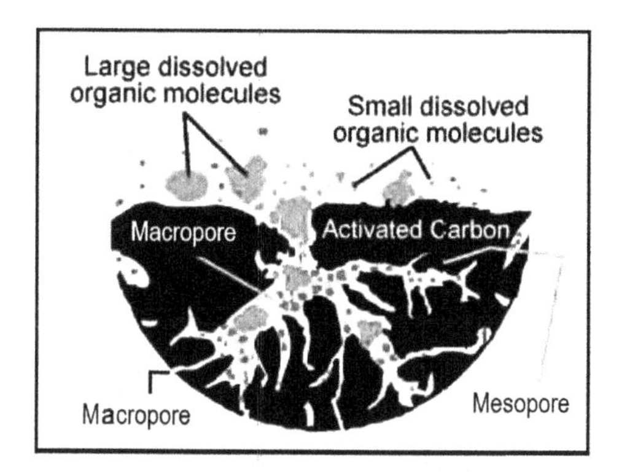

FIGURE 12.1 Porous structure of activated carbon (Selasa, 2018).

mentioned applications require different activated carbon porous structures. For the purification of air in inhabited spaces, highly microporous carbons are required to obtain greater adsorption at lower concentrations. In the case of activated carbons for control of air pollution, the pores must have higher adsorption capacity in the concentration range between 10 and 500 ppm. It is challenging to specify the pore diameters accurately, but generally, it is preferred to identify in the micro-and meso-range because they fill in this concentration range. Figure 12.1 gives a general idea of the porous structure of activated carbon materials (Bansal and Goyal, 2005).

Activated carbon has also been used as catalyst support in the gasification of coal and coke and for air pollution control. Metals are incorporated onto carbon by impregnation or adsorption. Since activated carbon has a high surface area (1,000 m^2/g) and a wide range of pore size distribution, nonuniformly distributed catalysts, which can offer superior conversion, selectivity, and durability under certain conditions, can be easily obtained by using activated carbon as the support. The large pores serve for the mass transfer and support the active element for the catalytic oxidation of large organics into intermediate products. Most active metals are dispersed into the small pores to help further oxidize the intermediate products into final products (carbon dioxide and water) (Hu et al., 1999).

12.2.1 Activated Carbon for Filtration Purposes

Activated carbon filters are trivial pieces of carbon, characteristically in granular or powdered block form, that has been treated to be exceedingly porous in nature. It is so cavernous, that merely 1 g of activated carbon can easily have a surface area of about 500 m^2 or higher. Enormous surface area enables these carbon filters to adsorb contaminants and allergens more exponentially than traditional carbon materials. Through the process of adsorption, organic compounds in the air or water chemically react with the activated carbon, causing them to adhere to the filter. The more porous the activated carbon is the more contaminants will it capture. These filters are remarkably used to eliminate hazardous compounds in air purification systems. There are various benefits associated with using activated carbon filters. These purifiers can be used to free your air of undesirable or harmful contaminants that can pose a hazard to your health. Although a filtration system that uses chemical reactions to cleanse air may sound complicated, it is relatively simple. It is then incorporated into the main purifying device, such as an air purifier or HEPA air filter system. Contaminated air enters the filtration system, moves through the active carbon, undergoes adsorption, and purified air exits the filter. If used in combination with a HEPA filter, the activated carbon prevents larger-sized particles like lint and dust from reaching it, assisting the HEPA filter to perform better and last longer. Each particle/granule of carbon provides a large surface area/pore structure, letting contaminants the maximum potential exposure to the active sites within the filter media. Activated carbon filters are most effective at eliminating chlorine, sediment, volatile organic compounds (VOCs), taste, and odor from water. They are not effective at removing minerals, salts, and any dissolved inorganic compounds. Typical particle sizes that can be removed by carbon filters range between 0.5 and 50 μm. The efficacy of a carbon filter is also dependent on the flow rate regulation

12.2.2 Activated Carbon for Effective Face Masks

Activated carbon, also known as activated charcoal, plays a very vital function in respiratory protection. It was primarily intended for military use since protective masks had become a very important part of personal protective equipment for the military. N95 masks and other respirators have a two-stage inhalation protection process regarding inhaling pollutants from the air. Activated carbon filters are more commonly found in the higher-end N95 and similar other types of masks and is highly efficient in removing gaseous pollutants in comparison to filters that are incapable of the same. An activated carbon filter can also efficiently filter bacteria and viruses present in the air. The most important benefits of using activated carbon incorporated in masks are as follows:

- It prevents the cause of long-term effects of pollution in our body. It can thoroughly disable the inhalation of polluted air that can cause airborne allergies and other lung diseases.
- Helps increase our lung capacity, is another edge. Moreover, having a carbon filtered mask still helps you inhale pure air and makes breathing easier.
- Finally, face masks with activated carbon filters protect us against bacteria, viruses, and fungus, that can be trapped in the air, thereby preventing several diseases and allergic reactions.

A study shows that the activated carbon combination mask model group has a greater ability than other types of masks, in absorbing CO_x. The average results of investigation of exposure to CO_x exhaust emissions without using a mask (control) were higher (3.40 ppm) than when using masks, namely, cotton cloth material (3.39 ppm), spunbond and meltblown (PPE) material (1.13 ppm), and a mask of activated carbon combination spunbond and meltblown material (0.10 ppm). Table 12.1 shows that the results of the analysis of variance showed a significant difference between the decrease in CO_x exposure between control and when using masks ($p \leq 0.001$). There was a significant difference between the use of spunbond and meltblown

TABLE 12.1
Effectiveness of the Use of Mask and Exposure to CO_x

		n	Mean (ppm)	CI 95%		p[a]
				Min	Max	
Co_x not Adsorbed	Control	7	3.40	2.16	4.80	≤0.001*
	Cotton	7	3.39	2.12	4.48	
	Spunbond and Meltblown	7	1.13	0.48	1.76	
	Combination Active Carbon, Spunbond and Meltblown	7	0.10	0.04	0.24	

masks and active carbon combination with spunbond and meltblown ($p = 0.021$) (Khayan et al., 2019).

The combination mask made from spunbond and meltblown has advantages and is a unification of several mask ingredients. Combination masks are commonly used in hospitals, as surgical masks. This model is for filtering dust and bacteria. Besides that, the combination mask also implemented the N95 type which can filter bacteria and dust size in the range of 1–10 μm up to 95% efficiency, while having other functions such as a respirator mask that can be used to filter and absorb toxic particles such as heavy metals or poisonous gases such as NO_x, CO_x, and SO_x. Therefore, masks made from activated carbon incorporating spunbond and meltblown are very meaningful for filtration and absorbing toxic gas material inhaled ($p < 0.05$) compared to other types of markers. Activated carbon masks absorb gases that cannot be absorbed or filtered with just ordinary masks or surgical masks (Loeb et al., 2009; Khayan et al., 2019; Atrie and Worster, 2012).

12.3 EFFECTIVE PROPERTIES INFLUENCING THE ADSORPTION BEHAVIOR OF ACTIVATED CARBON

The surface of activated carbon has a unique character. Its porous structure determines its adsorption capacity, its chemical structure influences its interaction with polar and nonpolar adsorbates, and its active sites in the form of edges, disruptions, and discontinuities determine its reactions with other atoms. Thus, the adsorption behavior of activated carbon cannot be taken into consideration based on surface area and pore size distribution alone. Activated carbons with identical surface area but prepared by different methods or given different activation treatments display different adsorption properties. The determination of an accurate model for adsorption on activated carbon with a complex chemical structure is, therefore, a complicated issue. A model must take into consideration both the chemical as well as the porous structure of the activated carbon, which includes the nature and concentration of the surface chemical groups, the polarity of the surface, surface area, and pore size distribution. In the case of adsorption from solutions, the concentration of the solution and its pH are also critical added factors (Bansal and Goyal, 2005). The heterogeneous surface of activated carbon is typically characterized into three main zones: the carbon basal planes, heterogeneous surface groups (mainly oxygen-containing groups), and inorganic ash. The bulk of the adsorption sites for liquid organics are on the basal planes, which constitutes more than 90% of the carbon surface. However, the much higher activity of the heterogeneous groups can result in significant effects on the total adsorption capacity. Surface oxygen groups, particularly carboxylic groups, are believed to adsorb water molecules, creating water clusters through H-bonding, which lowers the accessibility and affinity for aromatic adsorbates and therefore reduces the adsorption capacity. Oxygen groups, on the other hand, can increase the adsorption capacity in the deficiency of water, by forming H-bonds with the aromatics (Franz et al., 2000).

12.4 FORMS OF ACTIVATED CARBON

Activated carbons can be used in various forms such as powdered form, granulated form, and lately the fibrous form. Powdered activated carbons (PACs) typically have a finer particle size of about 44 μm, which allows faster adsorption. The granular activated carbon (GAC) has granules ranging from 0.6 to 4.0 mm and are hard, abrasion-resistant, and comparatively dense to endure operating conditions. Activated carbon fibers (ACFs) have the advantage of having the ability to be molded easily into the shape of the adsorption system and produce low hydrodynamic resistance to flow (Bansal and Goyal, 2005).

12.4.1 Granular Activated Carbon (GAC), Powdered Activated Carbon (PAC), and Super-Powdered Activated Carbon (S-PAC)

GAC is considered as a supporting medium due to its high chemical stability, mechanical robustness, large specific surface area, and easy availability. Furthermore, GAC is well known to have abundant micropores, macropores, and a high adsorption capacity for organic materials. Decreasing the particle size of activated carbon increases the rate of adsorption, causes the travel distance for intraparticle radial diffusion to be reduced, and the specific surface area per adsorbent mass to be increased. The problems of slow adsorption kinetics can be overcome by pulverizing activated carbon, but the PAC particle size was previously limited to about 5 μm. Recent developments in nanotechnology now enable pulverization down to submicron or nanometer size ranges at a reasonable expense, yielding super-powdered activated carbon (S-PAC). A study revealed that the S-PACs, which could remove a virus effectively, had a rough surface with numerous mesopores with diameters of ~23 nm, where a virus cannot pass through pores smaller than this. The closer the diameter of an adsorbate molecule to the pore size of an adsorbent, the greater is the adsorption (Matsushita et al., 2013). Over the years, some studies have shown that GAC is used in combination with ozonation for removing by-products derived from the oxidative decomposition of organic matter. PAC is occasionally used for removing chemicals with a musty odor and pesticides. It has also shown a positive effect when tested for the removal of the virus. Experiments conducted with a GAC-loaded (20 × 50 mesh, equivalent to 297–853 μm) column-type reactor removed only 24%–50% of poliovirus. Inferior, GAC filtration did not remove bacteriophage MS2. These results indicate that GAC is not suitable for substantial removal of the virus within the permitted contact time for drinking water purification treatment, possibly on account of a low rate of adsorption of the virus. Indeed, only 70% of bacteriophage T4 was eliminated by activated carbon (300–425 μm) after a contact time of 2 h. Accordingly, effective removal of the virus by activated carbon will require a longer contact time and a higher dosage of activated carbon, or both (Goddard and Butler, 2015; Hijnen et al., 2010). The reduction in the surface charge (by addition of Ca^2) increases the hydrophobicity of the carbon surface because the charge reduction allows negatively charged adsorbates to move nearer to the graphite structure on the carbon. Complementing this reduction in the electrophoretic repulsive force, the apparent increase in hydrophobicity of the surface most likely contributed to the high virus removal. Likewise, the hydrophobicity of the virus also possibly contributed to the higher removal: the virus having a hydrophobic surface was removed more significantly with the presence of activated carbons (Matsushita et al., 2013). PAC not only removes the target contaminant but also the natural organic matter. The PAC dose required to mitigate a particular organic contamination problem in the water supply generally depends on the adsorptive capacity of the carbon and the rate of adsorption of the target contaminant (Matsui et al., 2003).

12.4.2 Multiwalled Carbon Nanotubes (MWCNTs) and Graphene

Multiwalled carbon nanotubes (MWCNTs) consist of layered tubes that have diameters in the nanoscale and lengths of several micrometers. MWCNT aggregation is widespread, and the voids created by the MWCNT aggregates can act as pores for adsorption. Consequently, adsorption may happen on the external surface of MWCNTs, on the inner wall of the center tube, or in the pores formed due to aggregation. Aggregation is accompanied by a reduction in surface area, yet, described as an unfavorable phenomenon. Adsorption between the walls of a multiwalled tube is unlikely because the openings are too minute for contaminant's entrance. Like MWCNTs, graphene is a hydrophobic material derived from graphite; it comprises carbon atoms arranged hexagonally into sheets of single-atom thickness. While it has potential applications in electronics, its utility as an adsorbent has also received some attention. Adsorption mechanisms on the graphene surface should be comparable to those on MWCNTs, except that the available surface area and pore structure formed through aggregation is different due to its sheet-like primary morphology. In a comparative analysis, the average particle diameters for PACs and S-PACs were determined to be 37 μm and 0.24 μm, respectively, and were calculated based on volume percentage. The MWCNTs were reported to have a length of 10–20 μm, and the outer diameter was 29 ± 13 nm, measured using transmission electron microscopy. The nanographene platelets (NGPs) had dimensions of <5μm diameter and <1nm thickness. Surface area, pore-volume, pore morphology, and pH are shown for all the adsorbents in Table 12.2. MWCNTs with the lowest surface area, followed by NGPs, S-PACs, and PACs were noted. The total surface area of S-PAC was lower than PAC, its precursor, although S-PAC had a smaller particle size (Ellerie et al., 2013).

TABLE 12.2
Adsorbent Characteristics

Adsorbent	SSA_{BET}[a] (m^2/g)	V_T[b] (m^2/g)	Pore Volume Fractions[c] Micro (<2 nm)	Meso (2–50 nm)	Macro (>50 nm)	pH_{pzc}[d] pH units
MWCNTs	174	0.70	0.01	0.51	0.48	3.9
NGPs	624	2.69	0	0.59	0.41	~9.5
PAC	900	0.46	0.7	0.01	0.29	6.1
S-PAC	773	1.01	0.27	0.29	0.44	5.7

12.4.3 Activated Carbon Fibers (ACFs)

ACFs are highly microporous carbon materials, which are commercially available in the form of fiber tows, fabrics, papers, mats, and felts. ACFs have a larger micropore volume and a more uniform micropore size distribution than GACs and thus, are taken to have a larger adsorption capacity and higher rates of adsorption and desorption. The ACF may be packed to fit almost any geometry, for almost any catalytic application, and fulfills the requirements for high catalyst effectiveness and low-pressure drop for finely distributed catalysts, but avoids the technical issues concerned with powders. A study showed that with copper particles deposited onto the ACF, the NO removal efficiency was tested at four different deposition times (5, 10, 15, and 20 min). At all deposition times, the NO removal efficiency increased with increasing reaction temperature well up to 673 K. The highest NO removal efficiency was when the amount of Cu/ACF was 110 mg/g (with deposition time of 5 min). The pore size distributions of all the samples were concentrated at pore diameters lesser than 30 Å. Pores are typically classified as micropores (<20 Å), mesopores (20–500 Å), and macropores (>500 Å), according to the classification approved by the IUPAC. Finally, based on observations it was confirmed that the specific areas, average pore diameters, total pore volumes, and micropore volumes decreased with increasing deposition time (Byeon et al., 2008).

12.5 MODIFIED ACTIVATED CARBON

Activated carbon in general has a relatively low adsorption capacity and affinity for metals. Modifying activated carbon with suitable additives has been investigated to enhance its adsorption efficiency. Some studies based on modified activated carbon include palladium, silver, copper sulfide, titanium dioxide, manganese oxide, and zinc oxide nanoparticles loaded with activated carbon. Composite adsorbents of this type include Fe^{2+}-treated activated carbon, Fe-coated sand, granular iron oxide, Fe-coated GAC, and Mn oxide-coated sand. Experiments and evaluated results from these studies suggest that composite adsorbents can be synthesized by creating an oxide surface coating on another solid and these composite adsorbents can easily be separated from an aqueous solution after the adsorption process. In comparison to GAC, oxides have higher metal affinities and adsorption capacities. Furthermore, oxides can remove metals to lower concentrations and the metals that are absorbed can be recovered and reused (Fan and Anderson, 2005; Lee et al., 2015). Some experimental studies carried out over the past years in the lines of modified activated carbon for improved and more effective adsorptive properties have been mentioned and discussed as follows.

Manganese oxides. Manganese oxides are very important scavengers of aqueous trace metals in soil, sediments, and rocks because of their dominant absorptive behavior. The Cd(II), Pb(II), and Cu(II) adsorption capacities of GAC were significantly improved by Mn oxide coating. In comparison, the GAC and MnGAC had pHs of about 8.5 and 5.8, respectively, which shows that it has the potential to be an efficient way to remove and recover metals from metal-contaminated wastewater (Fan, Anderson, 2005; Lee et al., 2015).

Copper(II) and iron compounds. Since copper nitrate has proved to the best catalyst in the catalytic oxidation of dyeing and printing, it will be selected as the active element to be deposited onto the porous adsorbent. Copper ions form bonds with glycoproteins and may potentially influence capturing the virus–host cells. The effective action of copper is because of the formation of Cu^{+} and/or Cu^{2+} ions (Wu et al., 2010). The pretreatment of activated carbon with Cu(II) leads to a 30% increase in the removal of arsenic. Impregnation of activated carbon with $FeCl_3$ leads to an increase in iron content on the surface of activated carbon (Ghanizadeh et al., 2010; Hu et al., 1999). The techniques applied for modification and impregnation to increase AC surface adsorption and removal capacity and to add selectivity to carbon investigations have demonstrated that the presence of oxygenated groups on the surface of AC aids metal adsorption through the formation of metal complexes. Chemical oxidation of AC by nitric acid is one of the most effective means to provide acidic carboxylic and phenolic groups to the AC surface. AC oxidation with HNO_3 leads to a major increase in the number of surface acid groups with the acid dissociation constant (Ka) distributed over a wide range. These modifications do not relate to any evident change on specific surface area and porosity values but improve the adsorption capacity of copper ions on oxidized AC. The application of Langmuir isotherms to experimental data indicates that copper adsorption increases 3.5 times for modified AC as compared with data for regular AC. Experiments show that the maximum copper adsorption capacity depends on the ionic strength and the predominant copper species in the aqueous solution (de Mesquita et al., 2006).

Titanium dioxide. Activated carbon can adsorb pollutants and then release them onto the surface of TiO_2. Firstly, a higher concentration of pollutants in the vicinity of the TiO_2 surface than that in the bulk solution is created leading to an increase in the degradation rate of the pollutants. Secondly, the charge transfer between TiO_2 and activated carbon causes acidification of surface hydroxylic groups in the TiO_2. This will augment the interaction between some pollutants and TiO_2 to further promote the degradation process. Moreover, the ability of TiO_2 to absorb visible light is also enhanced. Thirdly, the intermediates produced through degradation can also be adsorbed by activated carbon and then oxidized further on. Therefore, secondary pollution of the intermediates can be avoided. The initial increase in the loading of TiO_2 results in a significant increase in the catalytic efficiency obviously due to the increase of active sites. When the loading of TiO_2 is 12 wt%, maximum catalytic efficiency is

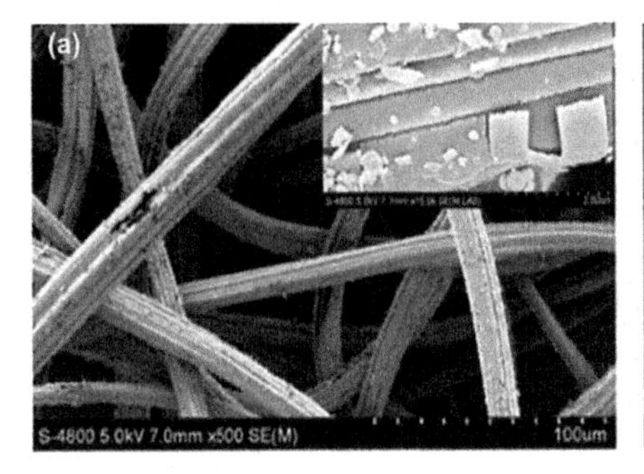

FIGURE 12.2 SEM images of samples: (a) S1 and (b) S2; (Reprinted from Shi et al., 2012).

reached. Also, TiO_2/AC has two practical advantages over the powdered TiO_2 photocatalyst: one being that it can simply be separated and recovered from water; another is that it has a high capacity to mineralize the target pollutant and the intermediates, indicating that secondary pollution can be avoided (Zhang et al., 2005). There are typically two kinds of coating methods without binding materials, dip-coating and hydrothermal treatment. These methods are applied to immobilize TiO_2 films on ACFs for obtaining the photocatalysts with high photocatalytic activity and good adhesion to the substrate. From the SEM images of titanium tetraisopropoxide treated ACF felts mentioned as samples S1 and S2 (Figure 12.2a and b), it can be observed that the smooth surface of ACFs have been covered by rough titania precursor, which can be confirmed from the magnified SEM image shown in the inset in Figure 2.1a, where titania precursors in the form of films anchored onto the fibers are clearly observed, and titania precursors in the form of larger particles or clusters can also be found. From the magnified image in 12.2b, it can be observed that smooth surfaces were preserved, implying that the titania precursor was uniformly anchored onto ACFs in the form of smooth films (Shi et al., 2012).

In another study, pitch-based ACFs in the form of nonwoven fabrics, with an average diameter of 4 μm was used to assess adsorption isotherm, kinetics, and mechanism of some substituted phenols on ACFs. Characterizations of ACFs demonstrated the narrow distribution of the low-size pores and neutral nature of the surface. Redlich–Peterson model gives the best fitting for the adsorption isotherms in most cases, while the Langmuir model is reasonably relevant in all cases. In kinetic studies, ACFs show a high adsorption rate because of their open pore structure. The pseudo-second-order model gives an acceptable fitting, and the intraparticle diffusion model describes the adsorption process well. Figure 12.3 demonstrates the nitrogen adsorption isotherm of the ACFs and the pore size distribution based on the Horvath-Kawazoe method, showing a sharp growth at the low-pressure range and a plateau at the high-pressure range demonstrating an H-type isotherm, indicating the microporous nature of the ACFs. The BET surface area and the micropore volume are 920.3 m^2/g and 0.422 cm^3/g, respectively, which are moderate for ACFs. The pore size distribution curve illustrates that the distribution is quite narrow and the pore size is centered at about 8.8 Å. The amounts of the acidic and basic groups on ACF's surface are measured as 0.12 and 0.15 mol/g, respectively, which are remarkably

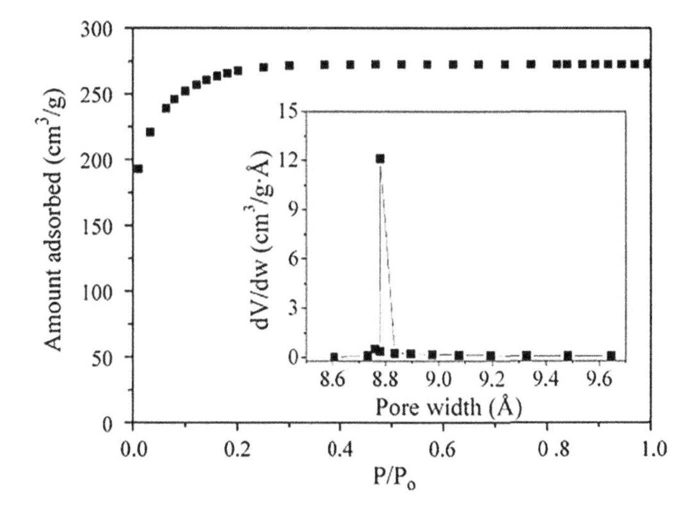

FIGURE 12.3 Nitrogen adsorption isotherm (77 K) and pore size distribution of the ACFs.

lower than the values reported by other studies. The pH value was measured as 6.9, which suggests a neutral surface (Liu et al., 2010).

12.6 CONCLUSION

Using a membrane coating rather than adding the adsorbent to a stirred vessel resulted in greater initial contaminant removal, as shown by lower permeate concentrations for the membrane coating setup at the beginning of the filtration experiments. When used as a membrane coating, S-PAC was more effective than its parent PAC material for methylene blue and atrazine removal. S-PAC also performed better than the graphene and MWCNT materials used in this study. The fast adsorption kinetics seen with S-PAC coatings can be advantageous in a scenario in which an immediate contaminant removal is required. However, the superior adsorption capability of S-PAC compared with PAC and the other adsorbents is coupled with flux resistance resulting from the small particle size, which could lead to increased energy requirements. Using S-PAC with a slightly larger particle size may be advisable for achieving a lower energy operation while still harnessing the enhanced adsorption capabilities of small particles. A carbonaceous coating can potentially serve as a secondary membrane, effectively preventing foulants such as organic matter from reaching the microfiltration or ultrafiltration membrane and restricting flow. However, as shown with the S-PAC results, the adsorbent coating itself may contribute to flow restrictions, depending on particle size and compatibility of the membrane and adsorbent. A balance among flux resistance, adsorption capacity, and adsorption kinetics is desirable for optimizing filtration efficiency. With a smaller particle size, S-PAC has a greater external adsorption area and a lower number of internal adsorption sites compared to PAC. This results in high immediate contaminant removal, but a steep breakthrough curve because of the limited internal adsorption sites. For PAC, pores are deeper because of the higher particle volume, so pore diffusion progresses throughout the filtration, and the breakthrough curve is more gradual. This is consistent with previous research showing that the resistance to solute mass transfer decreases with decreasing particle size It is noteworthy that the benefit of the coating technique was seen for all four materials, which have different surface areas, a range of pH, and varying pore morphologies. This points to the possibility that the coating approach is advantageous over the stirred vessel approach regardless of adsorbent properties. Most of the porous substrates used for adsorption are still in the form of granules, and the problem of separation and recovery of the photocatalysts from the reaction media still exists. Undoubtedly, ACFs in the form of felt possess prominent predominance in comparison with these granular porous substrates.

ACKNOWLEDGEMENT

This work was supported by the Ministry of Education, Youth and Sports of the Czech Republic, European Union – European Structural and Investment Funds in the Frames of Operational Programme Research, Development and Education – project Hybrid Materials for Hierarchical Structures (HyHi, Reg. No. CZ.02.1.01/0.0/0.0/16 _019/0000843), the Ministry of Education, Youth and Sports in the frames of support for researcher mobility (VES19China-mobility, Czech-Chinese cooperation) Design of multilayer micro/nano fibrous structures for air filters applications, Reg. No. 8JCH1064., and project "Intelligent thermoregulatory fibers and functional textile coatings based on temperature resistant encapsulated PCM" SMARTTHERM (Project No. TF06000048).

REFERENCES

Selasa, 2018. *Activated Carbon Untuk Industri Emas dan Perak*. Inov. Biomasa - Renew. Ind. Enable BioEconomy. URL http://inovasibiomasa.blogspot.com/2018/07/activated-carbon-untuk-industri-emas.html (accessed 4.9.20).

Atrie, D., Worster, A., 2012. Surgical mask versus N95 respirator for preventing influenza among health care workers: a randomized trial. *CJEM* 14, 50–52. doi: 10.2310/8000.2011.110362

Bansal, R.C., Goyal, M., 2005. *Activated carbon adsorption*. Taylor & Francis, Boca Raton.

Byeon, J.H., Yoon, H.S., Yoon, K.Y., Ryu, S.K., Hwang, J., 2008. Electroless copper deposition on a pitch-based activated carbon fiber and an application for NO removal. *Surf. Coat. Technol.* 202, 3571–3578. doi: 10.1016/j.surfcoat.2007.12.032

Ellerie, J.R., Apul, O.G., Karanfil, T., Ladner, D.A., 2013. Comparing graphene, carbon nanotubes, and superfine powdered activated carbon as adsorptive coating materials for microfiltration membranes. *J. Hazard. Mater.* 261, 91–98. doi: 10.1016/j.jhazmat.2013.07.009

Fan, H., Anderson, P., 2005. Copper and cadmium removal by Mn oxide-coated granular activated carbon. *Sep. Purif. Technol.* 45, 61–67. doi: 10.1016/j.seppur.2005.02.009

Franz, M., Arafat, H.A., Pinto, N.G., 2000. Effect of chemical surface heterogeneity on the adsorption mechanism of dissolved aromatics on activated carbon. *Carbon* 38, 1807–1819. doi: 10.1016/S0008-6223(00)00012-9

Ghanizadeh, G., Ehrampoush, M.H., Ghaneian, M.T., 2010. Application of iron impregnated activated carbon for removal of arsenic from water. *J. Environ. Health Sci. Eng.* 7, 12.

Goddard, M., Butler, M., 2015. *Viruses and Wastewater Treatment, Proceedings of the International Symposium on Viruses and Wastewater Treatment*, Held at the University of Surrey, Guildford, 15–17 September 1980. Elsevier.

Hijnen, W.A.M., Suylen, G.M.H., Bahlman, J.A., Brouwer-Hanzens, A., Medema, G.J., 2010. GAC adsorption filters as barriers for viruses, bacteria and protozoan (oo)cysts in water treatment. *Water Res.* 44, 1224–1234. doi: 10.1016/j.watres.2009.10.011

Hu, X., Lei, L., Chu, H.P., Yue, P.L., 1999. Copper/activated carbon as catalyst for organic wastewater treatment. *Carbon* 37, 631–637. doi: 10.1016/S0008-6223(98)00235-8

Khayan, K., Anwar, T., Wardoyo, S., Lakshmi Puspita, W., 2019. Active carbon respiratory masks as the adsorbent of toxic gases in ambient air. *J. Toxicol.* 2019, 1–7. doi: 10.1155/2019/5283971

Lee, M.-E., Park, J.H., Chung, J.W., Lee, C.-Y., Kang, S., 2015. Removal of Pb and Cu ions from aqueous solution by Mn_3O_4-coated activated carbon. *J. Ind. Eng. Chem.* 21, 470–475. doi: 10.1016/j.jiec.2014.03.006

Liu, Q.-S., Zheng, T., Wang, P., Jiang, J.-P., Li, N., 2010. Adsorption isotherm, kinetic and mechanism studies of some substituted phenols on activated carbon fibers. *Chem. Eng. J.* 157, 348–356. doi: 10.1016/j.cej.2009.11.013

Loeb, M., Dafoe, N., Mahony, J., John, M., Sarabia, A., Glavin, V., Webby, R., Smieja, M., Earn, D.J.D., Chong, S., Webb, A., Walter, S.D., 2009. Surgical mask vs n95 respirator for preventing influenza among health care workers: a randomized trial. *JAMA* 302, 1865–1871. doi: 10.1001/jama.2009.1466

Matsui, Y., Fukuda, Y., Inoue, T., Matsushita, T., 2003. Effect of natural organic matter on powdered activated carbon adsorption of trace contaminants: characteristics and mechanism of competitive adsorption. *Water Res.* 37, 4413–4424. doi: 10.1016/S0043-1354(03)00423-8

Matsushita, T., Suzuki, H., Shirasaki, N., Matsui, Y., Ohno, K., 2013. Adsorptive virus removal with super-powdered activated carbon. *Sep. Purif. Technol.* 107, 79–84. doi: 10.1016/j.seppur.2013.01.017

de Mesquita, J.P., Martelli, P.B., Gorgulho, H. de F., 2006. Characterization of copper adsorption on oxidized activated carbon. *J. Braz. Chem. Soc.* 17, 1133–1143. doi: 10.1590/S0103-50532006000600010

Shi, J.-W., Cui, H.-J., Chen, J.-W., Fu, M.-L., Xu, B., Luo, H.-Y., Ye, Z.-L., 2012. TiO2/activated carbon fibers photocatalyst: Effects of coating procedures on the microstructure, adhesion property, and photocatalytic ability. *J. Colloid Interface Sci.* 388, 201–208. doi: 10.1016/j.jcis.2012.08.038

Wu, Z., Fernandez-Lima, F.A., Russell, D.H., 2010. Amino acid influence on copper binding to peptides: Cysteine versus arginine. *J. Am. Soc. Mass Spectrom.* 21, 522–533. doi: 10.1016/j.jasms.2009.12.020

Xu, J., Gao, N., Deng, Y., Sui, M., Tang, Y., 2011. Perchlorate removal by granular activated carbon coated with cetyltrimethyl ammonium chloride. *Desalination* 275, 87–92. doi: 10.1016/j.desal.2011.02.036

Zhang, X., Zhou, M., Lei, L., 2005. Preparation of photocatalytic TiO2 coatings of nanosized particles on activated carbon by AP-MOCVD. *Carbon* 43, 1700–1708. doi: 10.1016/j.carbon.2005.02.013

13 Antiviral Finishes for Protection against SARS-CoV-2

Divan Coetzee, Mohanapriya Venkataraman and Jiri Militký
Technical University of Liberec, Czech Republic

CONTENTS

13.1 INTRODUCTION

To determine whether a compound could be used as an effective antiviral agent against SARS-CoV-2, the first step is to look at the compound's effectiveness against other viral pathogens with similarities. The SARS-CoV-2 virus exhibits a solar crown shape with beta linkages on the surface which is known to cause severe infection in humans that can be fatal. Like other coronaviruses, it has an enveloped structure. Transmission occurs through contact with bodily fluids from an infected person which could be when an infected person is still in the viral incubation period or asymptomatic (Ceccarelli et al., 2020). Transmission typically occurs when an infected person produces a virus-laden aerosol by coughing, sneezing, and breathing. The aerosol is then either inhaled by another person or accumulates on surfaces where the virus can remain active for days (Qu et al., 2020). It is therefore important to reduce transmission by incorporating antiviral agents on surfaces and respirators or face masks. As countries around the world are relaxing restrictions it is important to mitigate viral spread where social distancing is less effective to avoid large increases in infections (Xu and Li, 2020). Respirators can reduce the risk of viral spread significantly, however, the use of face coverings is not customary in many countries to be used by the general population. This could lead to improper care of the face coverings used thus reducing its protective effectiveness (MacIntyre et al., 2009). Health care providers are required to use N95 respirators which can prevent airborne particles up to 0.3 μm from being inhaled (Repici et al., 2020). Antiviral respirators or face masks could provide an additional layer of protection by disinfecting the viral molecules by contact. This would be possible due to the random movement of air molecules that collides with viral particles. As the viral particles move in the airstream, they encounter a fiber which contains an antiviral component. This mechanism of viral particle movement in the air was found valid for Adenovirus, Filoviruses which causes Ebola hemorrhagic fever, Orthomyxoviridae causing types A, B, and C influenza, and Coronaviruses such as the SARS type (Majchrzycka et al., 2019).

Due to global production constraints, the use of home-made cloth masks has become a cost-effective and more comfortable form of a respirator. The effectiveness of this can be questioned as these do not follow specific guidelines for effective protection in terms of particle filtration. These types of face masks are best suited to protect against splashes (Repici et al., 2020). Often these masks are reusable and without proper disinfection could promote viral transmission. The effectiveness of face masks can be improved by incorporating an antiviral finish on the cloth from which the face mask is produced, but this could also lead to production constraints given the large size of the population that would require it. For this purpose, this chapter also investigates natural antiviral finishes that could be easily applied at home. Nanoparticles have gained increasing attention in the field of medicine due to their potential antimicrobial effects. The antimicrobial effects of nanoparticles can be attributed to their small size which is typically below 100 nm (Coetzee et al., 2020). This makes them ideal to interact with viruses on a molecular level as indicated in Figure 13.1. There are mainly three mechanisms by which nanoparticles can act as antiviral agents. Firstly, nanoparticles can inhibit the virus directly by penetrating the viral cell. Some nanoparticles are effective for enveloped and some for nonenveloped viruses. Secondly, nanoparticles can bind to viruses preventing them from attaching to host cells when entering the body. Thirdly, nanoparticles can prevent replication of the virus (Chen and Liang, 2020). For finishing purposes, the first two mechanisms are important as the virus can be inactivated on contact with an antiviral surface or

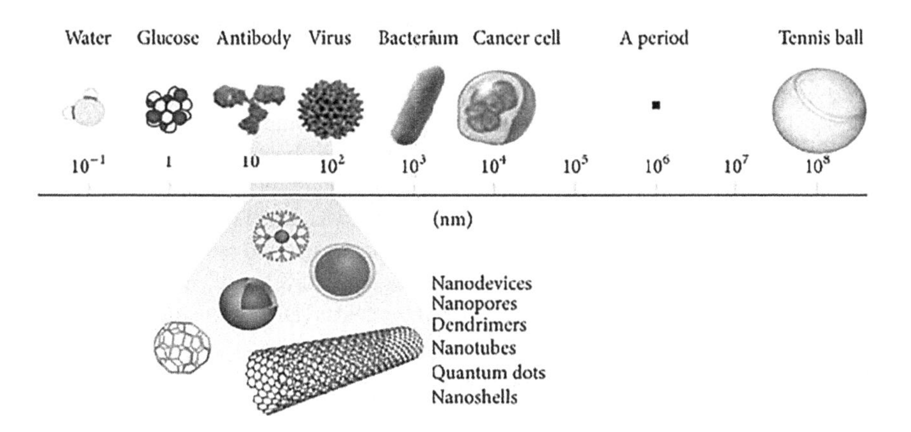

FIGURE 13.1 Size comparison between nanomaterials and cells, https://creativecommons.org/licenses/by/3.0/ (Amin, Alazba and Manzoor, 2014).

fiber. Cationic polymers have proven to have antiviral properties. Organic materials such as chitosan operate in a two-step mechanism. Firstly, the cationic chitosan binds with sialic acid in phospholipids which restrains the movement of the microbiological substance. Secondly, the chitosan molecules penetrate the cell wall where it inhibits cell growth by preventing DNA to RNA transformation. This action is attributed to the amino groups of chitosan (Singh et al., 2017). The amino groups in chitosan can be protonated by acidic medium and form salt linkages with sulfonate groups in dyes (Muthu, 2017). Polymers tend to be inert, however biodegradable polymers provide excellent eco-friendly alternatives with characteristics such as molecular penetrability which makes them ideal for disposable antiviral purposes. Chemical or natural oil extract-based systems typically operate by interfering with the viral cell wall destroying the cell. These are typically encapsulated or could be directly applied as a finish.

13.2 NANOPARTICLE-BASED ANTIVIRAL SYSTEMS

13.2.1 Quantum Dots (QDs)

Quantum dots (QDs) are semiconductor nanocrystals. This form of nanoparticles has been widely used for virus and cell labeling, image tracking, and viral detection. They exhibit distinguished luminescent properties which make QD's suitable for excitation, narrow, and bright emission spectroscopy. The antiviral properties of traditional QDs have been largely unknown, however, studies have shown that glutathione capped CdTe QDs exhibit antiviral properties. Systematic analysis using a one-step growth curve, MTT assay, and fluorescence colocalization was performed by Du et al. The study showed that the QDs altered the structure of the pseudorabies viral cell surface proteins. This inhibited the viral cell from entering host cells rendering it inactive. The release of Cd^{2+} ions from the QD's also inhibited the viral cell's ability to replicate. Results indicated that the viral binding of CdTe QDs was greater than that of Cd^{2+} ions and larger nanoparticles were more effective in preventing viral replication (Du et al., 2015; Chen and Liang, 2020). Carbon dots (CDs) exhibit lower toxicity compared to cadmium containing nanoparticles. The viricidal effects of CD's were found to differ based on their carbon precursors and synthesis conditions (Chen and Liang, 2020). Huang et al. fabricated a CD nanoparticle using benzoxazine monomer (BZM). It was found that the nanoparticles were able to bind with the virus rendering it unable to bind with host cells. The effectiveness of the nanoparticles was tested against enveloped viruses' Japanese encephalitis (JEV), Zika virus (ZIKA), and dengue virus (DENV) also, nonenveloped virus's porcine parvovirus (PPV), and adenovirus-associated virus (AAV). Results showed that the EC_{50} of BZM-CDs against the tested viruses was 18.63 μg/mL (JEV), 3.715 μg/mL (ZIKV), 37.49 μg/mL (DENV), 40.25 μg/mL (AAV), and 45.51 μg/mL (Huang et al., 2019). Łoczechin et al. prepared CDs from hydrothermal carbonization of ethylenediamine/citric acid precursors which was modified with boronic acid ligands. These were tested against Human Coronavirus HCoV-229E. CD's modified by hydrothermal carbonization of phenylboronic acid and 4-aminophenyl boronic acid proved to be the most effective viral inhibitors. CD's with $R{-}B(OH)_2$ and NH_2 functionalization proved to be effective in interfering with HCoV-229E entry receptors which prevented the viral cells from binding with human host cells at an EC_{50} of 5.2 ± 0.7 μg/mL. A reduction in viral replicability was also noted (Łoczechin et al., 2019).

13.2.2 Silver Nanoparticles (AgNPs)

Silver nanoparticles (AgNPs) have been used widely in industry and pharmaceutics due to their unique properties. They have been known for their antibacterial properties, however, their use as antivirals is still in the infancy stage. Recent studies conducted have attributed their antiviral activity mainly due to blocking binding sites in host cells to prevent a virus from attaching (Chen and Liang, 2020).

Liga et al. prepared a photocatalytic antiviral agent for water treatment by silver doping TiO_2 nanoparticles. The authors hypothesized a possible synergic mechanism occurring between silver and TiO_2. They found that Ag/TiO_2 greatly enhanced the photocatalytic inactivation of viruses primarily by increasing hydroxyl free radical production while slightly increasing virus adsorption. This could be in addition to the antimicrobial mechanism employed by AgNPs. It was suggested that the Ag/TiO_2 nanoparticles could also be activated by visible light through the silver surface. It was found that the viral inactivation rate was increased greater with silver content (Galdiero et al., 2011; Liga et al., 2011). Mohanapriya et al. have studied antimicrobial properties of silver nanotreated sewing thread samples with müeller-hinton agar. The growth of culture was determined by turbidity which read at 600 nm after 24 h. Growth was inhibited in tubes containing silver nanotreated thread. Here is an increase in the effectiveness of antimicrobial properties of the treated threads. This affects cellular metabolism and reduces cellular growth. As a result, it reduces the growth of microbes and in turn may result in a reduction of infections (Venkataraman et al., 2014). Castro-Mayorga et al. produced poly(3-hydroxybutyrate-co-3-hydroxy valerate) (PHBV) films with a coating of thermally post-processed electrospun PHBV18/AgNP fiber mats over compression-molded PHBV3 films. AgNPs were added into the polymer solution before the formation of the films or electrospinning. The antiviral effects of the PHBV18/AgNP film were studied using the feline calicivirus (FCV) and murine norovirus (MNV). Antiviral activity of the AgNPs remained constant for concentrations greater than 2.1 mg/L for the 150-day duration of the study. Most promising results were obtained using a suspension of AgNPs with a particle size of 30 nm at a concentration of 400 mg/L. This resulted in a 6 $\log_{10}$ reduction of MNV after exposure at 25°C for 6 h. Only AgNPs smaller than 10 nm were effective at reducing the FCV levels. The best overall results were found to be with AgNPs with sizes of 7 nm ± 3 nm at a concentration of 21 mg/L. This highlighted the size dependence of the ability for nanoparticles to interact with viruses. PHBV3/PHBV18/AgNP films which contained a total silver concentration of 270 ± 10 mg/kg demonstrated to have higher antiviral activity against FCV than comparable studies. This was suggested to be attributed to the release of silver ions from the immobilized AgNPs which could have contributed to an increased antiviral effect (Castro-Mayorga et al., 2017). Mori et al. studied the antiviral effects of an AgNP/chitosan composite against the H1N1 influenza A virus. AgNPs were added to a chitosan solution and precipitated to form a yellow powder. The composite was suspended in PBS for antiviral assays. It was suggested that the antiviral activity of AgNPs against several other types of viruses is due to direct binding of the AgNPs to viral envelope glycoproteins, thereby inhibiting viruses to penetrate host cells as was assumed for this study. No increase in antiviral activity was observed for composites which contained AgNP concentrations above 200 μg per 1 mg of chitosan. It was found that similar concentrations of AgNPs with smaller particle sizes exhibited greater antiviral activity. The same was observed as AgNP concentrations increased. It was found that pure chitosan had no antiviral effects on the H1N1 virus and the antiviral effect of the composite was therefore attributed solely to the antiviral effects of the AgNPs (Mori et al., 2013).

13.2.3 Copper Nanoparticles (CuNPs) and Copper Oxide Nanoparticles (CuONPs)

Sinclair et al. produced an antiviral membrane for the purification of drinking water. The authors used a stable covalent layer-by-layer (LBL) approach to create a multilayer film from polyethyleneimine (PEI). A terephthalaldehyde (TA) crosslinking agent was used to create crosslinked multilayers on the developed model surfaces and commercial polyethersulfone, (PES) MF membranes. The substrates were then coated with antiviral AgNPs and copper nanoparticles (CuNPs) which were stabilized with PEI. Results indicated a viral reduction on model surfaces of 4.0 $\log_{10}$-units of MS2 viral titer. This was found to be independent of the crosslinked PEI layer thickness. The crosslinked PEI and Ag/CuNPs-modified membranes reduced infectious MS2 bacteriophages by 4.5–5.0 $\log_{10}$-units. This occurred by both adsorption and inactivation of viral particles. The antiviral mechanism of cationic polymers such as PEI is still under investigation. PEI may act as merely an adsorbent by which negatively charged viral molecules are attracted to its highly positively charged surface (Sinclair et al., 2019). It was suggested that this electrostatic attraction could also cause structural and or genomic damage to viruses as investigated by previous studies (Milović et al., 2005; Larson et al., 2011). It was found by Rao et al. that both silver and CuNPs released free ions which were able to damage viral cells. AgNPs proved to be more effective at lower doses; however, the viricidal effects of both copper and AgNPs were increased with higher nanoparticle loading (Rao et al., 2016). It has been proven that CuCl nanoparticles can inactivate avian influenza viruses (Armstrong, Sobsey, and Casanova, 2017). Borkow et al. developed an antiviral respiratory face mask which contained two external layers made of spunbonded polypropylene fabric containing 2.2% (w/w) copper oxide nanoparticles (CuONPs). The internal mask structure consisted of a melt-blown polypropylene fabric which contained 2.0% (w/w) CuONPs contributing as the barrier layer that provides the physical filtration properties to the mask. This was placed behind the outer exterior layer. A second interior layer of plain polyester was placed behind the filtration layer which was designed to contribute to the mask shape. The final external layer is the layer in contact with the wearer's face. This layer is identical to the outer layer of the mask as illustrated in Figure 13.2. The mask's antiviral effectiveness was tested against human influenza virus A and avian influenza virus. Both produced masks and N95 control masks performed equally in terms of the amount of virus that was able to pass through them. By filtration, both test and control masks reduced the infectious viral molecules that passed through the masks by 3 logs for the human influenza A virus and by 4 logs for the avian influenza virus. Slightly better performance was obtained with the developed test mask. The key function of CuONP incorporation was to promote a reduction of viral molecules trapped in the mask. The infectious human influenza A viral titers in the test masks were reduced by $\geq 4.78 \pm 0.88$ log and avian influenza virus by $\geq 5.20 \pm 0.84$ log after 30 min of exposure, respectively. Direct contact tests with the control and the test masks indicated a reduction in

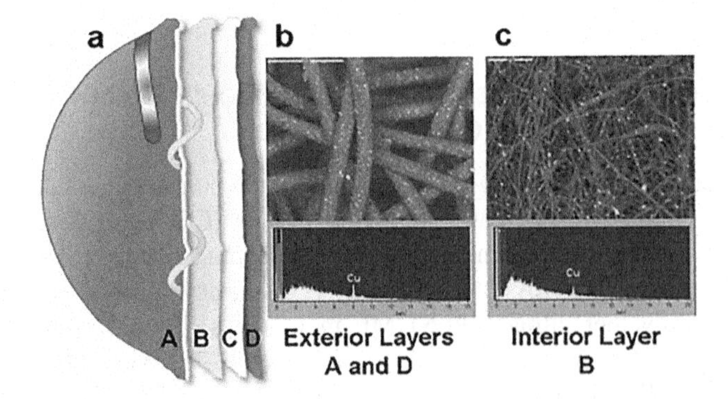

FIGURE 13.2 Copper oxide impregnated test mask composition. a) external spunbonded polypropylene layers (A and D) containing CuNPs, one internal melt-blown polypropylene layer (B) containing CuNPs, and one polyester layer containing no CuNPs (C). b) Scanning electronic microscope picture and X-ray analysis of external layer A. c) Scanning electronic microscope picture and X-ray photoelectron spectrum analysis of internal layer B, https://creativecommons.org/publicdomain/mark/1.0/ (Borkow et al., 2010).

infectious virus titers by 1.90 ± 1.03 log for human influenza A and 1.34 ± 0.84 log for avian influenza virus, respectively. Toxicity tests indicated that the Cu released from the mask was almost three times lower than the 20.8 µg/h minimum risk level for oral exposure for a person weighing 50 kg. Ingestion by respiration was more than 10^5 times lower than the permissible exposure limit. The viricidal mechanism of CuONPs was thought to be achieved via the interaction of copper ions with the virions which were entrapped in the mask or with direct contact with the CuONPs impregnated in the outer surfaces of the masks (Borkow et al., 2010).

13.2.4 Graphene Oxide Nanoparticles (GONPs)

Song et al. investigated the use of graphene oxide for use as an antiviral disinfectant by superficial bioreduction. Graphene oxide has proved to efficiently capture viruses, destroy their surface proteins, and extract viral RNAs in an aqueous environment. The authors tested their material against avian influenza (H7N9) and pathogenic agents of hand foot and mouth disease (EV71) (Song et al., 2015). The disinfection effects of GO was found to be weak at room temperature as viral capturing occurred. With an increase in temperature, the disinfection rate increased to a $\geq 1.0 \log_{10}$ reduction of viral titers. This led the authors to increase the temperature to a commonly used disinfection temperature of 56°C which resulted in complete disinfection of both virus types. It was found that at room temperature the viruses in the solution survived for 2 days on the GO before being completely disinfected. It was stated that potential interaction between GO and viruses could be attributed to hydrogen bonding and electrostatic interactions while FT-IR analysis indicated a redox reaction had occurred. It was found that the GO was reduced in localized areas due to the release of oxides on the surface which resulted in the capture and destruction of the viruses (Song et al., 2015). Graphene sheets have proven their ability to attract and destroy viral cells. Carbon nanohorns (CNHs) are classified as graphene nanostructures and have been receiving attention recently for their anticancer properties. They exhibit excellent biocompatibility and low toxicity. CNHs have large surface areas and total pore volumes which enables them to adsorb and store large quantities of guest molecules. This could make them potential carriers for disinfectants. CNHs are spherical aggregates with diameters of around 100 nm, making them ideal for passive tumor-targeting conditions. It should still be investigated whether these could be used for antiviral purposes. CNHs are easy to produce in large quantities of high purity without requiring metal catalysts. This makes CNH materials different from carbon nanotubes. CNHs absorb light in a wide spectrum of wavelengths. This ranges from infrared to ultraviolet light and converts it into heat energy, which could be beneficial for their utilization as photo-hyperthermia agents (Zhang and Yudasaka, 2016; Devereux et al., 2019).

13.2.5 Halloysite Nanotubes (HNTs) as Biocidal Carriers

Halloysite nanotubes (HNTs) are naturally occurring tubular clay nanomaterials that are biocompatible (Panchal et al., 2018). These nanomaterials are made of aluminosilicate kaolin ($Al_2Si_2O_5(OH)_4.nH_2O$) sheets with a hollow tubular structure and high aspect ratio. The aluminol and siloxane groups on the surface of HNT facilitate the formation of hydrogen bonding with biomaterials such as biocides onto its surface. The outer diameter of the nanotubes ranges between 40 and 70 nm with an inner diameter of 10–20 nm and a length of 500–1,500 nm. These properties render HNT very suitable for diverse drug delivery systems since they can act as carriers for antiviral agents (Satish, Tharmavaram, and Rawtani, 2019).

13.3 POLYMER-BASED ANTIVIRAL SYSTEMS

13.3.1 Polyethyleneimine

Cationic polymers in the past have demonstrated antiviral activity against several viruses and model viruses and are well suited for the modification of porous polymeric materials (Larson et al., 2011). PEI comprising of polycationic moieties has been widely used to modify various substrates. This is due to their long-term antimicrobial activity with no resistance development, the possibility for regeneration upon loss of activity, minimal cytotoxicity to mammalian cells, and biocidal and viricidal activity against a broad variety of pathogens in short contact times. It was suggested that cationic polymers provide a simple electrostatic adsorption mechanism that would remove negatively charged viruses from any permeating liquid (Sinclair et al., 2019). This could be a suitable antiviral mechanism when virus-containing moisture and liquids emitted by a person's respiratory system interact with the polymer; however, this would require further investigation.

13.3.2 Chitosan

Li et al. produced a hydrogel containing cationic dimethyl–decyl ammonium chitosan with high quaternization as an antiviral component. The mechanism of operation included the cationic chitosan attracting negatively charged microbes into the nanopores of the hydrogel which would bind with the microbial cell wall which resulted in cell death. The authors' experimental data showed that when a bacterial lipid bilayer was placed near the polymer chains the viral cell was significantly disturbed. Some of the anionic cell membrane components were pulled out of the bilayer and drawn into the polymer chain pores. The suction action induced by the charge difference between the polymer and bacterial cell resulted in the distortion of the cell walls until it started to disintegrate resulting in microbe death. Similar results were obtained using a Gram-positive bacterial-membrane model (Li et al., 2011). Based on this mechanism, previous research, and the charge similarities between viruses and bacterial cells, cationic chitosan polymers could have possible viricidal effects (Randazzo et al., 2018).

13.3.3 Polymers with Regenerative Oxygen Species (ROS)

Si et al. produced daylight rechargeable, antibacterial, and antiviral bioprotective nanofibrous membranes (RNMs) which effectively produced biocidal reactive oxygen species (ROS). The photoactive RNMs could store the biocidal activity under light irradiation which readily released ROS to provide biocidal functions under dim light or dark conditions. The RNMs exhibited integrated properties of fast ROS production, ease of ROS activity storing, excellent durability and breathability, high biocidal efficacy, and the ability to intercept fine particles. The RNMs could be incorporated as a biocidal protective layer on PPE. The produced film achieved promising contact killing against pathogens in aerosol and liquid forms. Benzophenones, 4-benzoyl benzoic acid (BA), benzophenone tetracarboxylic dianhydride (BD), and polyphenol chlorogenic acid (CA) were chosen by the authors to be used as photo-biocides. Poly(vinyl alcohol-co-ethylene) was used as a polymer precursor to construct the nanofibrous networks. PVA-co-PE nanofiber membranes were electrospun and grafted with THF solution containing the various photosensitizers. The esterification reactions between the membrane-OH groups and COOH groups of the photo-biocides were catalyzed by carbonyl diimidazole. The samples were rinsed, and further grafting was performed on one BA sample with CA to produce a BACA-functionalized RNM. UV-vis analysis was used to estimate the delight absorbance coefficients of the materials which indicated 21.62%, 35.33%, 66.04%, and 45.41% for the BA, BD, CA, and BDCA-RNMs, respectively. The results indicated a robust utilization of daylight source. It was found that the BDCA-RNM exhibited the highest recharging capacity. OH, radicals, and H_2O_2 were released at amounts of 2,332 and 670 μg/g by 1 h of daylight charging which corresponded to the charging rates of 38.86 and 11.16 μg/g.min, respectively. This indicated a 70% energy conversion efficiency compared to the irradiation tests. After a 30-day storage test, it was found that the BDCA-RNM retained more than 55% of its light-absorbing transient structure which supported the durability claim. The viricidal properties of the materials were tested against T7 phage which is a nonenveloped double-stranded DNA virus with a single proteinaceous capsid. This modal virus was known to be less sensitive to photodynamic destruction than enveloped or RNA-based viruses. The BACA-RNM sample achieved a viral reduction of 5.0 log PFU under daylight conditions after 5 min of exposure and the same viral reduction after 30 min under dark conditions (Si et al., 2018).

13.3.4 Polymers as Antiviral Carriers

Polyhydroxyalkanoates, such as poly(3-hydroxybutyrate-co-3-hydroxy valerate), are considered as alternatives to petroleum-based polymers. The polymer's permeability to low molecular weight compounds was one of its disadvantages. However, this is considered an advantage in the field of controlled released drug delivery. The incorporation of AgNPs resulted in an antimicrobial composite which was effective even after 7 months (Castro-Mayorga, Fabra and Lagaron, 2016; Randazzo et al., 2018). The same properties were observed using polylactic acid (PLA) which exhibits the same permeable properties as PHBV and is biodegradable. These were also incorporated with AgNPs to produce biodegradable antiviral packaging (Martínez-Abad et al., 2013).

13.4 CHEMICAL/ESSENTIAL OIL EXTRACT-BASED ANTIVIRAL SYSTEMS

13.4.1 Didecyldimethylammonium Chloride

Majchrzycka et al. used didecyldimethylammonium chloride as biocide loaded into HNTs which was used as carrier nanoparticles. These were then incorporated into a polypropylene/polyester melt-blown nonwoven material from which a face mask was produced. In addition to the biocidal HNTs, superabsorbent polymer nanoparticles (SAPs) were added. SAPs fall under a class of hydrogel polymers which absorb water through physical adsorption by means of water retention in micropores, hydrogen bonding of functional groups in polymer chains. Solvation is limited by crosslinking and thus the expansion of polymer chains results in a polymer volume increase. The study found that the produced masks were most effective at reducing *Staphylococcus aureus* and *Candida albicans* bacterium after 24 h incubation. The authors' inclusion of SAPs to reduce moisture content assisted in reducing the bacterium's ability to grow. The loss of active substance did not exceed 0.02% after testing the mask under extreme conditions (Majchrzycka et al., 2019). Didecyldimethylammonium chloride is an amphiphilic quaternary ammonium compound. Their ability to adsorb to negatively charged surfaces makes these compounds ideal for use as disinfectants. Their biocidal mechanism is based on the extraction of lipids from the viral cell membrane which is driven by complexation (Leclercq et al., 2016).

13.4.2 Oregano Essential Oil

Gilling et al. investigated the antiviral properties of oregano essential oil (OEO) against the MNV. The primary active component in OEO is carvacrol which was found to bind with the capsid or block the epitopes which were required for the virus to bind to host cells contributing to a loss in viral infectivity. After binding with the viral capsid, the carvacrol would start breaking it down and subsequently the viral RNA. The results indicated that the OEO produced statistically significant reductions in virus infectivity within 15 min of exposure reducing viral titers by 1.0 $\log_{10}$. After 24 h of exposure, the viral infectivity reduction remained stable for OEO however, pure carvacrol was far more effective, resulting in a 3.87 $\log_{10}$ viral reduction after 1 h of exposure (Gilling et al., 2014). Pilau et al. investigated the antiviral effects of OEO and carvacrol against enveloped virus's acyclovir-resistant herpes simplex virus type 1 (ACVR-HHV-1), acyclovir-sensitive (HHV-1), human respiratory syncytial virus (HRSV), bovine herpesvirus types 1, 2, and 5 (BoHV-1/2/5), bovine viral diarrhea virus (BVDV) and nonenveloped human rotavirus (RV), respectively. It was noted that for BVDV, the OEO was more effective when applied after inoculation with the virus, but against all other viruses tested OEO was more effective before inoculation, unlike pure carvacrol. OEO was found to have no antiviral effects on BoHV-1 and 5 however it did inhibit BoHV-2 very well. Carvacrol was ineffective against BoHV-2, but effective against human RV. Both OEO and its carvacrol were more effective against ACVR-HHV-1 than acyclovir sensitive HHV-1. Both were also successful in inhibiting HRSV with adequate inhibition of other viruses to which it was effective. In general, OEO was more effective in inhibiting viral infection than pure carvacrol. Differences in antiviral activity between the effectiveness of the OEO and its pure active ingredient carvacrol were attributed to the possibility of synergistic effects from the organic components in the oil (Pilau et al., 2011). Fraj et al. were able to successfully encapsulate OEO in polycaprolactone nano- and microspheres to produce an antiviral wound dressing. The capsules had a greater affinity for polyamide than for cotton fibers used due to greater capsule-polymer bonding. The authors successfully produced a finished material with encapsulated OEO; however, the antiviral properties were not tested (Fraj et al., 2018).

13.4.3 Sodium Pentaborate Pentahydrate (SPP)

Iyigundogdu et al. produced an antiviral cotton fabric that incorporated sodium pentaborate pentahydrate (SPP) as the main antiviral ingredient. Cotton samples were modified with 3.0% SSP and 0.03% Triclosan which was emulsified with Glucapon 215 CS UP and tested against poliovirus type 1 (PV-1) and adenovirus type 5 (AV-5). The authors performed an ICP-MS analysis to determine the boron ion contents of the treated textile samples. This was measured to be 0.70 ± 0.050% (w/w) which was equivalent to 3.8 ± 0.272% SPP. The antiviral mechanism of boron and boron-containing compounds such as SPP is still not fully understood. Previous research in the medicinal field has indicated the following possible mechanisms for boron-containing compounds. Potential mechanisms include the inhibition of numerous enzymatic processes and influencing Ca^{2+} receptors, by inhibiting cell division, nuclear receptor binding mimicry, and the induction of apoptosis which all lead to viral inhibition (Scorei and Popa, 2012; Das et al., 2013; Soriano-Ursúa, Das and Trujillo-Ferrara, 2014). These mechanisms have been reported for boron compounds mainly in the form of antiviral drugs, however, Iyigundogdu et al.'s study proved that boron-containing compounds also exhibit antiviral properties as a textile finish. Antiviral test results indicated that the viral titers were reduced by 3 $\log_{10}$ for treated cotton textiles, while untreated cotton textiles exhibited no viral reduction. In total, the viral titers were reduced by 60% for both PV-1 and AV-5 after 72 h of passing the virus titers through the textiles. The authors stated that based on the effectiveness of the finished samples these should be effective against enveloped and nonenveloped RNA viruses such as HIV, HCV, Ebola, MERS, and SARS (Iyigundogdu et al., 2017).

13.4.4 Triclosan

Triclosan is a synthetic, nonionic, chlorinated bisphenol, and broad-spectrum antimicrobial agent. Triclosan possesses mostly antibacterial properties with some antifungal and antiviral properties (Orhan, Kut, and Gunesoglu, 2007; Malmsten, 2011). A study by Iyigundogdu et al. has reported that triclosan acts as an antiviral compound by blocking lipid biosynthesis and inhibiting microbial growth. Together with SPP, the antiviral effects of the compounds were enhanced (Iyigundogdu et al., 2017).

13.4.5 Hypothiocyanite

Hypothiocyanite ($OSCN^-$) is produced by the oxidation of the thiocyanate anion (SCN^-) with a lactoperoxidase catalyst. The antiviral effects of hypothiocyanite are attributed to the formation of its conjugate acid, hypothiocanous acid, which can oxidize viral protein sulfhydryl groups (Wu et al., 2017). In a recent paper, it was stated that hypothiocyanite is an effective biocide against a vast range of viruses and microorganisms for which its mechanism of action is not directed at specific proteins. The biocidal complex $LPO/H_2O_2/OSCN^-$ had proven to be effective against various Gram-positive and Gram-negative bacteria and viruses. These include *C. albicans* and *C. krusei* fungi, HIV, herpes simplex virus (HSV-1), adenovirus (AV), echovirus (EV), and respiratory syncytial virus (RSV). Given the compound's effective mechanism it could be a promising antiviral candidate against SARS-CoV-2 (Cegolon, 2020).

13.4.6 Carrageenan

Carrageenan is a polysaccharide that is derived from red seaweed and is typically used in edible coatings. Different types of alginate polysaccharides display various viral inhibiting mechanisms. These mechanisms mainly pertain to the polymer binding with host cells; however, it was stated that the ι-carrageenan polysaccharide could inhibit influenza A virus infection by binding with the viral particles (Wang et al., 2011; Besednova et al., 2019). Carrageenan's are naturally occurring polysaccharides with antiviral activities correlated with their molecular weights and the existence of sulfonate groups in their molecular structure. Studies have shown that carrageenan can inhibit influenza virus, avian leukosis virus type J, DENV, HSV-1, HSV-2, HPV, HRV, and HIV viral infection (Ahmadi et al., 2015; Sun et al., 2018).

13.4.7 Green tea extract (GTE)

Green tea extracts viricidal ability is mainly attributed to their polyphenol content referred to as catechins. The catechins include epicatechin (EC), epigallocatechin (EGC), epicatechin gallate (ECG), and epigallocatechin-3-gallate (EGCG). EGCG contributes to around 50% of the GTE viricidal ability. The viricidal efficiency of GTE is pH dependent and has proved to be more effective at a neutral pH in previous studies. However, for their study the authors Falcó et al. obtained greater antiviral results at pH 5.5 which was possibly attributed to catechin degradation at pH 7 during analysis. The authors prepared alginate films with GTE at pHs 5.5 and 7.0. Results showed that alginate films containing 0.75 g GTE/g alginate which was prepared at pH 5.5 reduced MNV and hepatitis A (HAV) infectivity by 1.97 and 1.25 log after ON incubation at 37°C. Complete inactivation was observed for both viruses after overnight storage at 25°C (Falcó et al., 2019). The antiviral mechanism of EGCG is attributed to its interference with the viral lipid envelope resulting in cell destruction. This mechanism had proven to be effective against HSV, HIV, DENV, JEV, TBEV, and ZIKV (Xu, Xu, and Zheng, 2017).

13.5 CONCLUSION

PHBV18/AgNP film coatings and polymers with ROS could be promising for use in public spaces to limit the lifespan of viruses on surfaces. It has been shown that the use of reactive dyes on cotton textiles increased the absorption of Ag/TiO_2 nanoparticles which resulted in an increase in antimicrobial activity and UV protective properties (Mavrić, Tomšič and Simončič, 2018). Gold nanoparticles (GNPs) were considered in this study, however, research indicated that GNPs mainly influence host cells and other functions inside the body. For this reason, they were not further investigated for this study (Kerry et al., 2019). CNHs are a class of graphene nanoparticles which have antiviral potential. Studies have mainly investigated their use inside a host for antiviral purposes; however, their mechanism could prove useful when used as an antiviral finish with further investigation. It was suggested that cationic polymers provide a simple electrostatic adsorption mechanism that would remove negatively charged viruses from any permeating liquid. It could be a suitable antiviral mechanism for when virus-containing moisture and liquids emitted by a person's respiratory system interact with the polymer; however, this would require further investigation (Sinclair et al., 2019). ROS exhibit excellent antiviral properties by destroying viral cells with radicals and hydrogen peroxide. The authors tested the membrane on PPE and found that it had excellent breathability. This would make the BACA-RNM antiviral finish very useful in increasing the effectivity and time for which PPE could be used (Si et al., 2018). Majchrzycka et al.'s inclusion of SAPs to reduce moisture content which would typically assist in reducing a bacterium's ability to grow was justified in their experiment, however, further investigation should be performed on the inclusion of SAPs on materials for antiviral purposes (Majchrzycka et al., 2019). Didecyldimethylammonium chloride is stated as an effective biocide, however, their antiviral ability requires further investigation as they have proved effective against bacteria. The compound's mechanism of disinfection could potentially be effective against viruses. OEO was more effective in inhibiting viral infection than pure carvacrol. Differences in antiviral activity between the effectiveness of the OEO and its pure active ingredient carvacrol were attributed to the possibility of synergistic effects from the organic components in the oil (Pilau et al., 2011). For many studies, it was found that synergistic antiviral effects were observed by utilizing multiple antiviral components. It was also found that nanoparticles had greater antiviral effects than the free ions of the metals they were produced from. However, this could be due to the ability of nanoparticles to provide a controlled release of free ions at a higher dosage. The antiviral effectiveness of AgNPs was found to be more effective at particle sizes below 10 nm depending on the type of virus it was tested against. Nanoparticle systems prove to increase in cost as particle size decreases due to manufacturing complexity and thus costs can vary depending on the particle size (Khan, Saeed, and Khan, 2019). In general, chemical antiviral finishes are easier to produce on large scale and are therefore less expensive compared to nanoparticle systems. The use of essential oils could prove to be particularly useful as the environmental effects of these natural compounds are lesser than that of inorganic chemical-based antiviral systems. These can also be easily purchased by the general population to improve the effectiveness of nonmedical grade PPE.

ACKNOWLEDGEMENT

This work was supported by the Ministry of Education, Youth and Sports of the Czech Republic and the European Union – European Structural and Investment Funds in the frames of Operational Programme Research, Development and Education under project Hybrid Materials for Hierarchical Structures [HyHi, Reg. No. CZ.02.1.01/0.0/0.0/16_019/0000843].

REFERENCES

Ahmadi, A. et al. (2015) 'Antiviral potential of algae polysaccharides isolated from marine sources: A review', *BioMed Research International*. Hindawi Publishing Corporation, 2015, pp. 1–10. doi: 10.1155/2015/825203.

Amin, M. ., Alazba, A. . and Manzoor, U. (2014) 'A review of removal of pollutants from water/wastewater using different types of nanomaterials', *Advances in Materials Science and Engineering*, 1, p. 24. doi: doi:10.1155/2014/825910.

Armstrong, A. M., Sobsey, M. D. and Casanova, L. M. (2017) 'Disinfection of bacteriophage MS2 by copper in water', *Applied Microbiology and Biotechnology*, 101(18), pp. 6891–6897. doi: 10.1007/s00253-017-8419-x.

Besednova, N. et al. (2019) 'Metabolites of seaweeds as potential agents for the prevention and therapy of influenza infection', *Marine Drugs*, 17(6), pp. 1–21. doi: 10.3390/md17060373.

Borkow, G. et al. (2010) 'A novel anti-influenza copper oxide containing respiratory face mask', *PLoS ONE*, 5(6). doi: 10.1371/journal.pone.0011295.

Castro-Mayorga, J. L., Fabra, M. J. and Lagaron, J. M. (2016) 'Stabilized nanosilver based antimicrobial poly(3-hydroxybutyrate-co-3-hydroxyvalerate) nanocomposites of interest in active food packaging', *Innovative Food Science and Emerging Technologies*. Elsevier Ltd, 33, pp. 524–533. doi: 10.1016/j.ifset.2015.10.019.

Castro-Mayorga, J. L. et al. (2017) 'Antiviral properties of silver nanoparticles against norovirus surrogates and their efficacy in coated polyhydroxyalkanoates systems', *LWT – Food Science and Technology*, 79, pp. 503–510. doi: 10.1016/j.lwt.2017.01.065.

Ceccarelli, M. et al. (2020) 'Editorial – Differences and similarities between Severe Acute Respiratory Syndrome (SARS)-CoronaVirus (CoV) and SARS-CoV-2. Would a rose by another name smell as sweet?', *European Review for Medical and Pharmacological Sciences*, 24(5), pp. 2781–2783. doi: 10.26355/eurrev_202003_20551.

Cegolon, L. (2020) 'Investigating hypothiocyanite against SARS-CoV-2', *International Journal of Hygiene and Environmental Health*, 227, pp. 1–8. doi: 10.1016/j.ijheh.2020.113520.

Chen, L. and Liang, J. (2020) 'An overview of functional nanoparticles as novel emerging antiviral therapeutic agents', *Materials Science and Engineering: C. Elsevier*, 112(January 2019), p. 110924. doi:10.1016/j.msec.2020.110924.

Coetzee, D. et al. (2020) 'Influence of nanoparticles on thermal and electrical conductivity of composites', *Polymers*, 12(4), p. 742. doi: 10.3390/polym12040742.

Das, B. C. et al. (2013) 'Boron chemicals in diagnosis and therapeutics', *Future Medicinal Chemistry*, 5(6), pp. 653–676. doi: 10.4155/fmc.13.38.

Devereux, S. J. et al. (2019) 'Spectroscopic study of the loading of cationic porphyrins by carbon nanohorns as high capacity carriers of photoactive molecules to cells', *Journal of Materials Chemistry B*. Royal Society of Chemistry, 7(23), pp. 3670–3678. doi: 10.1039/c9tb00217k.

Du, T. et al. (2015) 'Probing the interactions of CdTe quantum dots with pseudorabies virus', *Scientific Reports*. Nature Publishing Group, 5, pp. 1–10. doi: 10.1038/srep16403.

Falcó, I. et al. (2019) 'Antiviral activity of alginate-oleic acid based coatings incorporating green tea extract on strawberries and raspberries', *Food Hydrocolloids*. Elsevier Ltd, 87(August 2018), pp. 611–618. doi: 10.1016/j.foodhyd.2018.08.055.

Fraj, A. et al. (2018) *Antimicrobial finishing of cotton and polyamide with nano–microparticles, International Conference of Applied Research on Textile, CIRAT-8*, Monastir, Tunisia, Researchgate.

Galdiero, S. et al. (2011) 'Silver nanoparticles as potential antiviral agents', *Molecules*, 16(10), pp. 8894–8918. doi: 10.3390/molecules16108894.

Gilling, D. H. et al. (2014) 'Antiviral efficacy and mechanisms of action of oregano essential oil and its primary component carvacrol against murine norovirus', *Journal of Applied Microbiology*, 116(5), pp. 1149–1163. doi: 10.1111/jam.12453.

Huang, S. et al. (2019) 'Benzoxazine monomer derived carbon dots as a broad-spectrum agent to block viral infectivity', *Journal of Colloid and Interface Science*, 542, pp. 198–206. doi: 10.1016/j.jcis.2019.02.010.

Iyigundogdu, Z. U. et al. (2017) 'Developing novel antimicrobial and antiviral textile products', *Applied Biochemistry and Biotechnology*, 181(3), pp. 1155–1166. doi: 10.1007/s12010-016-2275-5.

Kerry, R. G. et al. (2019) 'Nano-based approach to combat emerging viral (NIPAH virus) infection', *Nanomedicine: Nanotechnology, Biology, and Medicine* Elsevier Inc., 18, pp. 196–220. doi: 10.1016/j.nano.2019.03.004.

Khan, Ibrahim, Saeed, K. and Khan, Idrees (2019) 'Nanoparticles: Properties, applications and toxicities', *Arabian Journal of Chemistry*. The Authors, 12(7), pp. 908–931. doi: 10.1016/j.arabjc.2017.05.011.

Larson, A. M. et al. (2011) 'Hydrophobic polycationic coatings disinfect poliovirus and rotavirus solutions', *Biotechnology and Bioengineering*, 108(3), pp. 720–723. doi: 10.1002/bit.22967.

Leclercq, L. et al. (2016) 'Supramolecular assistance between cyclodextrins and didecyldimethylammonium chloride against enveloped viruses: Toward eco-biocidal formulations', *International Journal of Pharmaceutics*, 512(1), pp. 273–281. doi: 10.1016/j.ijpharm.2016.08.057.

Li, P. et al. (2011) 'A polycationic antimicrobial and biocompatible hydrogel with microbe membrane suctioning ability', *Nature Materials*. Nature Publishing Group, 10(2), pp. 149–156. doi: 10.1038/nmat2915.

Liga, M. V. et al. (2011) 'Virus inactivation by silver doped titanium dioxide nanoparticles for drinking water treatment', *Water Research*. Elsevier Ltd, 45(2), pp. 535–544. doi: 10.1016/j.watres.2010.09.012.

Łoczechin, A. et al. (2019) 'Functional carbon quantum dots as medical countermeasures to human coronavirus', *ACS Applied Materials and Interfaces*, 11(46), pp. 42964–42974. doi: 10.1021/acsami.9b15032.

MacIntyre, C. R. et al. (2009) 'Face mask use and control of respiratory virus transmission in households', *Emerging Infectious Diseases*, 15(2), pp. 233–241. doi: 10.3201/eid1502.081167.

Majchrzycka, K. et al. (2019) 'Application of biocides and super-absorbing polymers to enhance the efficiency of filtering materials', *Molecules*, 24(18), pp. 1–14. doi: 10.3390/molecules24183339.

Malmsten, M. (2011) 'Antimicrobial and antiviral hydrogels', *Soft Matter*, 7(19), pp. 8725–8736. doi: 10.1039/c1sm05809f.

Martínez-Abad, A. et al. (2013) 'Evaluation of silver-infused polylactide films for inactivation of Salmonella and feline calicivirus in vitro and on fresh-cut vegetables', *International Journal of Food Microbiology*. Elsevier B.V., 162(1), pp. 89–94. doi: 10.1016/j.ijfoodmicro.2012.12.024.

Mavrić, Z., Tomšič, B. and Simončič, B. (2018) 'Recent advances in the ultraviolet protection finishing of textiles', *Tekstilec*, 61(3), pp. 201–220. doi: 10.14502/tekstilec2018.61.201-220.

Milović, N. M. et al. (2005) 'Immobilized N-alkylated polyethylenimine avidly kills bacteria by rupturing cell membranes with no resistance developed', *Biotechnology and Bioengineering*, 90(6), pp. 715–722. doi: 10.1002/bit.20454.

Mori, Y. et al. (2013) 'Antiviral activity of silver nanoparticle/chitosan composites against H1N1 influenza A virus', *Nanoscale Research Letters*, 8(1), p. 93. doi: 10.1186/1556-276x-8-93.

Muthu, S. S. (2017) Textiles and Clothing Sustainability: Nanotextiles and Sustainability, *Textile Science and Clothing Technology*. Edited by M. S. Senthilkannan. Hong Kong: Springer Nature. doi: 10.1007/978-981-10-2182-4_2.

Orhan, M., Kut, D. and Gunesoglu, C. (2007) 'Use of triclosan as antibacterial agent in textiles', *Indian Journal of Fibre and Textile Research*, 32(1), pp. 114–118.

Panchal, A. et al. (2018) 'Self-assembly of clay nanotubes on hair surface for medical and cosmetic formulations', *Nanoscale*, 10(38), pp. 18205–18216. doi: 10.1039/c8nr05949g.

Pilau, M. R. et al. (2011) 'Antiviral activity of the Lippia graveolens (Mexican oregano) essential oil and its main compound carvacrol against human and animal viruses', *Brazilian Journal of Microbiology*, 42(4), pp. 1616–1624. doi: 10.1590/S1517-83822011000400049.

Qu, G. et al. (2020) 'An imperative need for research on the role of environmental factors in transmission of novel coronavirus (COVID-19)', *Environmental Science and Technology*, pp. 3730–3732. doi: 10.1021/acs.est.0c01102.

Randazzo, W. et al. (2018) 'Polymers and biopolymers with antiviral activity: potential applications for improving food safety', *Comprehensive Reviews in Food Science and Food Safety*, 17(3), pp. 754–768. doi: 10.1111/1541-4337.12349.

Rao, G. et al. (2016) 'Enhanced disinfection of Escherichia coli and bacteriophage MS2 in water using a copper and silver loaded titanium dioxide nanowire membrane', *Frontiers of Environmental Science and Engineering*, 10(4). doi: 10.1007/s11783-016-0854-x.

Repici, A. et al. (2020) 'Coronavirus (COVID-19) outbreak: what the department of endoscopy should know', *Gastrointestinal Endoscopy*. American Society for Gastrointestinal Endoscopy, pp. 1–6. doi: 10.1016/j.gie.2020.03.019.

Satish, S., Tharmavaram, M. and Rawtani, D. (2019) 'Halloysite nanotubes as a nature's boon for biomedical applications', *BJGP Open*, 6, pp. 1–16. doi: 10.1177/1849543519863625.

Scorei, R.I. and Popa, R. (2012) 'Boron-containing compounds as preventive and chemotherapeutic agents for cancer', *Anti-Cancer Agents in Medicinal Chemistry*, 10(4), pp. 346–351. doi: 10.2174/187152010791162289.

Si, Y. et al. (2018) 'Daylight-driven rechargeable antibacterial and antiviral nanofibrous membranes for bioprotective applications', *Science Advances*, 4(3). doi: 10.1126/sciadv.aar5931.

Sinclair, T. R. et al. (2019) 'Cationically modified membranes using covalent layer-by-layer assembly for antiviral applications in drinking water', *Journal of Membrane Science*. Elsevier B.V., 570–571, pp. 494–503. doi: 10.1016/j.memsci.2018.10.081.

Singh, N. et al. (2017) 'Sustainable fragrance cum antimicrobial finishing on cotton: Indigenous essential oil', *Sustainable Chemistry and Pharmacy*, 5(February), pp. 22–29. doi: 10.1016/j.scp.2017.01.003.

Song, Z. et al. (2015) 'Virus capture and destruction by label-free graphene oxide for detection and disinfection applications', *Small*, 11(9–10), pp. 1771–1776. doi: 10.1002/smll.201401706.

Soriano-Ursúa, M. A., Das, B. C. and Trujillo-Ferrara, J. G. (2014) 'Boron-containing compounds: Chemico-biological properties and expanding medicinal potential in prevention, diagnosis and therapy', *Expert Opinion on Therapeutic Patents*, 24(5), pp. 485–500. doi: 10.1517/13543776.2014.881472.

Sun, Y. et al. (2018) 'Antiviral activity against avian leucosis virus subgroup j of degraded polysaccharides from ulva pertusa', *BioMed Research International*, 2018, pp. 1–11. doi: 10.1155/2018/9415965.

Venkataraman, M., Mishra, R., Subramaniam, V. et al. (2014) 'Application of silver nanoparticles to industrial sewing threads: Effects on physico-functional properties & seam efficiency.' *Fibers Polym* 15, pp. 510–518

Wang, W. et al. (2011) 'In vitro inhibitory effect of carrageenan oligosaccharide on influenza A H1N1 virus', *Antiviral Research*. Elsevier B.V., 92(2), pp. 237–246. doi: 10.1016/j.antiviral.2011.08.010.

Wu, X. et al. (2017) 'Biocatalytic nanocomposites for combating bacterial pathogens', *Annual Review of Chemical and Biomolecular Engineering*, 8(1), pp. 87–113. doi: 10.1146/annurev-chembioeng-060816-101612.

Xu, S. and Li, Y. (2020) 'Comment beware of the second wave of COVID-19', *The Lancet*. Elsevier Ltd, 2019(20), pp. 2019–2020. doi: 10.1016/S0140-6736(20)30845-X.

Xu, J., Xu, Z. and Zheng, W. (2017) 'A review of the antiviral role of green tea catechins', *Molecules*, 22(8), pp. 1–18. doi: 10.3390/molecules22081337.

Zhang, M. and Yudasaka, M. (2016) Carbon Nanoparticles and Nanostructures *Carbon Nanohorns and Their High Potential in Biological Applications*. Edited by N. Yang, X. Jiang, and D.-W. Pang. Springer, Cham. doi: 10.1007/978-3-319-28782-9.

14 Fundamental Principles for Moisture Harvesting System and Its Design of Fabric

Kai Yang, Mohanapriya Venkataraman and Jiri Militký
Technical University of Liberec, Czech Republic

CONTENTS

14.1 INTRODUCTION

Fibrous face masks were raised in the last century and were proved to protect the human body from PM 2.5, bacteria, and so on (Davis 1991). Most studies focused on the filtration in the air and water system (Teo and Ramakrishna 2006). However, the microenvironment between the mouth and face masks was much more complicated, where there was an amount of moisture content and heat. Besides, the last report said that the virus could be adhered alive to the substrate of the face masks with a suitable humid environment (Van Doremalen et al. 2020). To control the microenvironment between skin or mouth and the face mask was essential. The humidity, airspeed, and temperature difference in the microenvironment has been proved to affect the moisture harvesting by the face mask. So, how to lead the harvested water in the face mask also remained to be solved reasonably. It was a pity that there were few studies related to this complicated topic. Besides, the stuffiness increased with the increased adsorbed water content by the mask and the decreased interspace between the fibers, which was initiated by various researches. It was noticed that the previous work was strong with the relatively open objectives, while the microenvironment between the mouth and the mask is strongly limited. Namely, the water transfer rate should be enhanced. Besides, nanofibers as the major part of the masks may contribute to different effects because the thermal and chemical properties of the material is significantly changed due to the nanosize (Bergman, Incropera, DeWitt, and Lavine, 2011). Therefore, how to keep the mask-wearing comfortable still required to be discussed. The review separated the continuous phenomena in the microenvironment into moisture harvesting and water transfer. The aim was to obtain ideas to design the suitable fibrous materials for the microenvironment between mouth and face mask.

14.2 MOISTURE HARVESTING ON THE FABRIC

Moisture harvesting is usually a dynamic cooling or isotherm process for the water droplet to form on the solid surface with the saturation state (RH = 100%) of the mixture of air and water vapor under or below the dew-point temperature at a constant pressure. The dewing was controlled by the pressure and RH, which means that the dewing could happen in the microenvironment between the mouth and face mask followed by dropwise condensation. The main factors affecting the moisture harvesting included the surface chemistry (e.g., topology) and thermodynamic equilibrium (e.g., airspeed). Most studies started to harvest moisture in the saturated water vapor environment, namely via dewing (Lee et al. 2012). The substrates for moisture harvesting including hydrophobic, hydrophilic, and hydrophobic–hydrophilic surfaces were studied. It was proposed that the hydrophilic coatings promote the heterogeneous nucleation and lower the thermal resistance by forming a thin liquid film on the surface (film-wise condensation) (Kim, Lee, and Webb 2002). The less fan power was requested for the same cooling duty with fixed frontal velocity when the plasma-treated heat exchanger was proposed. Besides, it was able to have higher cooling capacity by using higher air velocity. The thermal resistance was considered as lower when the hydrophobic surfaces were used because the condensed water drops on the hydrophobic surfaces were easy to be removed via gravity and the air streams (Narhe and Beysens 2007). In addition, the dropwise condensation on the hydrophobic surface was modeled in a few studies (Rose 2002). One of the earliest dropwise condensation models was proposed by Le Fevre and Rose (1996), who combined a calculation for the heat transfer through a single drop with a calculation for the drop size distribution. In the heat transfer model, four factors were taken into consideration, including the conduction resistance, the vapor–liquid interfacial matter transfer, the promoter layer resistance, and the resistance due to the convex liquid surface. The concept of the surface area fraction occupied by drops equal to or larger than the drop size was proposed for the drop size distribution. According to this model, various advanced models have been developed by having more precise physical and mathematic expressions for the growth of the small water drops. Tanaka (1975) proposed a population balance theory to investigate the transient change of local drop size distribution, which was based on the two mechanisms of small drop's growth: the direct condensation onto the drop and the coalescence of the neighboring drops.

Although the work provided a more accurate prediction of drop size distribution, especially for small drops, the measurement was much more difficult to experimentally examine. It was still noticeable that the advancement made from Tanaka's model was remarkable because it is true and suitable for most of the heat transfer that takes place through small drops (Wen and Jer 1976). Then, the population balance theory has been used to estimate the population of drops of a given size. Wen and Jer (1976) proposed the drop size distribution of small drops with the assumption of steady size distribution by using the population balance concept. In addition, the surface texture was also a factor for the drop distribution. In the study (Lee et al. 2012), hydrophobic, hydrophilic, and hydrophobic–hydrophilic surfaces were studied separately. The results proposed hydrophobic–hydrophilic surface system patterns for the higher water collection performance although the hydrophilic or hydrophobic surface system patterns were well established for the water collection. Based on the aforementioned mechanism, the transfer from the moisture to the water droplet and from the water droplet to the water film was revealed theoretically. However, it was not enough for effective water collection or moisture harvesting. The fog collector (Ju et al. 2012) based on cactus (Figure 14.1a–c) was proposed and provided an amount of information for such aim. It was found that there were three different spins in the cactus including oriented barbs, gradient grooves, and belt-structure trichomes, which are shown in Figure 14.1d. The structure of the three spins was also different as shown in Figure 14.1e–f, which provided the various pressures. When the water condenses on the surface of the cactus, "Deposition" initially occurs on the barb and the spine, with the water drops moving directionally along them, which is shown in Figure 14.1i.

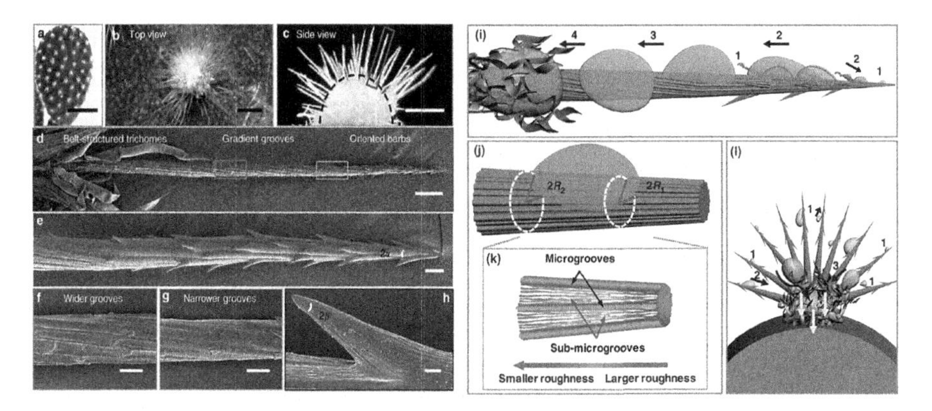

FIGURE 14.1 Structural images of the cactus and the mechanism of the fog collection on the cactus (Reprinted from Nat. Commun., vol. 3, J. Ju, H. Bai, Y. Zheng, T. Zhao, R. Fang, and L. Jiang, A multistructural and multifunctional integrated fog collection system in cactus, pp. 1–6, Copyright (2012)) (Ju et al. 2012).

When the deposition starts to proceed, the water drops tend to coalesce, and these drops increase in size to leave from the tip side of the spine, which is labeled as the "Collection" process. Then, bigger drops are transported along the gradient grooves which is labeled as "Transportation." Finally, the water drops are absorbed through the trichomes at the base of the spines, which is labeled as "Absorption." The essential point was the control of the water drop transportation. In this case, the gradient of the Laplace pressure arising from the conical shape of the spine and the gradient of the surface-free energy arising from the gradient of the surface roughness along the spine accounts for the directional movement of the water drops, which is schemed in Figure 14.1j,k in the right side. As a result, fog collection was proposed especially for moisture harvesting. However, it seems not strongly to connect with the fabric and there was no relative reference for the moisture harvesting by fabrics.

14.3 WATER TRANSFER THROUGH THE FABRIC

In Section 14.2, the principle for moisture harvesting is demonstrated and the basic theory for the water transfer is illustrated. However, the fabric is different from the fog collector and the texture is much more complicated. To reduce the harvested water on the surface of the fabric, the harvested water should be required to be faster transferred from inside to outside. For the single-layer fabric, more water adsorption and reduction of interspace between fibers with time in the microenvironment happened. Namely, the wettability and wickability of the fabric were focused on in this section.

14.3.1 WETTING

Wetting is a phenomenon which is caused by the cohesion and adhesion between the liquid and the solid surface. The interaction between the forces of cohesion (within the liquid) and the forces of adhesion (between the fibers and the liquid) determines whether wetting takes place or not and the spreading and absorption of the liquid on the surface of the textile material. From this point of view, wettability is defined as the first impression of fabric when the fabric is kept in touch with the liquid (Kissa 1996). Wetting is characterized by observing the displacement of the fiber–vapor interface to the fiber–liquid interface. The factors for the wettability of the fibrous materials include the surface chemical property, the geometry of fiber, and the surface roughness of the fabric.

14.3.2 CONTACT ANGLE

The measurement of contact angles is usually used as the primary data to characterize the wettability studies, which indicates the degree of wetting when a solid and liquid interact. The Young's Equation (14.1) was used to characterize the wettability (Figure 14.2a). The hydrophobic surface can be determined when $\theta > 90°$ and the surface is considered as hydrophilic when $\theta < 90°$.

$$\gamma_{sv} = \gamma_{sl} + \gamma_{lv} \cos\theta \tag{14.1}$$

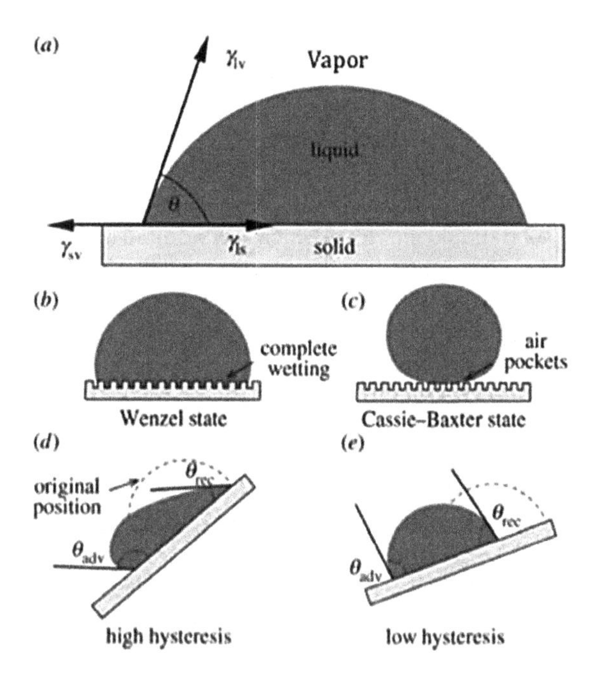

FIGURE 14.2 Model of water contact angle and water contact hysteresis (Reprinted from *Philosophical Transactions of the Royal Society A: Mathematical, Physical and Engineering Sciences*. Royal Society of London., vol. 2073, Philip S. Brown and Bharat Bhushan, Bioinspired Materials for Water Supply and Management: Water Collection, Water Purification, and Separation of Water from Oi, pp. 1–40, Copyright (2016)) (Brown and Bhushan 2016).

Where γ_{sv} (dyn/cm) is solid–vapor surface tension, sl is solid–liquid surface tension, and γ_{sv} is the liquid–vapor surface tension. Besides, $\gamma_{lv}\cos\theta$ is to characterize wetting energy.

Furthermore, the spreading parameter (S_{eq}) is evaluated by Equation (14.2). It is found that the S_{eq} is strongly related to γ_{lv} and θ. When S_{eq} was smaller than 0, it meant that the hydrophobic surface and the water could roll on the surface.

$$S_{eq} = \gamma_{sv} - (\gamma_{sl} + \gamma_{lv}) = \gamma_{lv}(\cos\theta_{eq} - 1) \quad (14.2)$$

14.3.3 Roughness on Wettability

Surface roughness is also important to affect dynamic wetting behavior, which can improve the surface hydrophobicity. It is of great significance to have a better understanding of roughness effects from both theoretical and practical perspectives. Generally, there are two models to characterize the roughness including the Wenzel model and Cassie model, which are shown in Figure 14.2b,c. Wenzel's work is to investigate the relationship between roughness and wetting. He stated that the apparent contact angle (θ^*) was strongly related to the surface roughness. Namely, the θ^*

as a function of the surface roughness of solid (R_g) and the contact angle obtained in the case of the ideal surface of the same solid (θ) (Equation 14.3).

$$\cos\theta^* = R_g \cos\theta \tag{14.3}$$

In the Wenzel model, the increased surface roughness of a hydrophobic surface can form a larger solid–liquid interface, which will result in a higher apparent contact angle on the hydrophobic surface. It is also assumed that the droplet size is sufficiently large compared to the roughness scale and that the liquid completely penetrates the rough grooves on the solid surface. When the surface roughness increases and the roughness factor is higher than one, the hydrophilic solid surfaces ($\theta < 90°$) become more hydrophilic while the hydrophobic surfaces ($\theta > 90°$) show increased hydrophobicity (Quéré 2008). It is noticeable that the equilibrium contact angle will increase in the Wenzel state for a hydrophobic surface, while the measured contact angle hysteresis (CAH) is typically very large. The reason is that the contact line is pinned at each wetted feature as it recedes (Lafuma and Quéré 2003). Besides, it has been shown that the surface with the roughness at a small scale cannot sufficiently "trap" air, and hysteresis is increased (Kim, Kavehpour, and Rothstein 2015). Cassie (Cassie and Baxter 1944) developed a model for a heterogeneous surface following Equation (14.4), where the θ^* was the apparent contact angle for the area fraction and θ was the contact angle obtained in the case of the ideal surface of the same solid, respectively. The subscripts 1 and 2 in the Equation (14.4) indicate two different surface chemistries. In the Cassie model, the reason for the increased hydrophobic solid surface roughness resulting in the increased hydrophobicity is caused by the air pockets formed within the peaks of the surface roughness. The presence of the combined partial liquid–air interface and the partial solid–liquid interface was found. For heterogeneous surface cases, the composite surface consists of two different areas, including solid and trapped air (Cheng et al. 2016).

$$\cos\theta^* = w_1 \cos\theta_1 + w_2 \cos\theta_2 \tag{14.4}$$

If the second area is air instead of surfaces having different chemistries, Equation (14.5) can then be written as:

$$\cos\theta^* = w_1 \cos\theta_1 + w_1 - 1 \tag{14.5}$$

14.3.4 Water Contact Hysteresis

CAH is also one of the most important and classic elements of wetting of liquid droplets in systems, which is shown in Figure 14.2d,e. The CAH could be evaluated by the difference between θ_{adv} and $\theta_{rec.}$ Most of the observation for the CAH was to look at a water droplet resting on the window. The static water droplet was governed by the gravity force which pulled the water droplet and the CAH which resists the movement. Once the state was a table, the droplet tended to be more asymmetric and static and there was a difference between the top and bottom of the droplet in the shape. The former one was thin with a smaller contact angle and the latter one was thick with a higher contact angle. It is assumed that the droplet slides down in an asymmetric shape if a certain size is obtained. Then, the difference between its front (in the direction of the driving force in this case gravity) and back contact angles (in the direction opposing the driving force) is called the CAH. The CAH is essential in various fields like the coating processes (dynamic hysteresis), digital microfluidics, and evaporation of droplets (leading to the well-known coffee stain). Especially, the industrial applications which are governed by the CAH include immersion lithography, fiber coatings, and inkjet printing. In these cases, it is clear that hysteresis is a problem (immersion lithography) while in others, it is essential (dip-coating). Determining and controlling CAH are critical for the operation of these industrially relevant systems (Bonn et al. 2009). The CAH is strongly relative to the capillary number, which provides a difference consisting of two parts. Firstly, a jump in the contact angle at zero velocity was observed (Eral, Mannetje, and Oh 2013). This jump was known or labeled as the CAH in some other research works. The jump is the difference in the contact angle which is induced by surface effects including roughness and heterogeneity. The difference between the maximum (advancing) and minimum (receding) contact angles are given by the minimum Gibbs free energy (Marmur and Bittoun 2009). It is revealed that the angles including the advancing and receding angles are not stable because of the dynamic process involved in reaching the minimum free energy. The lowest energy barrier should be large enough to observe the CAH. There is also the lowest energy barrier of the theoretical advancing and receding angles. Then, a slight vibration possibly shifts the apparent contact angle to a minimum closer to the global minimum in energy. As a result, the measured CAH may be smaller than the theoretical CAH. It is noticed that static hysteresis is extremely stable. For example, a droplet can be deposited in a state with a stable contact angle between the two limiting angles with the assumption that there is no motion (assuming no evaporation occurs) or relaxation. However, a retention time dependence of CAH was observed, which was caused by the deformations of the substrates, and the balance between the surface normal components of surface tension was affected (Tadmor et al. 2008). The static CAH is purely calculated according to the surface directly underneath the contact line. Therefore, CAH can be estimated on the various surfaces by using an average over the solid–liquid surface. Furthermore, the interaction between the liquid and the solid surface is for the dynamic component of hysteresis and even the liquid only sticks to the solid surface. It was found that the velocity is also essential for the CAH. The static hysteresis will dominate if a drop moved slowly on a rough surface, while the dynamic hysteresis becomes extremely important when high velocities or low static CAH surfaces were used. The example of the former one is mica/graphene, and the example of the latter one is liquid-soaked solids. Also, the contact line is invisible in the static hysteresis while being clear in the dynamic hysteresis. As a result, the hysteresis is strong relative to the measurement (Eral, Mannetje, and Oh 2013).

14.3.5 Measurement of Wetting

From the aforementioned demonstration, the water contact angle was used to characterize the wetting. To measure the contact angle (including advancing and receding water contact angle), there are several methods: tilted plate method (Tadmor et al. 2009), sessile drop method or the captive bubble method (Yildirim Erbil et al. 1999), and Wilhelmy method. There is no clear consensus as to which method is better, and indeed both have advantages and disadvantages. The biggest advantage of the Wilhelmy method is to cover large surfaces and measure wettability quickly. The tilted plate method is performed by using a drop on the surface and then observing the sliding situation when the surface is inclined. However, there are some minor concerns to be considered in the tilted plate method: 1) The droplet is assumed to move at a finite velocity, and it is not avoidable to have problems for the measurement in optical measurements. 2) The uncontrollable change in pressure between the front and back of the droplet could possibly cause a strong or weak curvature. The sessile drop method based on the optical equipment is popular in characterizing the contact angles because of high convenience and good visual clarity. In contrast to the Wilhelmy method, the sessile drop method is suitable for the various situations and also for a larger surface, which is performed by depositing a small droplet on the surface. Besides, the sessile drop method is performed by directly measuring the contact angle while the Wilhelmy method is carried out by measuring the force resulting from this contact angle. The difference is that the actual length of the contact line between the liquid and surface is difficult to measure when a rough surface is used. In the sessile drop method, the apparent contact angle is measured while it may be different from the actual contact angle. Therefore, the sessile drop method is very suitable for a single surface also, while the Wilhelmy method assumed that each place on the surface is the same.

14.3.6 Wicking

Wicking is to characterize the sustained capillary motion in various porous materials, especially fabrics. It occurs when fibers with capillary spaces in between them are wetted by a liquid, which are caused by the capillary force. Therefore, the capillary pressure and permeability of the fabric are two essential parameters affecting the wicking. The water transport through the fabric was simply described by Darcy's law (Amico and Lekakou 2000) (Equation 14.6). However, the air permeability was difficult to quantify by applying Darcy's Law. Then, the water transport through the fabric was described by Lucas–Washburn kinetic (Kamath et al. 1994) (Equation 14.7). It was noticed that the effective radius of the applied capillary tube and the effective contact angle was determined by fitting the experimental data. Therefore, the suitable wicking model for fabrics still require to be modified.

$$Q = -K\Delta \mathrm{P} / \mathrm{L}_0 \tag{14.6}$$

$$\mathrm{dL} / \mathrm{dt} = \left(\mathrm{r}^2 \Delta \mathrm{P}\right) / \left(8\mu L\right) \tag{14.7}$$

Where Q is the rate of flow, K is a constant, ΔP is the pressure drop across the material, L_0 is the length of the sample, dL/dt is the velocity of the water, μ is the water velocity, r is the radius of the pipe and L is the wetting length.

14.3.7 Fabric with Fast Wetting and Quick Drying

From the aforementioned demonstration, the capillary force and the surface chemistry determined the wetting and wicking rate. By combing the principle with the fabric, it was similar to assuming that the material components in the fabric could be the surface chemistry, and both the structure of the fiber and texture of the fabric could account for the capillary force. On the one hand, blending yarns of the synthetic fibers and natural fibers were suggested by constructing the heterogeneous surface of the fabric, which was a simple and basic way. However, the blending yarns system only provided a limitation wetting and wicking rate. On the other hand, increasing the capillary force to enhance the water transfer through the fabric has been more and more significant, effective, and feasible especially with the development of nanomaterials. In this part, several methods suitable for fast wetting and quick-drying are discussed.

14.3.8 Increasing the Specific Surface Area of Fiber

The increased specific surface area of the fiber could enhance the capillary force, which could be obtained by modifying the shape of the fiber, changing the fiber structure, reducing the fiber size, and so on.

- **The shape of the fiber**. From various research work, various shapes of the fibers including the normal circle "O" type, "+"type, "Y" type, "U" type, trilobal, and triangle "Δ"were used for the wetting and wicking (Hasan et al. 2008). It was found that the modified type could enhance the wicking rate.
- **Interior fiber structure**. By comparing the normal fibers, the porous fibers were considered to have a higher wicking rate. For example, Watanabe et al. used the porous membranes to store the electrolyte and found better wickability of the porous membranes (Watanabe, Satoh, and Shimura 2019).
- **Fiber size**. In the work, De Schoenmaker et al. (2011) compared the wickability of the PA nanofibers (nanofibrous membranes) with PA microfibers (normal nonwoven fabrics). The results showed that the wicking height of nanofibrous structures is significantly higher than the height of spin-bound nonwovens. The wicking rate was optimized by using nanofibrous membranes with a larger nanoscale diameter and porosity. During the initial phase of the wicking experiment, the capillary forces determined mainly by the fiber diameter, establish the wicking rate. Besides, the wicking rate of the normal fabrics with microscale fibers was reduced by using a similar adjustment of the diameter and the porosity.

14.3.9 Introducing Functional Group into the Fiber

The chemical modification of the fiber usually based on the grafting is for the shift between hydrophobicity and hydrophilicity. For example, Liu et al. (2014) prepared the PNIPAM-grafted cotton fabric for a thermoresponsive system, and the wickability was enhanced. El Messiry, El Ouffy, and Issa (2015) grafted the microcellulose particles on the surface of the PET and PET/cotton fabric. More content of the microcellulose particles resulted in the reduced water contact angle and higher wicking height.

14.3.10 Blending Yarns System

By comparing with the aforementioned two methods, the blending yarns systems generally focused on the comfortability of the final fabric. For example, Öztürk et al. investigated wicking properties of the knitted fabrics based on cotton–acrylic rotor yarns (Öztürk, Nergis, and Candan 2011). The wicking height of the cotton/acrylic yarns was enhanced by comparing with the cotton yarns. The water movement and absorption occurred only on the surface of the acrylic fibers, while the moisture absorption of cotton fiber was higher. Besides, the smaller acrylic yarns resulted in better wicking property.

14.4 ONE-WAY DIRECTIONAL WATER TRANSFER THROUGH THE FABRIC

From Section 14.3, the main principle for fast wetting and quick-drying was shown. However, the water absorption or transfer in the fabric was unidirectional (Tang, Kan, and Fan 2014). To design the mask, it is important to control the water transportation along with the expected direction in practice (one-way water motion), which was guided by the structure and/or surface feature of the substrate. It was found that the directional water transportation in the fabric across the thickness was caused by two main mechanisms: i) creation of a gradient across the fabric thickness from hydrophobicity to hydrophilicity and ii) combination of a hydrophobic fibrous layer with a hydrophilic fibrous layer. Wang et al. (2010) found the directional water transfer through fabrics by adjusting the asymmetric wettability of the fabrics was based on the imbalanced surface tension of the front and back sides of the fabric. Furthermore, Wang et al. then observed the unidirectional water transport from the superhydrophobic to the hydrophilic side and from the hydrophilic to the superhydrophobic side with external force (Wang, Wang, and Lin 2013). In the following studies, Zhou et al. reported that a fibrous-based thin porous media was realized to have the directional fluid-transport ability for both water and oil fluids, which was based on the aforementioned mechanism (Zhou et al. 2013) (Figure 14.3). The details included the heating temperature and reirradiation on the samples with UV light, which were labeled as (a) and (b) in Figure 14.3. The deposition of the water, soybean oil, and hexadecane droplets on the surface of the treated samples were shown in Figure 14.3c–e. Based on the surface tension of the different liquids, the three types of water penetration through the irradiated fabric were found, which was shown in Figure 14.3f–h. As a result, the realization switch of the super-hydrophobicity and hydrophilicity was included and schemed in Figure 14.3i. It was also realized that the water could transfer against the gravity from the optimized hydrophobic side of the fabric while the adsorption was found when the water started to transfer from the hydrophilic sides of the fabric, which was shown in Figure 14.4a

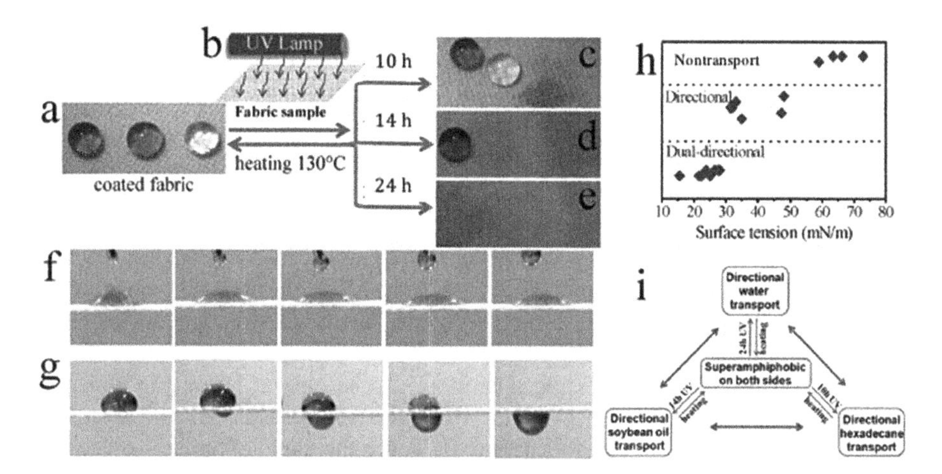

FIGURE 14.3 The scheme of the superphobicity/philicity janus fabric (Reprinted from Sci. Rep, vol. 3, H. Zhou, H. Wang, H. Niu, and T. Lin, Superphobicity/philicity janus fabrics with switchable, spontaneous, directional transportability to water and oil fluids, pp. 1–6, Copyright (2013)) (Zhou et al. 2013).

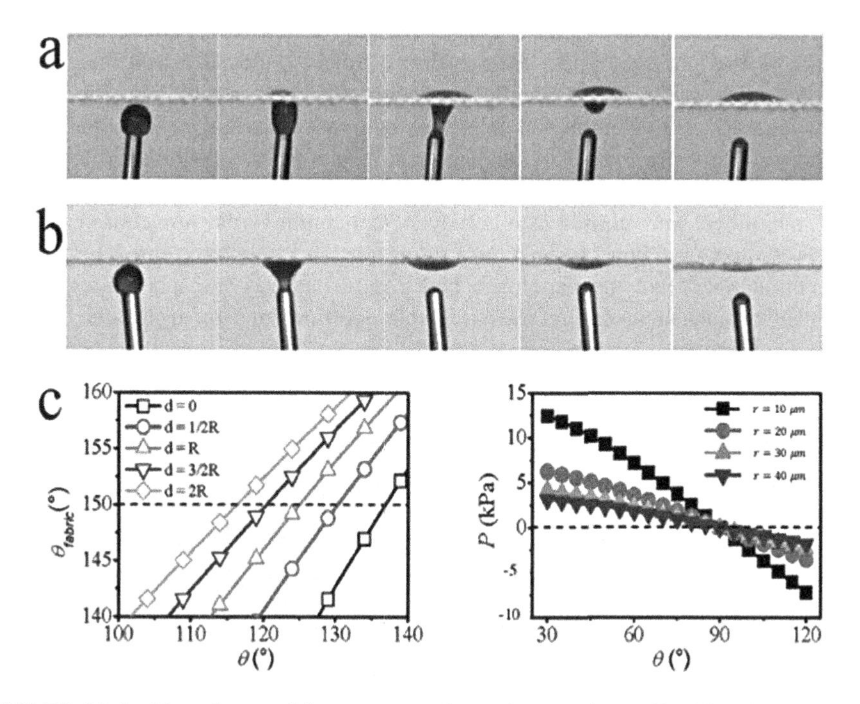

FIGURE 14.4 The scheme of the water transfer against gravity realized by UV-irradiated fabric (Reprinted from *Sci. Rep.*, vol. 3, H. Zhou, H. Wang, H. Niu, and T. Lin, Superphobicity/philicity janus fabrics with switchable, spontaneous, directional transportability to water and oil fluids, pp. 1–6, Copyright (2013)) (Zhou et al. 2013).

and b, respectively. The Young–Laplace capillary pressure (P) was considered for the difference and was proved by estimating the contact angle (θ_{fabric}) between the hydrophobic fabric and the water and the contact angle (θ) between the water drop and the capillary wall, which was shown in Figure 14.4c,d. From the aforementioned previous work, the water transfer was governed by the wettability, surface roughness, and capillary pressure in the hydrophobic area across the thickness. To control the water transfer from the hydrophilic side to the hydrophobic side, the capillary pressure was adjusted to be negative for the water transfer, which was caused by the high contact angle zone inside of the hydrophobic area. Besides, the control of the selectivity of the different liquids was mostly controlled by the surface tension, which was based on the theory that smaller surface tension resulted in a smaller contact angle. It was the same for the capillary pressure. Therefore, if one type of liquid can penetrate through a porous material, other liquids with lower surface tension are also able to permeate because of more wettable nature. However, the liquid of higher surface tension may be forbidden to penetrate if the contact angle on the same surface is large enough. Apart from the realization of the directional water transfer based on asymmetric wettability across the fabric thickness, the nanofibrous membranes also showed directional water transport by having a similar structure. Wu et al. (2012) and Wang, Wang, and Lin (2013) successfully prepared the nanofibrous membranes-based composites with directional water transfer, where the water can penetrate from the hydrophobic side rather than from the hydrophilic side.

It was noticed that the capillary pressure may be decreased with more water transfer across the fabric thickness from both controllable sides. To decrease capillary pressure in the hydrophobic area during water transfer, Mao et al. developed novel composite 1D fiber assembly via an electrospun yarn technology (Mao et al. 2020), which was shown in Figure 14.5a. The main idea is to modify the capillary of the yarn structure, which was similar to the tree structure shown in Figure 14.5b. The PAN nanofibers were chosen as outer layers and cotton yarns were chosen as inner layers as shown in Figure 14.5c,d. As a result, the fabrics with the core yarns having the cotton fibers and the nanofibers had a high one-way transport index, which revealed enhancement of the water transport performance by using the tree-like yarns.

14.5 OUTLOOK

For the face masks, the comfortability is governed by the moisture harvesting and water transfer property. Although there were various mechanisms for moisture harvesting, the practice for the face masks is not realized yet. Besides, the directional water transfer across the fabric thickness has been studied for decades. Most studies focused on these topics separately, however, they should be considered together in practice. Apart from the surface chemistry and topology, the capillary force across the fabric thickness is now assumed as the most important factor to govern the directional water transfer. Besides, the capillary force is affected significantly when there is water content inside the fabric, which lowers down the efficiency of the water transfer. Besides, the microenvironment between the mouth and face masks was much more complicated. How moisture harvesting happens, and the collected water

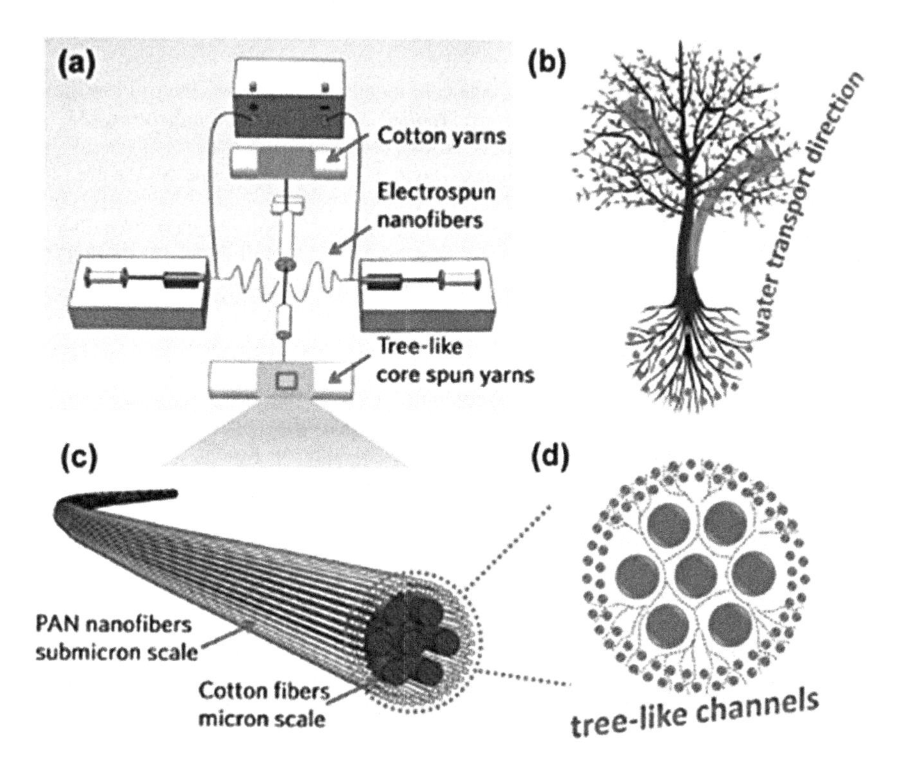

FIGURE 14.5 The scheme of the nanofiber-incorporated yarns (Reprinted from *Mater. Des.*, vol. 186r, Mao et al., tree-like structure driven water transfer in 1D fiber assemblies for Functional Moisture-Wicking Fabric, p. 108305., Copyright (2020), with permission from Elsevier) (Mao et al. 2020).

transfers out remains to be revealed, although the suggestions could be obtained from the aforementioned basic and novel research works.

ACKNOWLEDGEMENT

This work was supported by the research project of Student Grant Competition of the Technical University of Liberec no. 21406/2020 granted by Ministry of Education, Youth and Sports of Czech Republic and the Ministry of Education, Youth and Sports of the Czech Republic, European Union – European Structural and Investment Funds in the Frames of Operational Programme Research, Development and Education – project Hybrid Materials for Hierarchical Structures (HyHi, Reg. No. CZ.02.1.01/0.0/0.0/16_019/0000843), the Ministry of Education, Youth and Sports in the frames of support for researcher mobility (VES19China-mobility, Czech-Chinese cooperation) Design of multilayer micro-/nanofibrous structures for air filters applications, Reg. No. 8JCH1064., and project "Intelligent thermoregulatory fibers and functional textile coatings based on temperature resistant encapsulated PCM" SMARTTHERM (Project No. TF06000048).

REFERENCES

Amico, S., and C. Lekakou. 2000. "Mathematical Modelling of Capillary Micro-Flow through Woven Fabrics." *Composites Part A: Applied Science and Manufacturing* 31 (12): 1331–1344. doi:10.1016/S1359-835X(00)00033-6.

Theodore L. Bergman, Frank P. Incropera, David P. DeWitt, Adrienne S. Lavine. 2011. *Fundamentals of Heat and Mass Transfer*. Wiley.

Bonn, Daniel, Jens Eggers, Joseph Indekeu, and Jacques Meunier. 2009. "Wetting and Spreading." *Reviews of Modern Physics* 81 (2): 739–805. doi:10.1103/RevModPhys.81.739.

Brown, Philip S., and Bharat Bhushan. 2016. "Bioinspired Materials for Water Supply and Management: Water Collection, Water Purification and Separation of Water from Oil." *Philosophical Transactions of the Royal Society A: Mathematical, Physical and Engineering Sciences*. 374 (2073): 1–40.doi:10.1098/rsta.2016.0135.

Cassie, A. B.D., and S. Baxter. 1944. "Wettability of Porous Surfaces." *Transactions of the Faraday Society* 40: 546–551. doi:10.1039/tf9444000546.

Cheng, By Kuok, Blake Naccarato, Kwang J. Kim, and Anupam Kumar. 2016. "Theoretical Consideration of Contact Angle Hysteresis Using Surface-Energy-Minimization Methods." *International Journal of Heat and Mass Transfer* 102 (11): 154–161. doi:10.1016/j.ijheatmasstransfer.2016.06.014.

Davis, Wayne T. 1991. "Filtration Efficiency of Surgical Face Masks: The Need for More Meaningful Standards." *AJIC: American Journal of Infection Control* 19 (1): 16–18. doi:10.1016/0196-6553(91)90156-7.

De Schoenmaker, Bert, Lien Van der Schueren, Sander De Vrieze, Philippe Westbroek, and Karen De Clerck. 2011. "Wicking Properties of Various Polyamide Nanofibrous Structures with an Optimized Method." *Journal of Applied Polymer Science* 120 (1): 305–310. doi:10.1002/app.33117.

El Messiry, Magdi, Affaf El Ouffy, and Marwa Issa. 2015. "Microcellulose Particles for Surface Modification to Enhance Moisture Management Properties of Polyester, and Polyester/Cotton Blend Fabrics." *Alexandria Engineering Journal* 54 (2): 127–140. doi:10.1016/j.aej.2015.03.001.

Eral, H. B., T Mannetje, and J. M. Oh. 2013. "Contact Angle Hysteresis: A Review of Fundamentals and Applications." *Colloid and Polymer Science* 291: 247–260 doi:10.1007/s00396-012-2796-6.

Hasan, M. M.B., A. Calvimontes, A. Synytska, and V. Dutschk. 2008. "Effects of Topographic Structure on Wettability of Differently Woven Fabrics." *Textile Research Journal* 78 (11): 996–1003. doi:10.1177/0040517507087851.

Ju, Jie, Hao Bai, Yongmei Zheng, Tianyi Zhao, Ruochen Fang, and Lei Jiang. 2012. "A Multi-Structural and Multi-Functional Integrated Fog Collection System in Cactus." *Nature Communications* 3: 1–6. doi:10.1038/ncomms2253.

Kamath, Y.K., S.B. Hornby, H.-D. Weigmann, and M.F. Wilde. 1994. "Wicking of Spin Finishes and Related Liquids into Continuous Filament Yarns." *Textile Research Journal* 64 (1): 33–40. doi:10.1177/004051759406400104.

Kim, G rak, Hyunuk Lee, and Ralph L. Webb. 2002. "Plasma Hydrophilic Surface Treatment for Dehumidifying Heat Exchangers." *Experimental Thermal and Fluid Science* 27 (1): 1–10. doi:10.1016/S0894-1777(02)00219-4.

Kim, Jeong Hyun, Pirouz Kavehpour, and Jonathan P. Rothstein. 2015. "Dynamic Contact Angle Measurements on Superhydrophobic Surfaces." *Physics of Fluids* 27 (3): 032107. doi:10.1063/1.4915112.

Kissa, Erik. 1996. "Wetting and Wicking." *Textile Research Journal* 66 (10): 660–668. doi:10.1177/004051759606601008.

Lafuma, Aurélie, and David Quéré. 2003. "Superhydrophobic States." *Nature Materials*: 475–480. doi:10.1038/nmat924.

Le Fevre, E. J., and John W. Rose. 1996. "A Theory of Heat Transfer by Dropwise Condensation." *International Heat Transfer Conference* 3: 362–375. doi:10.1615/ihtc3.180.

Lee, Anna, Myoung Woon Moon, Hyuneui Lim, Wan Doo Kim, and Ho Young Kim. 2012. "Water Harvest via Dewing." *Langmuir* 28 (27): 10183–10191. doi:10.1021/la3013987.

Liu, Xuqing, Yi Li, Junyan Hu, Jiao Jiao, and Jiashen Li. 2014. "Smart Moisture Management and Thermoregulation Properties of Stimuli-Responsive Cotton Modified With Polymer Brushes." *The Royal Society of Chemistry*, 63691–63695. doi:10.1039/C4RA11080C.

Mao, Ning, Jiao Ye, Zhenzhen Quan, Hongnan Zhang, Dequn Wu, Xiaohong Qin, Rongwu Wang, and Jianyong Yu. 2020. "Tree-like Structure Driven Water Transfer in 1D Fiber Assemblies for Functional Moisture-Wicking Fabrics." *Materials and Design* 186: 108305. doi:10.1016/j.matdes.2019.108305.

Marmur, Abraham, and Eyal Bittoun. 2009. "When Wenzel and Cassie Are Right: Reconciling Local and Global Considerations." *Langmuir* 25 (3): 1277–1281. doi:10.1021/la802667b.

Narhe, R. D., and D. A. Beysens. 2007. "Growth Dynamics of Water Drops on a Square-Pattern Rough Hydrophobic Surface." *Langmuir* 23 (12): 6486–6489. doi:10.1021/la062021y.

Öztürk, Merve Küçükali, Banu Nergis, and Cevza Candan. 2011. "A Study of Wicking Properties of Cotton–Acrylic Yarns and Knitted Fabrics." *Textile Research Journal* 81 (3): 324–328. doi:10.1177/0040517510383611.

Quéré, David. 2008. "Wetting and Roughness." *Annualreviews.Org* 38: 71–99. doi:10.1146/annurev.matsci.38.060407.132434.

Rose, J. W. 2002. "Dropwise Condensation Theory and Experiment: A Review." *Proceedings of the Institution of Mechanical Engineers, Part A: Journal of Power and Energy* 216 (2): 115–128. doi:10.1243/09576500260049034.

Tadmor, Rafael, Kumud Chaurasia, Preeti S. Yadav, Aisha Leh, Prashant Bahadur, Lan Dang, and Wesley R. Hoffer. 2008. "Drop Retention Force as a Function of Resting Time." *Langmuir* 24 (17): 9370–9374. doi:10.1021/la7040696.

Tadmor, Rafael, Prashant Bahadur, Aisha Leh, Hartmann E. N'Guessan, Rajiv Jaini, and Lan Dang. 2009. "Measurement of Lateral Adhesion Forces at the Interface between a Liquid Drop and a Substrate." *Physical Review Letters* 103 (26): 266101. doi:10.1103/PhysRevLett.103.266101.

Tanaka, Hiroaki. 1975. "A Theoretical Study of Dropwise Condensation." *Journal of Heat Transfer* 97 (1): 72–78. doi:10.1115/1.3450291.

Tang, Ka Po Maggie, Chi Wai Kan, and Jin Tu Fan. 2014. "Evaluation of Water Absorption and Transport Property of Fabrics." *Textile Progress* 46 (1): 1–132. doi:10.1080/00405167.2014.942582.

Teo, W. E., and S. Ramakrishna. 2006. "A Review on Electrospinning Design and Nanofibre Assemblies." *Nanotechnology* 17 (14). doi:10.1088/0957-4484/17/14/R01.

Van Doremalen, Neeltje, Trenton Bushmaker, Dylan H. Morris, Myndi G. Holbrook, Amandine Gamble, Brandi N. Williamson, Azaibi Tamin, et al. 2020. "Aerosol and Surface Stability of SARS-CoV-2 as Compared with SARS-CoV-1." *New England Journal of Medicine*. doi:10.1056/NEJMc2004973.

Wang, Hongxia, Jie Ding, Liming Dai, Xungai Wang, and Tong Lin. 2010. "Directional Water-Transfer through Fabrics Induced by Asymmetric Wettability." *Journal of Materials Chemistry* 20 (37): 7938–7940. doi:10.1039/c0jm02364g.

Wang, Hongxia, Xungai Wang, and Tong Lin. 2013. "Unidirectional Water Transfer Effect from Fabrics Having a Superhydrophobic-to-Hydrophilic Gradient." *Journal of Nanoscience and Nanotechnology* 13 (2): 839–842. doi:10.1166/jnn.2013.6008.

Watanabe, Masahiro, Yasutaka Satoh, and Chiyoka Shimura. 2019. "Management of the Water Content in Polymer Electrolyte Membranes with Porous Fiber Wicks." *Journal of the Electrochemical Society* 140 (11): 3190–3193. doi:10.1149/1.2221008.

Wen, Hai Wu, and Ru Maa Jer. 1976. "On the Heat Transfer in Dropwise Condensation." *The Chemical Engineering Journal* 12 (3): 225–231. doi:10.1016/0300-9467(76)87016-5.

Wu, Jing, Nü Wang, Li Wang, Hua Dong, Yong Zhao, and Lei Jiang. 2012. "Unidirectional Water-Penetration Composite Fibrous Film via Electrospinning." *Soft Matter* 8 (22): 5996–5999. doi:10.1039/c2sm25514f.

Yildirim Erbil, H., G. McHale, S. M. Rowan, and M. I. Newton. 1999. "Determination of the Receding Contact Angle of Sessile Drops on Polymer Surfaces by Evaporation." *Langmuir* 15 (21): 7378–7385. doi:10.1021/la9900831.

Zhou, Hua, Hongxia Wang, Haitao Niu, and Tong Lin. 2013. "Superphobicity/Philicity Janus Fabrics with Switchable, Spontaneous, Directional Transport Ability to Water and Oil Fluids." *Scientific Reports* 3: 1–6. doi:10.1038/srep02964.

Index

G

H

I

K

L

M

N

O

P

Printed in the United States
by Baker & Taylor Publisher Services